New Forms of Procurement

T0188199

The last three decades have seen the evolution of Public-Private Partnerships (PPPs) and Relational Contracting (RC) as alternative procurement approaches to traditional methods of delivering public infrastructure. The potential for growth in these new forms of procurement has led to an ongoing debate on the nature of requirements, particularly in terms of policy development, encouraging private investment and value for money. A key argument for governments to procure projects using PPPs and RC is that the process delivers better value for all the stakeholders, including the community and asset end-users. This wide-ranging study of such crucial procurement issues includes international historical context, collaboration and risk management, with a focus on sustainable procurement approaches. The international significance of PPPs and RC procurement is reinforced with case study examples from the UK, Europe, North America, South Africa and the Asia-Pacific. It features cutting-edge research from around the world on subjects such as:

- reviews and reflection of the PPP approach
- project alliancing
- implementation of RC in developing countries
- changes in procurement policy
- value for money, collaboration and stakeholder involvement
- growth and emergence of PPPs in Asia
- risk management.

Including contributions from some of the world's most prominent academics and practitioners in the field, this volume is a crucial guide to the strategic choices governments now face in the provision of infrastructure, between using 'public' or 'private' mechanisms, or a combination of the two.

Marcus Jefferies is a Senior Lecturer in the School of Architecture and Built Environment, University of Newcastle, Australia.

Steve Rowlinson is a Professor in the Department of Real Estate and Construction at the University of Hong Kong, SAR, China.

Spon Research

Spon Research publishes a stream of advanced books for built environment researchers and professionals from one of the world's leading publishers. The ISSN for the Spon Research programme is ISSN 1940-7653 and the ISSN for the Spon Research E-book programme is ISSN 1940-8005

Published

Free-Standing Tension Structures: From Tensegrity Systems to Cable-Strut Systems
978-0-415-33595-9
W. B. Bing

Performance-Based Optimization of Structures: Theory and applications
978-0-415-33594-2
Q. Q. Liang

Microstructure of Smectite Clays and Engineering Performance
978-0-415-36863-6
R. Pusch and R. Yong

Procurement in the Construction Industry: The Impact and Cost of Alternative Market and Supply Processes
978-0-415-39560-1
W. Hughes et al.

Communication in Construction Teams
978-0-415-36619-9
S. Emmitt and C. Gorse

Concurrent Engineering in Construction Projects
978-0-415-39488-8
C. Anumba, J. Kamara and A.-F. Cutting-Decelle

People and Culture in Construction
978-0-415-34870-6
A. Dainty, S. Green and B. Bagilhole

Very Large Floating Structures
978-0-415-41953-6
C. M. Wang, E. Watanabe and T. Utsunomiya

Tropical Urban Heat Islands: Climate, Buildings and Greenery
978-0-415-41104-2
N.H. Wong & C. Yu

Innovation in Small Construction Firms
978-0-415-39390-4
P. Barrett, M. Sexton and A. Lee

Construction Supply Chain Economics
978-0-415-40971-1
K. London

*Employee Resourcing in the
Construction Industry*
978-0-415-37163-6
A. Raiden, A. Dainty and
R. Neale

*Managing Knowledge in the
Construction Industry*
978-0-415-46344-7
A. Styhre

*Collaborative Information
Management in Construction*
978-0-415-48422-0
G. Shen, A. Baldwin and
P. Brandon

*Containment of High Level
Radioactive and Hazardous Solid
Wastes with Clay Barriers*
978-0-415-45820-7
R. N. Yong, R. Pusch and
M. Nakano

*Performance Improvement in
Construction Management*
978-0-415-54598-3
B. Atkin and J. Borgbrant

*Organisational Culture in the
Construction Industry*
978-0-415-42594-0
V. Coffey

*Relational Contracting for
Construction Excellence: Principles,
Practices and Partnering*
978-0-415-46669-1
A. Chan, D. Chan and J. Yeung

Soil Consolidation Analysis
978-0-415-67502-4
J. H. Yin and G. Zhu

*OHS Electronic Management
Systems for Construction*
978-0-415-55371-1
I. Kamardeen

*FRP-Strengthened Metallic
Structures*
978-0-415-46821-3
X. L. Zhao

*Leadership and Sustainability in the
Built Environment*
978-1-138-77842-9
A. Opoku and V. Ahmed

*The Soft Power of Construction
Contracting Organisations*
978-1-138-80528-6
S. O. Cheung, P. S. P. Wong and
T. W. Yiu

*Fall Prevention Through Design
in Construction: The Benefits of
Mobile Construction*
978-1-138-77915-0
I. Kamardeen

Trust in Construction Projects
978-1-138-81416-5
A. Ceric

*New Forms of Procurement: PPP
and Relational Contracting in the
21st Century*
978-1-138-79612-6
M. Jefferies and S. Rowlinson

New Forms of Procurement

PPP and Relational Contracting in
the 21st century

Edited by
Marcus Jefferies and Steve Rowlinson

Routledge
Taylor & Francis Group

LONDON AND NEW YORK

First published 2016
by Routledge

2 Park Square, Milton Park, Abingdon, Oxfordshire OX14 4RN
711 Third Avenue, New York, NY 10017

Routledge is an imprint of the Taylor & Francis Group, an informa business

First issued in paperback 2018

Copyright © 2016 selection and editorial material, Marcus Jefferies and Steve Rowlinson; individual chapters, the contributors.

The right of the editors to be identified as the authors of the editorial material, and of the authors for their individual chapters, has been asserted in accordance with sections 77 and 78 of the Copyright, Designs and Patents Act 1988.

All rights reserved. No part of this book may be reprinted or reproduced or utilised in any form or by any electronic, mechanical, or other means, now known or hereafter invented, including photocopying and recording, or in any information storage or retrieval system, without permission in writing from the publishers.

Notice:
Product or corporate names may be trademarks or registered trademarks, and are used only for identification and explanation without intent to infringe.

British Library Cataloguing-in-Publication Data
A catalogue record for this book is available from the British Library

Library of Congress Cataloging in Publication Data
Names: Jefferies, Marcus, editor. | Rowlinson, Stephen M., editor.
Title: New forms of procurement : PPP and relational contracting in the 21st
    century / edited by Marcus Jefferies and Steve Rowlinson.
Description: Abingdon, Oxon ; New York, NY : Routledge, 2016. | Includes
    bibliographical references and index.
Identifiers: LCCN 2015044956| ISBN 9781138796126 (hardback : alk. paper) |
    ISBN 9781315758053 (ebook : alk. paper)
Subjects: LCSH: Public works. | Construction projects. | Industrial procurement. |
    Public-private sector cooperation.
Classification: LCC HD3850 .N49 2016 | DDC 658.7/2—dc23
LC record available at http://lccn.loc.gov/2015044956

ISBN: 978-1-138-79612-6 (hbk)
ISBN: 978-0-367-00118-6 (pbk)

Typeset in Sabon
by Swales & Willis Ltd, Exeter, Devon, UK

Contents

Figures

Tables

Contributors

Co-editors and chapter authors/co-authors

Marcus Jefferies (co-editor of the book, co-author of the Preface and co-author of Chapters 1, 2, 5, 9 and 12) is a Senior Lecturer in the School of Architecture and Built Environment at the University of Newcastle, Australia. His PhD investigated innovative risk management in social infrastructure Public-Private Partnerships (PPPs) and involved a multi-stage case study approach at both organisation and project levels. Marcus has been a Chief Investigator for two Australian Research Council (ARC) Linkage Projects and prior to this was involved in three Cooperative Research Centre (CRC) projects. He teaches contract administration, procurement and risk management at both Bachelor and Master's levels and is a Member of the Australian Institute of Building (AIB) and the Chartered Institute of Building (CIOB). He is also part of the Technical Committee of Working Commission W092 Procurement Systems of the International Council for Research and Innovation in Building and Construction (CIB) and is a reviewer for several international journals including *Construction Management and Economics*, *Engineering, Construction and Architectural Management* and *International Journal of Managing Projects in Business*. Prior to entering academia he worked in the construction industry for over 10 years as a site engineer in both Hong Kong and the UK. He is a keen follower of English and Australian sporting teams, and a willing participant in sport when time and injuries allow, and is an avid collector of Australian red wine.

Steve Rowlinson (co-editor of the book, co-author of the Preface, co-author of Chapter 1, and author of Chapter 10) is a Professor in the Department of Real Estate and Construction at Hong Kong University and is involved actively in research and PhD supervision in the areas of procurement systems, project management, occupational health and safety and ICT. He has been Coordinator of the CIBW 092 Working Commission on Procurement Systems for over 15 years and has co-organised numerous conferences and symposia in this capacity. Steve has authored and co-authored more than 10 books and over 100 peer-reviewed papers.

He is an Adjunct Professor at Queensland University of Technology where he has a particular interest in international construction innovation and has been involved in several funded research projects. Steve has acted as a consultant to the Hong Kong Works Bureau, Hong Kong Housing Authority and Queensland Department of Main Roads, in the process producing numerous consultancy and expert reports. He is a Member of the Institution of Engineers (HK), the Institution of Civil Engineers (UK) and is a Fellow of the Royal Institution of Chartered Surveyors. In his spare time he is a keen golfer, skier, snowboarder and ex-footballer.

Chapter authors and co-authors

Hanna Boodhun (co-author of Chapter 17) is a graduate of the University of Cape Town. She is currently working in the construction industry as a Candidate Quantity Surveyor for a local contractor in Mauritius and is pursuing further research in the field of urban facilities management.

Michelle Bunting (co-author of Chapter 17) is a Quantity Surveying graduate from the University of Cape Town. She currently works as an expat within the construction industry and specialises in energy-related projects. She is also engaged in community development programmes which strive to uplift underprivileged communities within the African countries in which she operates.

Le Chen (co-author of Chapter 7) is an Associated Lecturer at the School of Civil Engineering and Built Environment, Queensland University of Technology (QUT), Australia. Her research focuses on innovation, strategic management, performance evaluation and project governance.

Jacky Chung (co-author of Chapter 15) is an Assistant Professor with the Department of Building, National University of Singapore. He is an experienced researcher in the areas of collaborative team work, construction briefing, stakeholder engagement, Public-Private Partnership, value management and building information modelling. He has been presented six scholarly awards and has produced more than 30 research publications in leading academic journals and international conferences.

Peter Davis (author of Chapter 8) is the Chair of Construction Management at the University of Newcastle, Australia. He holds a PhD in Economics and Finance, studying the impact of relationship-based procurement, from the Royal Melbourne Institute of Technology (RMIT). He has a Master's Degree in Project Management from Curtin University of Technology, Perth, WA.

Myungsik Do (co-author of Chapter 13) received his PhD (Eng.) from Kyoto University and is currently a Professor in the Department of Urban Engineering, Hanbat National University, Korea. His research interests are transportation planning, SOC project evaluation and asset management of infrastructure facilities. He was awarded the 2001 KSCE Young Researcher Award and the 2014 Best Researcher Award of the Korean Society of ITS Engineers.

Jennifer Firmenich (co-author of Chapter 5) completed her PhD at the Federal Institute of Technology in Switzerland (ETH Zurich). It focused on PPP risk management and involved international construction group STRABAG as industry partner. Currently, she is deepening her expertise at PricewaterhouseCoopers (PwC) in Switzerland within the Real Estate Consulting team. Her responsibilities cover the development of financial, organisational and risk management solutions for real estate and infrastructure projects.

Mark Gannon (author of Chapter 4) is a Senior Lecturer in Project Management at Hertfordshire Business School, the University of Hertfordshire, UK. He has developed a wide range of project management, commercial and analytical expertise through working on some of the largest and most innovative Private Finance Initiative/Public-Private Partnership projects undertaken in the United Kingdom.

Yongjian Ke (co-author of Chapter 12) is currently a Lecturer in Construction Management at the University of Newcastle, Australia. He obtained his PhD in 2010 from Tsinghua University, China and on completion he joined the National University of Singapore as a Research Fellow. His research experience concentrates on two interrelated areas: Public-Private Partnerships (PPPs) and Relational Contracting on which he has built a track record of over 70 peer-reviewed publications which have attracted over 1,000 citations with an h-index of 16.

Mohan Kumaraswamy (co-author of Chapter 15) is an Honorary Professor of the University of Hong Kong and the T. N. Suba Rao Brigade Group Adjunct Chair Professor of IIT Madras. He is the Founding Director of the Centre for Innovation in Construction and Infrastructure Development and Editor-in-Chief of the *Journal of Built Environment Project and Asset Management*. He is one of the two Joint Co-ordinators of the CIB Task Group TG72 on Pubic-Private Partnership.

Sang Hyuk Lee (co-author of Chapter 13) is a Senior Researcher in the Korea Institute of Civil Engineering and Building Technology (KICT). His research interests are in transportation asset management with risk and uncertainty and transportation safety modelling.

Beverley Lloyd-Walker (co-author of Chapter 2) is a Research Fellow within the Centre for Integrated Project Solutions (CIPS) at RMIT University, Melbourne. Beverley's research interests include relationship-based project procurement, the professional skills required for successful collaborative project delivery, and career planning and development within the project management profession.

Kerry London (co-author of Chapter 11) is Professor of Construction Management at RMIT University, Melbourne, Australia. She is a national leader with many Australian firsts: first female Chair in Construction Management in 2008, course Director of the first online internationally accredited Master's of Construction Management programme, first academic from her discipline to be a member of the Australian Research Council College of Experts – Humanities and Creative Arts, only female academic appointed by the Federal Minister of the Department of Industry, Innovation, Science and Research to the Built Environment Industry Innovation Council, author of one of the seminal research books on construction supply chains and first female and academic President of the Chartered Institute of Building Australasian Council. She has held various senior executive management and leadership positions.

Denny McGeorge (co-author of Chapter 9) is an Emeritus Professor of the University of Newcastle, NSW and is a leading academic in the field of construction economics, with extensive international experience in the fields of benchmarking, construction costs and productivity. He served as an industry advisor to the Gyles and Cole Royal Commissions and assisted a Senate enquiry into the Australian construction industry. He is currently the Education Leader for the Cooperative Research Centre (CRC) in Low Carbon Living.

Karen Manley (co-author of Chapter 7) is a global thought leader in the area of innovation on infrastructure projects. She is currently an Associate Professor at the School of Civil Engineering and Built Environment, Queensland University of Technology (QUT), Australia. She specialises in the application of post-neoclassical approaches to the analysis of innovation and industry growth.

Kathy Michell (co-author of Chapter 17) is an Associate Professor in the Department of Construction Economics and Management and a senior researcher in the UCT-Nedbank Urban Real Estate Research Unit at the University of Cape Town. Her research interests lie in community-based and urban facilities management, the social reality intrinsic to facilities and the relationships that exist between the facility, the users and the wider urban precinct.

Giovanni Migliaccio (co-author of Chapter 6) is Associate Professor in the Department of Construction Management at the University of Washington and holds the P. D. Koon Endowed Professorship in Construction Management. His primary areas of research interest include innovative project contracting and delivery, sustainable management of construction workforce with a focus on ergonomics and the physiological demand of construction work, and innovative construction engineering and management education. As a researcher, he has been part of over $2.1 million in research grants and has published about a hundred manuscripts and reports, including two books, over 20 refereed journal publications and book chapters, and over 40 refereed proceedings.

Edward Minchin (co-author of Chapter 6) is Professor and Director of Master's Programs at the University of Florida's Rinker School of Construction Management. His primary areas of construction research interest include supply chain integrity and project delivery methods. As a researcher, he has been part of over $3.1 million in research grants and his 100-plus published manuscripts and reports include one book, over 30 refereed journal publications and book chapters, and over 30 refereed proceedings.

Wishnu Bagoes Oka (co-author of Chapter 14) is an economist for the Government of Ontario, Canada. His research publications focus on infrastructure development and public finance. Wishnu holds a Master's Degree in Public Administration from the University of Waterloo, Canada and an undergraduate degree in Regional and City Planning from Bandung Institute of Technology, Indonesia.

Pradono (co-author of Chapter 14) is a Professor in Infrastructure Planning at Institut Teknologi Bandung, Indonesia. His research interests are mainly in economics and the management of infrastructure and transportation. He holds a Doctor of Engineering from Tokyo University, a Master of Economics of Development from Australian National University, and a Bachelor of Economics from Gadjah Mada University, Yogyakarta, Indonesia.

Leila Rostom (co-author of Chapter 17) graduated from the University of Cape Town and joined a local registered professional quantity surveying firm. She has since qualified as a Chartered Quantity Surveyor with the Royal Institution of Chartered Surveyors and is broadening her career and experience by focusing on the green building sector.

P. D. Rwelamila (author of Chapter 16) is a Professor of Project Management at the Graduate School of Business Leadership, University of South Africa, past President of the South African Council for Project and Construction Management Professions, Joint Coordinator: CIB - W107: Construction in developing countries, and past Chairperson and Non-executive Director of MSINGI Construction Project Management (Pty) Ltd, a construction project management consulting firm based in Cape Town, South Africa. He has worked in a number of countries in Africa and Europe. He has more than 200 published and peer-reviewed journal papers, chapters in books and conference proceedings.

Sandra Schmidt (co-author of Chapter 3) is Research Programme Manager at the University of Manchester. Previously, she has worked in policy development and public affairs in the public and non-governmental sectors in the UK and internationally.

Jessica Siva (co-author of Chapter 11) is an Australian Postgraduate Award PhD scholarship holder at RMIT, Australia. She previously obtained an MPhil from the University of Newcastle where she was awarded the 2007 Postgraduate Research Prize. She has tutored and supervised in the programmes of Master's of Architecture and Master's of Construction Management, has worked on various nationally funded research projects, has over 50 research publications, and held the role of Publications Officer of the Chartered Institute of Building (Australasia) while editor of their professional journal. Recently she was awarded the RMIT European Union Centre Grant for Academic Excellence.

Derek Walker (co-author of Chapter 2) is Professor of Project Management, RMIT University and editor of the *International Journal of Managing Projects in Business*. He worked in project management roles the UK, Canada and Australia before commencing an academic career in 1986. His research interests centre on collaborative project delivery.

ShouQing Wang (co-author of Chapter 12) is a Professor at the Department of Construction Management, Deputy Director at the Institute of International Engineering Project Management and Director of the PPP Lab, Hang Lung Centre for Real Estate, all at Tsinghua University, Beijing, China. He has produced over 300 PPP-related publications and is one of the three key drafting members for PPP law in China.

Graham M. Winch (co-author of Chapter 3) is Professor of Project Management and Director for Social Responsibility at Manchester Business School. He has run construction projects and researched project organising across a wide variety of engineering sectors. He is the author of *Managing Construction Projects: An Information Processing Approach*, 2nd edn (Wiley-Blackwell, 2010).

Kelwin Wong (co-author of Chapter 15) is a Post-Doctoral Fellow in the Department of Civil Engineering at the University of Hong Kong. He has served as the Research Manager on consultancy projects for the Hong Kong Government on strategies for construction industry professional resources development and establishing key performance indicators for New Engineering Contract (NEC) projects. His research interests include stakeholder and knowledge management in construction, sustainable housing development, Public-Private Partnerships and sustainability education.

Jing Xie (co-author of Chapter 12) is a Senior Vice President and registered builder of Jurong Primewide Pte Ltd, a Singapore Government-linked international design and build contractor. She has over 20 years' experience in sizeable engineering projects worldwide, including seven years in China on various metro, highway and high-speed rail projects and also working on the Taiwan High Speed Rail project, which at the time was the largest PPP project in Asia. She takes particular interest in hands-on programme management and township urban development and is currently a PhD candidate at Tsinghua University focusing on PPPs.

Weiwu Zou (co-author of Chapter 15) is a cost control manager at one of the top 10 Chinese real estate developers. He is also an experienced researcher in the areas of collaborative team working, Public-Private Partnership, green buildings and building information modelling. He earned his PhD from the University of Hong Kong, after which he worked on and managed research projects there as a Post-Doctoral Fellow.

Preface

Chapter 1 sees Jefferies and Rowlinson set the scene for the key issues discussed in the book. This chapter touches on both the aim of the book and its broader coverage. They provide some key historical context, definitions, and discuss the links between infrastructure, innovation and procurement. Some suggestions for further research are also highlighted.

In Chapter 2, Walker, Lloyd-Walker and Jefferies discuss how collaborative forms of project procurement for building and infrastructure projects have been steadily evolving towards greater levels of inter-team collaboration. Two forms that embrace collaboration between the design, construction and operation of infrastructure assets are Public-Private Partnerships (PPPs) and project alliances. However, they identify that the client's role in asset delivery collaboration varies considerably. From an asset ownership perspective PPPs and alliances may be seen as being polar opposites. However, from an inter-team relationship perspective, these forms share many similarities. The chapter examines the question. are PPPs and alliances opposites or do they lie along a continuum?

Winch and Schmidt, in Chapter 3, provide a review of the UK's Private Finance Initiative (PFI) which has been a remarkable policy experiment over the last 20 years. PFI as an innovative form of fee-based Public-Private Partnership (PPP) has transformed the provision of social and economic infrastructure in the UK. Launched in 1992, PFI, at its peak, formed up to 10 per cent of UK public sector capital investment. By 2012 its volume had significantly diminished so it is timely to review this policy initiative and assess its contribution to the UK economy and polity with a view to drawing broader lessons for governments around the world. This chapter provides such a review.

Gannon analyses the abandoned London Underground PPP in Chapter 4. Public-Private Partnerships (PPPs) were introduced into the UK by the Conservative Government in their 1992 Autumn Statement. Since then, PPPs in the UK accounted for over £57bn worth of capital expenditure. The transport sector, whilst not having signed the most PPP deals, accounted for the largest capital expenditure of all sectors; approximately £12.1bn of which urban transit PPPs accounted for approximately £7bn. The application of an innovative PPP funding policy to upgrade the London Underground was shrouded in controversy since its announcement in March 1998. After an extended transaction period that resulted in significant transaction costs, three infrastructure contracts were signed in 2002/3. During the operational period significant problems began to emerge between London Underground (LU) and the consortiums. Four years into the first review period the Metronet consortium that won two contracts went into administration; and two years later the Tube Lines consortium's contract returned to the public sector ownership and operation. In sum the PPP model, which had been expected to deliver £15bn of investment into London Underground's ailing system was abandoned. The chapter examines the lessons learned from a public sector's perspective when implementing an innovative PPP policy to upgrade LU's ailing infrastructure.

Firmenich and Jefferies reflect on project risk management in the context of life-cycle-oriented projects, such as PPPs, in Chapter 5. After stressing the relevance of the subject, the current state of practice and research are explored. A review of literature mainly considers publications from the UK and Germany and based on these findings, the project risk management for PPP projects is outlined. The chapter concludes with recommendations for further research and industry practice as it moves from a 'reactive short-term cost minimization' approach to a 'proactive long-term profit maximization' one.

In Chapter 6, Migliaccio and Minchin identify Relationship Contracting (RC) as a process to establish and regulate contractual relationships while removing barriers that would normally impede success. RC is a fluid concept that can manifest itself in many different ways in construction. Contractually, a recent study has analysed standard contracts for different project delivery systems, and has found that RC principles often permeate, at various levels, any standard construction contract. Beyond contractual language, another example of the pervasive nature of RC is seen in the way construction entities cluster at both the project and the strategic level. While being often identified as RC practices, joint venturing and alliancing actually occur on a project-by-project basis independently from the project delivery system (and contract) adopted by the client, or strategically over time across projects that are delivered with different delivery systems.

Conversely to this fluid nature, there is a widespread understanding that RC strongly relies on the establishment of alignment among all parties involved, including the client, the builder and the design professionals. Whereas achieving alignment among all parties may be more effective during the design phase, very little is known in terms of how RC may manifest itself during this phase. This chapter analyses design management practices for infrastructure projects delivered through Design and Build or Construction Management/General Contracting in the United States. Their narration focuses on how project outcomes have been successfully achieved through effective design management and they provide a critical comparison of various RC approaches.

In Chapter 7, Chen and Manley address how construction firms that employ collaborative procurement approaches develop operating routines through joint learning so as to improve infrastructure project performance. The authors report on a study based on a survey sample of 320 construction practitioners involved in collaborative infrastructure delivery projects in Australia. The study developed valid and reliable scales for measuring collaborative learning capability (CLC), and used the scales to evaluate the CLC of contractor and consultant firms within the sample. The evaluation suggests that whilst these firms explore knowledge from both internal and external sources, transform both explicit and tacit knowledge, and apply and internalise new knowledge, they can improve the extent to which these routines are applied to optimise project performance.

The purpose of Chapter 8, by Davis, is to review some of the fundamental behaviours that would underpin a successful PPP-type project. It identifies how an understanding of relationship marketing thinking in PPP delivery benefits all stakeholders and delivers their goals however diverse they may be. This chapter shows, using deliberate analysis of relationship marketing and the development of relationship contracting models, how close integration of the team in the project process may enhance cost and quality efficiencies, reduce the time to reach the operational phase of the project and manage financial risks that are inevitably embedded within the complexity of the project. The chapter draws on the author's experience in industry and extensive postdoctoral research on procurement and most particularly collaborative and integrated delivery approaches.

Chapter 9 maps current approaches to successful risk management of social infrastructure Public-Private Partnerships (PPPs) via a case study of a recently completed Australian project. Jefferies and McGeorge focus on the Top Ryde PPP project in Sydney, which is an A$800m mixed-use development of retail, commercial, residential and public services. They discuss the evolution of PPPs as an alternative procurement method to traditional methods of delivering public infrastructure and address how the competing demands for public sector investment for new infrastructure has prompted governments to turn to the private sector to form partnerships to deliver and operate these assets.

This chapter identifies five features that appear crucial for the success of PPP projects: the delivery model, costs, standardisation, relationships and risk management. A key argument for governments to procure projects using PPP is that the process would deliver better overall value for all the stakeholders, including the broader community. Social infrastructure PPPs are now part of the procurement landscape in Australia, therefore, the authors make a valid point that continued research into PPPs is vital to ensure the development of sustainable procurement methods that offer greater rewards for both public and private sector stakeholders.

Rowlinson, in Chapter 10, lays out the pre-requisites for successful procurement strategies and applies them to concept of PPPs through addressing governance issues. Having addressed these two basic concepts the author then looks at infrastructure types and classifications and the debates over privatisation. The Asian approach to PPPs is then investigated through looking at practice in four culturally and economically different countries: Hong Kong, Japan, South Korea and China. These countries are compared and contrasted in terms of the use of PPPs and the governance structures in place to enable PPPs.

Siva and London view innovation as requiring a governance environment that supports creativity in Chapter 11. Their research examines the governance context on megaprojects focusing on the client's role. An analytical model based on cultural political economy theory and the concept of governmentality is developed to re-examine megaproject governance. They present the results of a case study of what is not only Singapore's first PPP, but also the largest sports facilities infrastructure project in the world. PPP projects have multiple 'clients'; in this study the client is a network of 'actors' comprising the sports government agency responsible for setting up the project framework and the various PPP consortium stakeholders. A narrative analysis is used to present actors' stories to explain how the various instruments or types of power can be used to manipulate relationships. Through the re-telling of stories of the narrators' success with or against power an understanding of how to disrupt social structure is achieved. The critique of disruption to social structure makes clearer how instruments of power can be used in other situations to achieve better project outcomes.

In Chapter 12, Xie, Wang, Jefferies and Ke begin by reviewing the application and development of PPPs in China. Their discussion focuses on the background, evolution, government policy and administration, implementation, model examples etc. and goes on to contextualise the topic within the area of risk management. The chapter provides an update as to the progress of PPPs in one of the world's largest construction markets, which is still subject to many problems and challenges, and goes on to summarise the current drivers and barriers of PPP application while reflecting on the lessons learnt from past projects. Finally, the chapter provides a simple, but effective, comparison of PPP practice between China and the West. This is concluded by making suggestions on how to improve the process of procuring PPPs in China as the government assesses future infrastructure development.

In Chapter 13, Lee and Do discuss how the rise of the Korean economy over the last two decades has led to a significant demand for infrastructure. This growth has not been without problems however, as a great deal of public investment was committed to implement both economic and social infrastructure projects before the Asian economic crisis. As a direct result of the economic downturn, the government has been faced with downsizing budgets for infrastructure projects and had to develop initiatives to increase private sector investment in public projects. This has led to legislative change, greater financial risk sharing and the introduction of systems such as 'minimum revenue guarantee' (MRG). Therefore, private investment in public infrastructure has grown to become the common approach. MRG is an agreement for supplementing operational deficits that the private sector encounters due to problems with demand forecasting. However, after enforcing MRG, the government has encountered difficulties with redeeming revenue to private sector partners. Additional initiatives are discussed, such as 'standard cost support' (SCS), which is an agreement for a financial reward when actual operating revenue does not meet operating expenses. The authors present government strategies such as MRG and SCS within the context of the current economic climate and private investment in Korean public infrastructure by integrating supporting case study projects into their discussion.

Pradono and Oka in Chapter 14 focus on PPP frameworks in Indonesia. The implementation of these frameworks has seen economic infrastructure projects emerged as a more favourable investment than social infrastructure projects. This policy approach resulted in increased government financial responsibility, risk liability and other long-term liabilities. They identify that the development towards a strong and established PPP market, as well as a mature construction and investment sector, has helped government alleviate some of these responsibilities and liabilities. Improvements in the PPP framework processes include greater emphasis on project selection, cost consistency and scheduling. In order to achieve the successful delivery of PPPs and to continue to attract private sector investment, the government needs to ensure that sufficient mechanisms are in place to implement better policies and processes.

Chapter 15 sees Kumaraswamy, Chung, Zou and Wong make a strong case for prioritising relationship management in PPPs. Their discussion centres on longer-term RIVANS (Relationally Integrated Value Networks) in PPPs and is based on a focus on achieving best overall value for the complex network of PPP project stakeholders. This chapter draws together parallel research streams to develop a framework and tools for 'relationally integrating' PPP stakeholders by simultaneously targeting previously identified critical success factors in relationship management in PPPs and also aiming at the common cause of overall long-term network value. Specifics of the chapter include the early involvement of asset management representatives as co-creators of value based on end-user needs and facilitating knowledge exchange between both the project management and asset management teams.

In Chapter 16, Rwelamila identifies the use of Public-Private Partnerships (PPPs) as a response to societal problems that were previously held as intractable. This chapter traces historical and contemporary developments on the need for PPPs and current public protests against PPP initiatives which are perceived as impositions to what the greater public actually want. The chapter provides a detailed understanding of what constitutes the first 'P' (the public) in a PPP construct and shows how dominant approaches in PPP project arrangements have failed to embrace the real first 'P' (an inclusive approach) and have been dominated by the artificial first 'P' (an exclusive approach). The chapter takes both longitudinal and latitudinal cross-sections through the e-tolling of Gauteng's freeway in South Africa in order to understand the way in which the project was conceived, planned and implemented. Practical recommendations are made with a clear theme aimed at a paradigm shift from exclusion of the public to inclusion, where a conceptual multi-stakeholder management framework is formulated as a mechanism of bringing the public to the centre of PPP project planning, construction and operation. Finally, the chapter presents some justification for the framework and the tools used in order to reduce negative conflicts and enhance effectiveness throughout the PPP project life-cycle.

Chapter 17, which is the final chapter in this book, investigates community-based facilities management as a form of relationship contacting. The authors, Michell, Boodhun, Bunting and Rostom, see community-based facilities management as an integrated approach to the maintenance, improvement and adaptation of the land and buildings within a community in order to support the social objectives of local government. It is evident from research undertaken in South Africa that the success of the implementation of community-based facilities management is contingent on the relationship between local government and the community. This chapter utilises the findings from a case study into the role of community-based facilities management in 'relationship contracting' in the context of the townships of South Africa. It is argued that a non-performance-based contract or relationship contract is essential for the successful management, maintenance and operation of infrastructure and public facilities within marginalised communities.

1 Public-Private Partnerships and relationship-based procurement approaches

An introduction

Marcus Jefferies and Steve Rowlinson

Chapter introduction

The last three decades have seen the evolution of Public-Private Partnerships (PPPs) and Relationship Contracting (RC) as alternative procurement approaches to traditional methods of delivering public infrastructure. Governments have turned to the private sector to form partnerships in the design, construction, finance, ownership and operation of infrastructure assets and the emergence of PPPs and RC approaches, such as Alliancing, provides alternative means for developing infrastructure using private sector expertise. There is considerable growth potential for new forms of procurement given that many governments have now developed policies to expand the application of PPPs beyond economic infrastructure projects such as toll-roads to include social infrastructure projects such as healthcare and schools. A key argument for governments to procure projects using PPPs and RC is that the process would deliver better overall value for all the stakeholders, including the community and asset end-users.

In terms of coverage, this book includes issues of international historical context, collaboration, policy, risk management and emerging markets and focuses on sustainable procurement approaches through several supporting international case studies. Governments will be increasingly faced with strategic choices whether to use 'public' or 'private' mechanisms, or a combination of the two, for the provision of infrastructure. The principles embodied in RC are now established worldwide as a significant means of developing public services.

The increase in PPP-type schemes

The PPP method is a current innovation in construction procurement and has been implemented globally within the construction industry. The rise in popularity is first mentioned by McDermott (1999), who states that a significant development in construction procurement has been the rapid increase in the use of PPP arrangements. Angeles and Walker (2000) identify a growing trend for governments and other clients in the construction industry to place major projects into the private sector. This notion is further supported by Hardcastle *et al.* (2005) who identify Public-Private Partnerships (PPPs) as being increasingly used to provide public facilities and services.

PPPs are a natural progression from both the Build-Own-Operate (BOO) and the Build-Own-Operate-Transfer (BOOT) contracts (Jefferies, 2006). In making a fine distinction between PFPs (publicly financed partnerships) and PFIs (partnerships involving private financing) Jones (2003) groups operating franchises such as BOOT projects under PFIs, and long-term service agreements, such as Alliances and DCM (design, construct and maintain) projects under PFPs. Overall, using Jones's terminology, PFIs still dominate the PPP sector in many of the developed countries.

Procurement definitions

In this first chapter of the book, it is important to offer a definition for the term 'procurement'. As defined by Turner (1990), procurement is the act of obtaining, acquiring or securing. As an activity in industry it has come to mean the tendering and selection systems required to obtain anything from paper clips to power stations with organisations now having facilities departments and/or procurement managers to look at these requirements. It is now common to work with a procurement officer or project manager in order to agree on the holistic requirements for given projects.

A simple, yet effective definition of 'building procurement' is offered by Masterman (1992) as the organisational structure adopted by the client for the management of the design and construction of a building project. Lenard and Mohsini (1998) go one step further in their offering of a procurement definition by identifying it as 'a strategy to satisfy client's development and/ or operational needs with respect to the provision of constructed facilities for a discrete life-cycle'.

In progressing on from the generic definition of building- (or construction-) related procurement, it is important to define the specific procurement approaches that are the subject of investigation in this thesis, i.e. PPP and Relationship Contracting. One of the problems with PPP is its very definition. Definitions tend to depend on a commentator's own particular perspective and range from the very general to the quite particular. A general and well cited definition is provided by Akintoye et al. (2003): 'Public Private Partnerships (PPPs) are defined as a long-term contractual arrangement between a public sector agency and a private sector concern whereby resources and risk are shared for the purpose of developing a public facility.'

PPPs are a means of public sector procurement using a mix of private sector finance and best practice. PPPs can involve design, construction, financing, operation and maintenance of public infrastructure and facilities, or the operation of services, to meet public needs. They are often privately financed and operated on the basis of revenues received for the delivery of the facility and/or services. Key to this is the ability of the private sector to provide more favourable long-term financing options than may be available to a government entity and to secure financing in a much quicker time frame (NSW Government, 2012). Such contracts are long-term in nature and typically 25–30 years.

Relationship Contracting is designed to break down the contractual and commercial walls between owners, contractors, designers and suppliers so that a trusting team is formed which shares the risks when something goes wrong and shares the savings when the team performs exceptionally well. Costs are expected to be reduced and outstanding results in key areas can be achieved.

Cheung *et al.* (2005) provide a definition of relationship contracting:

> Relationship contracting is based on a recognition of and striving for mutual benefits and win-win scenarios through more cooperative relationships between the parties. Relationship contracting embraces and underpins various approaches, such as Partnering, Alliancing, joint venturing, and other collaborative working arrangements and better risk sharing mechanisms. Relationship contracts are usually long-term, develop and change over time, and involve substantial relations between the parties and development of trust.

Successful relationship management requires trust, commitment, cooperation, open communication, goal alignment and joint problem solving (Howarth *et al.*, 1995; Hampson and Kwok, 1997; Rowlinson and Cheung, 2004). Trust between alliance partners creates an opportunity and willingness for further alignment (such as future job opportunities), reduces the need for continuous cross monitoring of one's behaviour, reduces the need for formal controls, and reduces the tensions created by short-term inequities. It allows the partners to focus on their long-term business development as well as cutting down cost and time outlays (Rowlinson *et al.*, 2006).

PPPs and Relationship Contracting methods, such as Alliancing, are current examples of procurement innovation involving both government and private industry. The private sector is playing an increasingly important role in this trend that has partly arisen out of a necessity for the development of infrastructure to be undertaken at a rate that maintains growth, reduces public spending and allows for successful risk management. This in turn has become a major challenge for many countries, and particularly so where it is evident that these provisions cannot be met by government alone, as they have typically been in the past.

Adopting the PPP and RC approach

With the PPP family of procurement options an alliance or joint venture group forms to provide a facility for a client for which the client makes a concession agreement to fund the facility until that facility's ownership is transferred to the client. In the past this arrangement has been more common for infrastructure projects than for buildings because the concession allows for tolls or other payments to be made by end-users to cover the cost of both procuring the facility and its operation (Walker *et al.*, 2000). However, procurement approaches, such as PPPs or Alliances, have been used more recently for building projects, such as the National Museum in Canberra (Walker and Hampson, 2003), and various social infrastructure projects such as hospitals, schools and prisons (Duffield, 2005; Jefferies and McGeorge, 2009).

Many countries have now embarked on infrastructure projects procured via PPP or the use of similar methods. The scheme is now widely practised and is spread among a diverse range of countries from Australia, Canada, Hong Kong, the UK and the United States to countries like India, Malaysia, Mexico, Thailand and the Philippines (Walker and Smith, 1995; Walker and Hampson, 2003; Grimsey and Lewis, 2004; Bult-Spiering and Dewulf, 2006; Jefferies and McGeorge, 2009; Jefferies, 2014). Most of these projects are financed on a limited recourse basis and built and operated as private ventures under project agreements involving the host government.

It became evident several decades ago that governments globally had major shortcomings in funding public works. The fundamental influences from these issues are what have developed the trends towards greater private sector involvement and more specifically infrastructure procurement strategies such as PPP and Alliance Contracting.

Infrastructure development challenge

There are connections between a country's economy and successful infrastructure development. For instance, according to the World Bank (1994), infrastructure can support economic growth and make development environmentally sustainable. A growing number of countries are now demanding alternatives, especially options involving the private sector (World Bank, 2005). The last three decades have seen the evolution of PPPs as an alternative procurement method to traditional methods of delivering public infrastructure. Competing demands for public sector investment for new infrastructure has prompted governments to turn increasingly to the private sector to form partnerships for infrastructure delivery. This has become a major challenge for both public and private sector stakeholders but the emergence of PPPs and RC provides an alternate means for developing infrastructure using private sector expertise.

This trend in public-private infrastructure partnerships is supported by McDermott (1999), Curnow *et al.* (2005), and Jefferies and McGeorge (2009), who see governments as having to obtain private sector finance for infrastructure projects which have previously been a drain on public sector finances. Therefore, the importance of infrastructure as a trigger for both economic and urban growth cannot be underestimated. Infrastructure can support economic growth, reduce poverty and make development environmentally sustainable. According to the World Bank (1995) there are current problems, with inefficient public sector monopolies widely blamed for the ineffective provision of infrastructure services, and a growing number of countries are now demanding alternatives, especially options involving the private sector.

Private sector participation in the procurement of public works

Private sector input takes many forms. The simplest is a service or management contract by which private contractors assume responsibility for operations and maintenance. Contracts typically are of short duration but offer the advantage of higher efficiency. The private partner can assume a longer term and greater responsibility through lease contracts, in which systems are leased and operated by private firms. The responsibility for investment and financing remains with the public sector in this case. In taking private sector participation a step further, fuller efficiency gains are possible through long-term concessions and 'PPP-type' schemes. Here the private partner not only finances new works but constructs and operates them for indefinite or specified periods of time (World Bank, 1995).

The World Bank has placed full support behind increasing private sector involvement in a country's infrastructure provision. They have made clear their intentions that with projects the World Bank plans to finance, the package shall be designed from the outset in such a way that will facilitate private sector participation whenever the government chooses (World Bank, 2012). Underlying these issues is the fact that procurement logic must change. According to Miller (1999), procurement strategy should recognise explicitly what generations of experience has already taught, that innovations enter the infrastructure portfolio through each of the individual segments in the procurement process (design, construction, finance, operation, maintenance) and through combinations of these segments (D&B [Design and Build], DBO [Design Build Operate], BOT [Build Operate Transfer], PPP, Alliancing, Partnering).

PPPs appear to be the vehicle for continued finance and construction of infrastructure. Measuring the success of this type of project to date and identifying the various risks involved in order to achieve optimum success is still somewhat uncharted. McDermott (1999), Rowlinson (1999), Salzman and Mohamed (1999a and 1999b), Jefferies and Gameson (2002), McGeorge *et al.* (2006), Jefferies (2006), Walker and Rowlinson (2008) and Love *et al.* (2011) all state that the key to successful PPP infrastructure projects is the successful identification, evaluation and allocation of the risks involved.

History of PPP projects

In order to fully understand the concept, it is important to have a very clear grasp of the history and development of public-private sector joint ventures such as PPP. This form of public-private partnership is clearly not a recent form of infrastructure procurement. History records that the Industrial Revolution began when Abraham Darby first smelted coke in 1709. Urbanisation and the need for associated infrastructure were to follow. Governments of the time had only rudimentary tax arrangements primarily to service heads of state. Infrastructure was therefore left to individuals to finance and build. The canals and railroads of Europe and later those in the Americas, China and Japan were also procured this way (Smith, 1999).

The wonder of the age, the 195 km Suez Canal opened for navigation on 17 November 1869 (Smith, 1999). The Suez Canal Company was empowered by the Egyptian government to build and operate the canal. Finance was provided by European capital with Egyptian financial support, and a concession to design, construct and operate this revenue-producing facility was expected (Levy, 1996). The project agreement for the procurement of the canal was based around a 99-year concession contract (Walker and Smith, 1995).

Industrial countries generally funded new infrastructure between the late nineteenth century and the 1970s from their respective fiscal resources. However, a series of influences emerged in the late 1970s which placed pressure on this established system for both developed and developing countries. The infrastructures of 'developed' countries such as those of Western Europe, North America, Japan and Australia experience strain from two principal influences. First, the existing and limited infrastructure's ability to keep pace with the growth of the country and second, the demand for health and welfare facilities due to an ageing population. The problems and challenges for 'newly industrialised countries' such as Malaysia, Hong Kong, Taiwan, Mexico and South Africa are caused by a population explosion placing heavy demand on an already limited infrastructure (Walker and Smith, 1995).

Some of the key historical developments in pubic-private sector joint venture infrastructure projects are highlighted below (adapted from Jefferies, 2014):

- Industrial Revolution (1709)
- Water Distribution – Perrier Bros, Paris (1782)
- Suez Canal (1869)
- Manchester Ship Canal (1894)
- Cross Harbour Tunnel, Hong Kong (1972)
- Toronto Airport (1980)
- Dartford Bridge, UK (1991)
- Sydney Harbour Tunnel (1992)
- Western Harbour Crossing, Hong Kong (1997)
- Sydney Olympic Stadium (1999)
- London Underground (2003)
- Ministry of Defence Building, London (2003)
- Urban Water Supply, China (2006)
- National Highways of India (2000–7)

- Southern Cross Railway Station, Melbourne (2006)
- Royal Women's Hospital, Melbourne (2008)
- Hong Kong Hospital Authority (2010)
- US Highways Project (2012)
- Singapore Sports Hub (2014)
- Light Rail Project, Shenzen, China (2015)
- Beijing Expressway (2015)
- Dumfries-Galloway Hospital, Scotland (2015).

The Suez Canal experience demonstrated that the concept of private sector participation in infrastructure provision is not a new idea. It is, however, only in the last two to three decades that PPP-type concepts have become high on many government agendas.

Infrastructure and innovation

There is an obvious link between innovation and procurement, hence the need for the continuous development of new models in order to deliver and maintain infrastructure growth in times of limited finance. This is supported by Katsanis *et al.* (1997), who identify procurement as the most important determinant of the outcome of the process. This is especially relevant as project complexity and market dynamics have made it necessary for owners to seek alternative methods, including the PPP approach.

The construction industry is now learning from other industries and indeed from the history of early infrastructure development in forming on-going partnerships between the public and private sectors. According to Miller (1997) procurement systems are the conduit through which innovative methods, techniques and technologies for infrastructure facilities have been introduced. He examines how procurement strategies for both project delivery and finance affect the nature and source of innovation. This is tested within the hypothesis that choice of project delivery and financial method has a profound impact on the existence and source of innovation in infrastructure systems by dramatically changing the nature and sequence of the elements associated with design, finance, construction, operation and maintenance of the project. The case studies (West Harbour Crossing in Hong Kong, Highway 407 in Toronto and US Grants project for Wastewater Treatment Plants) indicate that the relationship between the procurement method and innovation is strong. The cases indicate that early procurement planning plays a critical role in the way in which the government gains access to innovative ideas, methods and conduct in the Engineering Procurement Construction (EPC) sector, both public and private.

In the PPP procurement process, innovation is sought in fundamentally different ways and through different channels. Competition occurs later in the procurement process and covers a significantly enlarged scope. Tenders are required to compete for combined functions of design, finance, construction, maintenance and operations, a competition that expressly values innovation in each of these areas and in the integration of one or more of these areas. The opportunity to package basic project elements differently is a driving factor for innovation in the PPP approach (Miller, 1997, 1999).

If innovation is to produce fundamental improvements in quality, cost and time of the delivery of public infrastructure projects, then the procurement method will need to be structured to produce competition that focuses on these improvements. A procurement system which openly embraces innovation sends a powerful signal to the private sector that government recognises that continued public-private cooperation is essential to sustainable self-adjusting improvements in infrastructure facilities and services (Miller, 1997, 1999). According to Angeles and Walker (2000), one way of meeting future procurement demands will be to develop better project delivery systems. A growing view is that new and innovative systems will continuously be used. These include Design and Build schemes, extending their scope to Design-Build-Finance-Operate, which will be similar in many ways to PPP schemes.

Encouraging and rewarding innovation in delivery and long-term operation of infrastructure facilities and services is becoming a major goal of modern procurement systems, particularly as public budgets prove to be inadequate to meet current infrastructure needs (Jefferies, 2006; Jefferies and McGeorge, 2009).

Chapter summary

The future of construction over the next 20–30 years will be influenced by the restructuring and re-engineering of the various processes in the construction industry (Angeles and Walker, 2000; Jefferies, 2006; Jefferies, 2014). This will be aimed at achieving improved quality and performances through the integration of project processes and management of supply chains. The move away from traditional procurement systems will have significant effects on innovation. Because the traditional process, such as design-bid-build, does not allow for intellectual property and knowledge externalities by contractors in their tenders, there is a perverse disincentive to innovate (de Valence, 2001). With the increase of non-traditional procurement methods, such as PPP, this disincentive is removed and firms can appropriate the benefits of innovation and research and development. This situation will continue to lead to the use of new forms of procurement of major infrastructure projects such as relationship-based procurement schemes (Rowlinson *et al.*, 2006; McGeorge *et al.*, 2008; Jefferies and McGeorge, 2009; Brewer *et al.*, 2013).

Bridge *et al.* (2015) outline several potential research directions for PPPs and RC; for instance, given the dearth of empirical studies there appears to be much value in developing a wide-ranging global study and deep database that explore the many underlying positions that exist within all variations of this type of procurement method. More specifically, research should seek to enhance the efficiency of the conventional PPP model by building on its expected strengths in terms of containing risk and seeking to address its expected weaknesses by developing a more conducive environment to innovation. They go on to highlight that the most obvious avenue for investigation in the alliance-type models concerns establishing the right balance of incentives and seeking to optimise the drivers of this balance.

Reflections

1 Why have we seen the global emergence of procurement methods, such as PPP, Partnering and Alliancing for the delivery of infrastructure?
2 What are the current procurement trends in infrastructure provision?
3 How have the likes of PPPs evolved in your country?
4 Walker and Rowlinson, in their 2008 book entitled *Procurement Systems: A Cross Industry Project Management Perspective*, published by Taylor & Francis, not only provides an historical overview of procurement development but also includes practical discussion on unravelling the complexities of various procurement methods. The book provides an excellent grounding in both project management and construction procurement.
5 What do you see as the key aspects in the future delivery of public infrastructure and the procurement methods used to deliver this type of project?

References

Akintoye, A., Beck, M. and Hardcastle, C. (2003). *Public-Private Partnerships: Managing Risks and Opportunities*. Oxford: Blackwell Science.

Angeles, N.H. and Walker D. (2000). PPP Schemes: A Project Delivery System for a New Millennium. *Chartered Building Professional*, March: 21–3.

Brewer, G.J., Gajendran, T., Jefferies, M., McGeorge, D., Rowlinson, S. and Dainty, A. (2013). *Value through Innovation in Long-term Service Delivery: Facility Management in an Australian PPP. Built Environment Project and Asset Management*, 3(1): 74–88.

Bridge, A.J., Soh, Y.I. and Rowlinson, S. (2015) Models for Engaging PPPs in Civil Infrastructure Projects: A Case of 'Having Your Cake and Eating It Too'?. *Proceedings of the Engineering Project Organization Conference (EPOC)*, Edinburgh, 24–26 June.

Bult-Spiering, M. and Dewulf, G. (2006). *Strategic Issues in Public-Private Partnerships: An International Perspective*. Oxford: Blackwell Publishing.

Cheung, F.Y.K., Rowlinson, S., Jefferies, M. and Lau, E. (2005). Relationship Contracting in Australia. *Journal of Construction Procurement*, 11(2): 123–35.

Curnow, W., Jefferies, M.C. and Chen, S.E. (2005). Unsustainable Biddings Costs: A Critical Issue for Public Private Partnerships. In (ed.) Ng, Thomas S. *Public Private Partnerships: Opportunities and Challenges*. Hong Kong: Centre for Infrastructure and Construction Industry Development, University of Hong Kong, pp.35–43.

de Valence, G. (2001). Trends in Procurement and Implications for Innovation and Competitiveness of Australian Building and Constructions. In (ed.) Duncan, J. *Performance in Product and Practice, Proceedings of the CIB World Building Congress*, Wellington, New Zealand, 2–6 April.

Duffield, C.F. (2005). PPPs in Australia. In (ed.) Ng, Thomas S. *Public Private Partnerships: Opportunities and Challenges*. Hong Kong: Centre for Infrastructure and Construction Industry Development, University of Hong Kong, pp.5–14.

Grimsey, D. and M.K. Lewis (2004). *Public Private Partnerships: The Worldwide Revolution in Infrastructure Provision and Project Finance*. Cheltenham, UK, Edward Elgar Publishing.

Hampson, K. and Kwok, T. (1997). Strategic Alliances in Building Construction: A Tender Evaluation Tool for the Public Sector. *Journal of Construction Procurement*, 3(1): 28–41.

Hardcastle, C., Edwards, P.J., Akintoye, A. and Li, B. (2005). Critical Success Fcators for PPP/PFI Projects in the UK Construction Industry: A Factor Analysis Approach. In (ed.) Ng, Thomas S. *Public Private Partnerships: Opportunities and Challenges*. Hong Kong: Centre for Infrastructure and Construction Industry Development, University of Hong Kong, pp.75–83.

Howarth, C.S., Gillin, M. and Bailey, J. (1995). *Strategic Alliances: Resource-sharing Strategies for Smart Companies*. Australia: Pearson Professional (Australia).

Jefferies, M.C. (2006). Critical Success Factors of Public Private Sector Partnerships: A Case Study of the Sydney SuperDome. *Engineering, Construction and Architectural Management*, 13(5): 451–62.

Jefferies, M.C. (2014). *An Analysis of Risk Management in Social Infrastructure Public-Private Partnerships (PPPs)*. Unpublished PhD Thesis, University of Newcastle, Australia.

Jefferies, M.C. and Gameson, R. (2002). Stadium Australia: Reflecting on the Risk Factors of PPP Procurement. In (ed.) Lewis, T.M. *Procurement Systems and Technology Transfer, CIB W92 Procurement Systems Symposium*, The University of The West Indies, Trinidad and Tobago, 14–17 January, pp.355–68.

Jefferies, M. and McGeorge, W.D. (2009). Using Public-Private Partnerships (PPPs) to Procure Social Infrastructure in Australia. *Engineering, Construction and Architectural Management*, 16(5): 415–37.

Jones, D. (2003). Evaluating What Is New in the PPP Pipeline. *Building and Construction Law*, 19: 250–70.

Katsanis, C.J., Cahill, D.J. and Davidson, C.H. (1997). Networks and Learning Organisation. In (eds) Davidson, C.H. and Abdel Meguid, T.A. *Procurement – A Key to Innovation, Proceedings of CIB W92 International Symposium*, Montreal, pp.323–332.

Lenard, D. and Mohsini, R. (1998). Recommendations from the Organisational Workshop. In (ed.) Davidson, C.H. *Proceedings of CIB W92 Montreal Conference, Procurement – The Way Forward*, Montreal, 20–23 May 1997, pp.79–81.

Levy, S.M. (1996). *Build, Operate, Transfer: Paving the Way for Tomorrow's Infrastructure*. New York, NY: John Wiley & Sons.

Love, P.E.D., Davis, P.R., Chevis, R. and Edwards, D.J. (2011). Risk/Reward Compensation Model for Civil Engineering Alliance Projects. *ASCE Journal of Construction Engineering and Management*, 137(2): 127–36.

McDermott, P. (1999). Strategic and Emergent Issues in Construction Procurement. In (eds) Rowlinson, S. and McDermott, P. *Procurement Systems: A Guide to Best Practice*. London: E. & F.N. Spon, pp.3–26.

McGeorge, D., Cadman, K., Jefferies, M. and Chen S.E. (2006). Public Private Partnerships: Private Sector Developers Partnering with Government for the Provision of Social Infrastructure. *Governments and Communities in Partnership*, Centre for Public Policy, University of Melbourne, 25–27 September.

McGeorge, D., Jefferies, M., Cadman, K. and Chen, S.E. (2008). Implications for Design and Build Contractors Bidding in Public Private Partnership Consortiums: An Australian Perspective. In (ed.) Hughes, W.P. *Construction Management and Economics: Past, Present and Future, Proceedings of the CME 25 Conference*, University of Reading, pp.109–24.

Masterman, J.W.E. (1992). *An Introduction to Building Procurement Systems*. London: E. & F.N. Spon.

Miller, J.B. (1997). Procurement Strategies Which Encourage Innovation: The Fundamental Element of Sustainable Public Infrastructure Systems. In (eds) Davidson, C.H. and Abdel Meguid, T.A. *Procurement – A Key to Innovation, Proceedings of CIB W92 International Symposium*, Montreal, pp.443–52.

Miller, J.B. (1999). Applying Multiple Project Procurement Methods to a Portfolio of Infrastructure Projects. In (eds) Rowlinson, S. and McDermott, P. *Procurement Systems: A Guide to Best Practice*. London: E. & F.N. Spon, pp.209–27.

New South Wales Government (2012). *Public Private Partnership Guidelines*. Sydney: NSW Treasury.

Rowlinson, S. (1999). Selection Criteria. In (eds) Rowlinson, S. and McDermott, P. *Procurement Systems: A Guide to Best Practice*. London: E. & F.N. Spon, pp.276–99.

Rowlinson, S. and Cheung, F.Y.K. (2004). A Review of the Concepts and Definitions of the Various Forms of Relational Contracting. In (eds) Kalidindi, S.N. and Varghese, K. *Proceedings of the International Symposium of CIB W92 on Procurement Systems*, Chennai, 7–12 January, pp.227–36.

Rowlinson, S., Cheung, F.Y.K., Simons, R. and Rafferty, A. (2006). Alliancing in Australia: No Litigation Contracts; A Tautology? *ASCE Journal of Professional Issues in Engineering Education and Practice: Special Issue on 'Legal Aspects of Relational Contracting'*, 132(1): 77–81.

Salzmann, A. and Mohamed, S. (1999a). Risk Identification Frameworks for International PPP Projects. In (ed.) Ogunlana, S. *Profitable Partnering in Construction Procurement*, CIB W92 Proceedings Publication 224, pp.475–85.

Salzmann, A. and Mohamed, S. (1999b). Risk Identification and Interaction in International PPP Projects. In (ed.) Williams, P.H.M. *Australian Institute of Building Papers*, vol. 9, pp.101–14.

Smith, A.J. (1999). *Privatized Infrastructure: The Role of Government*. London: Thomas Telford.

Turner, A. (1990). *Building Procurement*. Basingstoke: Macmillan.

Walker, C. and Smith, A.J. (1995). *Privatized Infrastructure: The BOT Approach*. London: Thomas Telford.

Walker, D.H.T. and Hampson, K. (2003). *Procurement Strategies: A Relationship-based Approach*. Oxford: Blackwell Science.

Walker, D.H.T. and Rowlinson, S. (eds) (2008). *Procurement Systems: A Cross Industry Project Management Perspective*. London: Taylor & Francis.

Walker, D.H.T., Hampson, K. and Peters, R. (2000). *Relationship Based Procurement Strategies for the 21st Century*. Canberra: AusInfo.

World Bank (1994). *World Development Report: Infrastructure for Development*. New York, NY: The World Bank; Oxford University Press.

World Bank (1995). *Meeting the Infrastructure Challenge in Latin America and the Caribbean*. Washington, DC: The World Bank.

World Bank (2005). *Meeting Infrastructure Challenges*. Washington, DC: The World Bank.

World Bank (2012). *Private Participation in Infrastructure*. Washington, DC: The World Bank. http://ppi.worldbank.org, viewed 1 July 2012.

2 Public-Private Partnerships and alliances

Opposites or a continuum?

Derek Walker, Beverley Lloyd-Walker and Marcus Jefferies

Chapter introduction

The value of this chapter is that it compliments many of the others in this book, particularly Chapter 1 (Jefferies and Rowlinson), Chapter 5 (Firmenich and Jefferies), Chapter 7 (Chen and Manley) and Chapter 9 (Jefferies and McGeorge), in its discussion of collaboration. This chapter also builds on Chapters 1, 5, 7 and 9 as it adds an alliancing perspective to the view of collaboration in PPPs.

There has been a great deal of research undertaken into PPPs in terms of achieving value for money (Akintoye *et al.*, 2003; Raisbeck *et al.*, 2010) or questioning if or how value for money is being achieved (Hodge, 2004; Grimsey and Lewis, 2005) and also from a perspective of how PPPs' governance and routines operate in practice (Williams, 2010; Wilson *et al.*, 2010; Jooste *et al.*, 2011; Regan *et al.*, 2013).

A growing body of literature on project and programme alliances has recently been focused on explaining what alliancing is (Hutchinson and Gallagher, 2003; Ross, 2003; Walker and Hampson, 2003a; Hauck *et al.*, 2004; Department of Infrastructure and Transport, 2011) and how alliance risk and reward routines operate (Lahdenperä, 2009, 2010; Love *et al.*, 2011). Others have explored how alliances perform (Wood and Duffield, 2009; Walker *et al.*, 2013) and behavioural aspects of their operation such as the importance of trust and commitment (Cheung *et al.*, 2005; Lau and Rowlinson, 2009) and the no-blame culture (Rowlinson *et al.*, 2006; Walker *et al.*, 2014) they invoke. An emerging interest has also developed in aspects of the workplace culture in alliances (Rowlinson *et al.*, 2008; Reed and Loosemore, 2012; Walker and Lloyd-Walker, 2015) including the feeling or ambience encountered in alliances (Walker and Lloyd-Walker, 2014a). First it may be best to clarify what we mean by a project alliance.

An alliance is a more structured and formal relationship-based procurement (RBP) arrangement than, for example, partnering or even joint ventures. The Department of Finance and Treasury Victoria (2010, p.9) describes project alliancing as, a method of procuring . . . [where] All parties are required to work together in good faith, acting with integrity and making best-for-project decisions. Working as an integrated, collaborative team, they make unanimous decisions on all key project delivery issues. Alliance agreements are premised on joint management of risk for project delivery. All parties jointly manage that risk within the terms of an 'alliance agreement', and share the outcomes of the project.

Ross (2003, 2008) provides us with one of the most comprehensive available descriptions of how three key contractual 'limbs' to an alliance may contribute to its success. Limb 1 specifies the commercial arrangements for reimbursement of project specific construction costs and overheads. These form the basis of the development between the design team, contractor and project owner parties to the alliance of a validated and rigorous target outturn cost (TOC) that is independently validated. Limb 2 is the fee arrangement that covers overhead and what would otherwise be considered profit. This is negotiated between the project owner and each of the alliance non-owner participants to reflect a reasonable profit level based on the complexity of the project. Limb 3 provides the incentivised arrangements that link performance with a risk and reward agreement so that parties are engaged in a joint enterprise in which performance of the whole project determines the risk and reward quantum. Additionally, an alliance agreement has *specific* terms and conditions that identify participant behaviours and shared values that will support collaboration.

What becomes clear, from the examples within the cited literature above of interesting perspective on the topic, is that PPPs and alliances share important characteristics that can be explored, compared and contrasted. Relationship-based aspects of PPPs have not been widely discussed in the literature, although Clifton and Duffield (2006) argue, based on a survey of 15 public sector experts and auditor general reports, that alliance principles could be applied to PPP projects to enhance value and deal better with risk. Walker and Jacobsson (2014) reported on a case study of a multi-billion PPP project completed in 2012 with a AUD$600 million alliance component undertaken between the PPP special purpose vehicle (SPV) construction parties and a major services subcontractor. An extensive search of the literature has not revealed any other documented case study of an alliance within a PPP. Walker and Jacobsson (2014) report that the alliance was successful and that the seven senior executive-level interviewees for their study indicated that the working relationships, project delivery performance and collaboration were far more productive and fruitful in outcomes than would otherwise have been the case. The executives interviewed stated that the choice of an alliance within the PPP was considered a major factor in the consortium winning the PPP bid. This study presented useful insights, especially as generally alliances and PPP projects have been considered 'role opposites'. Clifton and Duffield (2006) and others who have written extensively about PPPs have stressed the risk allocation model as a defining feature that differentiates PPPs from other more relational approaches such as alliances.

The PPP approach generally ensures that risk is passed from the client, for example a government agency (road, rail, water or social infrastructure entities), to the PPP SPV. A main objective of a PPP is that governments can pass on *all* risk whereas when they participate in an alliance the government entity fully enters into a risk *sharing* model. Here the government takes on risks that it can best manage and other alliance parties accept and manage those risks that they can best manage. This is what Walker and Jacobsson (2014) refer to as the project owner (government agency) taking an alliance 'hands-on' approach that can be compared to their 'hands-off' approach with a PPP. They viewed the PPP SPV as 'the client' in their analysis of the case study they report upon, because that entity took over ownership of the project once the consortium was appointed. Two joint venture contractor SPV parties delivered the construction works under a design and construct (D&C) contract. Tunnel fit out work was delivered as an alliance between the two contractors and a services sub-contractor under an alliance agreement. The governance arrangement then followed the same path as that of many other Australian and New Zealand infrastructure project alliances. This has led to a growing knowledge base of how alliances can and should operate. This is helpful as it allows us to feel confident not only that an alliance and PPP are not necessarily opposite ends of a continuum but that they can be successfully combined.

Walker and Lloyd-Walker (2014b) developed a RBP Taxonomy that can be used to help us visualise elements of any form of project delivery and is particularly useful for both comparing and contrasting project procurement forms.

The RBP Taxonomy

During 2012–13 Walker and Lloyd-Walker conducted a study involving a desk-top review of a wide range of literature about RBP (relationship-based procurement). Based on the analysis of over 600 pages of transcripts from interviews with RBP subject matter experts (SMEs), 34 senior managers engaged in alliances and 14 academic experts in this area, the authors developed the RBP Taxonomy (Walker and Lloyd-Walker, 2014c; Walker and Lloyd-Walker, 2015) illustrated in Figure 2.1. Academic SMEs were able to discuss not only research results that they had published but also a wealth of insights that had not been published because of editorial word count limitations, referee comments and constraints on time to write up 'everything' learned through their studies. Similarly, practitioner SMEs were able to discuss what they directly knew about by giving consideration to their experience and also by speculating and reflecting on conversations they had been engaged in with their colleagues and peers. These interviews were mainly undertaken in Australia but they also included academics and practitioners from the United States, the UK, Sweden, Finland, the Netherlands, France and Hong Kong. This provided a strong global perspective.

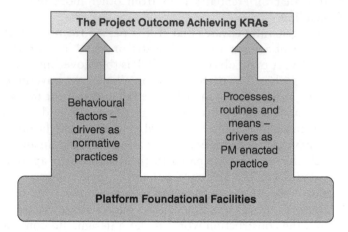

Figure 2.1 RBP Taxonomy: three main elements

The taxonomy model emerged from sensemaking based on analysis of the transcript data. Two principal theoretical concepts guided its development. The first was Nyström's (2005) application of Wittgenstein's family resemblance theory to categorise elements of project partnering. His taxonomy was developed using a similar approach to categorising family resemblance based on inherited features. This approach was also used by Yeung *et al.* (2007) to classify alliancing. More is now known about how alliancing operates since Yeung *et al.* (2007) undertook their research. The second principal theoretical concept to be applied to analysis of the data by Walker and Lloyd-Walker was drawn from a study by Jacobsson and Roth (2014) on the partnering approach used in Sweden. The report on their research conceptualised partnering as requiring a series of necessary building blocks for collaboration. They argued that 'Dialogue between parties involves a mutual interest to learn from and communicate with each other, and maintained that this involves more than merely listening to the customer, or in a partnering setting, the client' (p.423). They offered a perspective based on the nature of a co-created environment of partnering in terms of foundations, means and factors.

Figure 2.1 illustrates three elements of the RBP Taxonomy. Its logic assumes that a functioning alliance is severely limited without soundly functioning foundational facilities. Any project collaboration (whether delivered through a PPP, alliance or even by traditional design and construct means) requires certain behavioural features. These behavioural features may be supported or inhibited by the nature of the foundational facilities that promote and support a context for collaboration and in doing so ensure that the logic is sound. This element promotes and induces a tangible and intangible set of behaviours that drive social norms towards manifestly wanting to be, and succeeding in being, collaborative. These behaviours then need to be internalised through organisational routines and means that legitimise and entrench collaboration as the preferred way of working. Thus the logic of the RBP Taxonomy is based on identifying sub-elements from the three main elements and measures for these sub-elements so that we can compare and contrast various project delivery forms in terms of their levels of collaboration prescribed by the taxonomy sub-elements.

The sub-elements are now briefly illustrated and described (see Figures 2.2–2.4).

Element 1 comprises five sub-elements. Each sub-element has a course grained measurement indicator ranging from low to high. Assessment of the characteristics of each element for the project provides the means to rate the sub-element, for example using the five levels from low to high.

Sub-element 1 is about the client's – i.e. the project owner's – motivation and context that define the project as one that encourages collaboration. This in turn defines a substrate of circumstances that affects the potential degree of possible collaboration. Figure 2.2 illustrates the components of this dimension or sub-element. For example the main driver determining the environment for collaboration and therefore its resultant level could be attributable to a priority for best value. This could be predominantly focused on monetary value or on wider aspects that could be considered by participating parties as generating value. The main driver may be urgency to recover from an immediate disaster or emergency or it could be related to a vanguard-type project with a purposeful exploration of options and

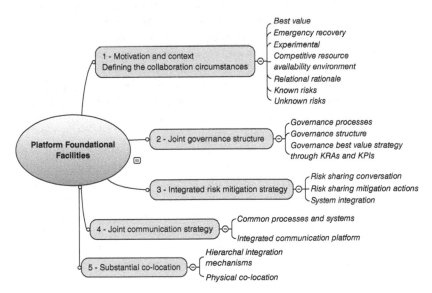

Figure 2.2 RBP Taxonomy: platform foundational facilities – Element 1

for which the ability to learn and reflect upon experience is desired. Other foci that could shape the level of motivation and mechanism to collaborate could be driven by external economic-cycle conditions that impose resource availability conditions. Similarly the driver may be a need to create relationships or to deal with known risks in complicated projects, or to deal with a realisation that the project presents many likely unknown conditions that make the project complex or chaotic to manage. The purpose of this perspective is to take an overall view in order to assess the extent to which the project context shapes the degree of collaboration needed and desirable for that context.

Sub-element 2 relates to the nature of the project governance structure. The aim is to have a unified way that each project delivery team party legitimises its actions through rules, standards and norms, values and coordination mechanisms. Organisational routines and the way that committees, liaison and hierarchy take a unified or complimentary approach to interacting impacts the quality of explicit understanding of how teams should collaborate and communicate. Components of this sub-element that influence a judgement on the nature of the governance structure would relate to how common assumptions and ways of working influence project governance processes and rules, how the level of flexibility/ rigidity, power and influence and communication symmetry directly influence the workplace culture and the strategy deployed to define, measure and assess success through key results areas (KRAs) and key performance indicators (KPIs).

Sub-element 3 relates to the extent to which an integrated risk mitigation strategy is organised for all parties as part of the client's proactive risk management system. The quality of explicit understanding of how to manage risk and uncertainty collaboratively and how potentially to gain advantage from a project-wide insurance policy is shaped by the conversation about risk sharing; who takes responsibility for any class of, or particular, risk. It is also shaped by the nature of the ways of managing risk, ranging from collective to individual, that are agreed upon to decide how to mitigate those risks and how the project will be structured and managed to provide a platform that is based upon the participant's philosophical stance about relationships between teams.

Sub-element 4 relates to the level of joint communication strategy platforms such as integrated processes and ICT groupware including building information modelling (BIM) and other electronic forms of communication. Two aspects of this sub-element were defined as shaping the assessment of the sub-element. The first is the extent to which project participants were assessed as needing to share a common way of working, and a common language and communication approach. This is to avoid misunderstanding that can undermine trust and commitment and consequently effective decision making and action. The second is evidence of a common ICT platform, including, for example, common BIM tools that can minimise the risk of poor coordination and communication and misunderstandings between participants.

Sub-element 5 pertains to the extent to which project teams are substantially co-located within easy physical reach of each other. Two aspects of this were identified. The first is the extent of evidence that is present of leaders inspiring motivation towards unity of purpose by physically interacting with individuals at the various levels of an organisation through site visits, meetings held on site and other ritualistic or practical events that are held in the workplace. These can be very important as a platform for integrated joint action. The second is the degree and quality of co-location being present that establishes a well-considered and conducive environment that facilitates positive interaction.

The behavioural factors element comprises five sub-elements. These include Sub-element 6: the degree of authentic leadership that is demonstrated by the project leadership team and team members. In general authentic leadership is characterised by reflectiveness, pragmatism, appreciativeness, resilience, wisdom, spirit to challenge the status quo, and authenticity (aligning action with rhetoric). A more detailed description of authentic leadership characteristics as adopted for the RBP Taxonomy can be found in several publications (Lloyd-Walker and Walker, 2011; Walker and Lloyd-Walker, 2015).

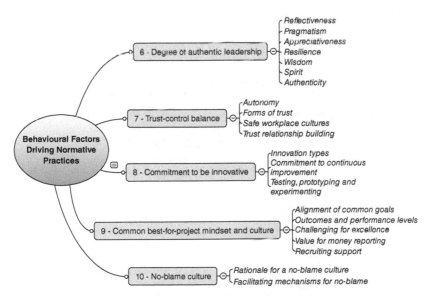

Figure 2.3 RBP Taxonomy: behavioural factors driving normative practices – Element 2

Sub-element 7 relates to the balance that has been struck between trust and control. This is described in terms of the autonomy of parties, the forms of trust exhibited, the workplace environment style and the nature of behaviours that build trusting relationships.

Sub-element 8 relates to a commitment to be innovative. This is defined by the types of innovation committed to and exemplified by a commitment to continuous improvement and a willingness and ability to undertake testing, prototyping and experimentation.

Sub-element 9 behaviours are evidenced by the degree of a genuine best-for-project culture. This is demonstrated by the degree to which common goals of teams are aligned for project success. This includes the level of achieved project performance outcomes, commitment to challenging 'business as usual' mindsets to strive for excellence in project outcomes, openness about reporting what value for money could and should be achieved and the ability and willingness to recruit support for taking decisions and committing resources for a best-for-project outcome.

Sub-element 10 behaviours relate to realising a no-blame culture so that problems are quickly identified and dealt with rather than people hiding them and engaging in blame-shifting tactics (Walker *et al.*, 2014). This behaviour relies on clarity about the need for and desirability of a no-blame culture as well as supporting and embracing mechanisms that develop and maintain such a culture.

All these behaviours, even if well supported by an underpinning set of facilities and platform infrastructure, cannot be completely effective without agreed routines and means (Element 3) to support and realise the required behaviours. Figure 2.4 illustrates this third element and the six sub-elements it comprises.

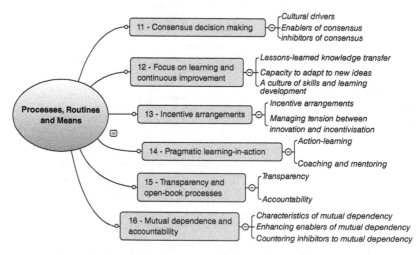

Figure 2.4 RBP Taxonomy: processes, routines and means – Element 3

Sub-element 11 relates to the level of consensus decision making. This can be assessed through the nature of the processes, routines and means that drive cultural behaviours, enable and maximise consensus decision making and minimise inhibitors of consensus being achieved. This sub-element's processes, routines and means link to Sub-element 2 – the joint governance structure – and the level to which it supports or hinders it with processes, facilities and intangible signals that impact upon how decisions are made and communicated.

Sub-element 12 relates to the focus placed on continuous learning and improvement. Maintaining an innovations register or having protocols to celebrate successfully changed practices provides the crucial link between espoused governance arrangements that require continuous improvement and the behavioural indicators of its practice. As with all sub-elements within this element the focus is on *how* behaviours and the underpinning infrastructure for collaboration are achieved.

Sub-element 13 relates to incentive arrangements and how they effectively ensure that any pain/gain agreement operates. This includes the characteristics of the incentive system and how it is operationalised as well as managing the tension between innovation and incentivisation. Innovation may not effectively yield the intended results or may even have adverse unforeseen consequences. While innovation is to be applauded, its effectiveness needs to be linked to the risk/reward agreements.

Sub-element 14 relates to the means that enable how pragmatic learning-in-action takes place. This sub-element is closely related to innovativeness and no-blame because action-learning is learning from experimentation, reflection and re-calibration of action to navigate towards a desired outcome. It also relates to the routines and means demonstrated in coaching, mentoring and providing support for learning.

Sub-element 15 relates to the processes and means that ensure transparency and openness. This is assessed by evidence of the level of transparency and open-book accountability. This clearly differentiates the requirements for various RBP approach forms. For example the SPV takes on all risk from the client for a PPP, therefore it would be undesirable and inappropriate that the client demand means to enable open-book access to the SPV; whereas for an alliance, open-book access would be expected and routines and means should be clearly in place to ensure transparency.

Sub-element 16 is related to the degree of mutual dependence and accountability. Routines may be established to specifically create a 'we all sink or swim together' mindset for alliances. However, within PPPs for example, each organisation within the SPV may have good intentions and even routines to act as 'one-team' but the individual participant firms may all focus, as an overriding priority, on ensuring that they achieve a good result for their shareholders. This sub-element is assessed by the way the routines and means express the characteristics of mutual dependency. This is shown by evaluating how these routines enable mutual dependency, a one-team style integration and adherence to project-driven common goals.

Having described the taxonomy elements and sub-elements we now illustrate, for one sub-element, how it might be assessed within a specific context. We do this using a relevant RBP Collaboration Taxonomy Sub-element, for example Sub-element 1 'Motivation and context'. This, as explained above, relates to the circumstances impacting upon a procurement choice and is described as defining a substrate of circumstances that affects the potential degree of possible collaboration. Its measure falls between two anchor points: Low = 1 and High = 5. The taxonomy specifies low and high anchor points for each of the 16 elements. An illustration of the anchor points for Sub-element 1 is provided below.

Low levels would be related to a *hostile environment* for collaboration. This may be due to a lack of conviction on the part of project participants in the value of collaboration within this project's context.

High levels would relate to the procurement choice solution being driven by the acceptance by project participants of the logic of a clear advantage being gained by adopting a focus on a *supportive and collaborative* approach to delivering benefits that align with the values of participants.

The assessment for Sub-element 1 – motivation and context for collaboration – is based on an overriding sense that one or more aspects of the sub-element influence the assessment. The dominant reason may be acute emergency because a deadline is unmovable. This occurred in the case of the National Museum of Australia NMA alliance project (Walker and Hampson, 2003b). Or it may be related to a realisation based on past experience that facing both known and unknown risks in a critical aspect of a project makes the only viable choice an alliance or alliance-type arrangement. We saw an example of this when a part of a PPP project was delivered via an alliance (Walker and Jacobsson, 2014). The most salient pressure point on the sub-element's seven identified components may steer the judgement of the assessment for the sub-element towards say 5 as very high, or the contextual motivation for a relatively straightforward PPP may be quite low. Naturally, each individual case must be judged on full information.

In this way a radar diagram visualisation may be prepared to illustrate the assessment for each of the 16 sub-elements. This visualisation tool can be used to illustrate the characteristics of the project in a parsimonious snapshot. The snapshot can then be used to compare one project to another to stimulate a broader discussion of the two projects' features and their influences on levels of collaboration. It could also be used to benchmark a single project at various intervals and this may trigger revision of the chosen approach and details of the level of infrastructure of Element 1, or the behaviours expressed in Element 2, or the routines and means as observed in Element 3. The visualisation tool could trigger a systematic way to review those aspects that are often difficult to assess such as culture or collaboration levels. This tool could also be used to *design* the project procurement and delivery approach by pegging values on the radar diagram and then considering how these will be achieved. For example if a client/ project owner is averse to an alliance approach but wishes to retain as many aspects of an alliance as possible, then designing the system to meet desired levels expressed on the radar diagram can be engineered.

We now illustrate the RBP Taxonomy by comparing a PPP and a project alliance with which we are familiar. The values along the dimensions for each of the 16 sub-elements have been assessed as an illustration of how the RBP Taxonomy can be used in practice and to show that PPPs and an alliance form can be compared rigorously using this approach.

Comparison of a PPP and alliance using the RBP Taxonomy

The nature and characteristics of PPPs are discussed in Chapter 3 (Winch and Schmidt), Chapter 9 (Jefferies and McGeorge) and Chapter 12 (Xie *et al.*). Historically PPPs have been used for the design, construction, financing, operation and maintenance of public infrastructure and facilities. They are often privately financed and operated on the basis of revenues received for the delivery of the facility and/or services. In Chapter 3, Winch and Schmidt state that the innovative nature of the UK's Public Finance Initiative (PFI) approach to PPPs sees a fee-based form of PPP where concessionaires are reimbursed directly by the sponsoring government body. This allows PPPs to be used for a much wider range of public projects, such as rebuilding hospitals, schools or prisons, rather than just 'classic' infrastructure such as road, tunnel and bridge tollway projects. Further, Jefferies and McGeorge (Chapter 9) identify that the application of the PPP approach to social infrastructure, such as education or healthcare projects, is a relatively recent trend and many of the projects currently under consideration by governments worldwide are for social PPP projects as well as the more traditional 'economic' forms of infrastructure projects such as toll-roads. Either way, whether infrastructure is deemed social or economic, several authors continue to cite the work of Akintoye *et al.* (2003) in providing a general definition of a PPP as a long-term contractual arrangement between a public sector agency and a private sector concern whereby resources and risk are shared for the purpose of developing a public facility.

Opposites or a continuum? Asset ownership identity

Looking at PPPs and alliancing we see one major and stark difference between them in relation to the asset *creation and ownership* identity during the concession period.

- In a PPP, the owner of the asset is the SPV until the end of the concession period at which point the asset ownership is passed to the 'client' that originally developed a brief and called for the PPP proposals. A fee for a *service* is paid (tolls or other arrangements such as payment for hospital patients for example) for use of the asset during the concession period and the client only gains ownership of the asset after that concession period ends.

- In an alliance, the alliance team works with the client during the design stage, sometimes helping to re-define elements of the brief in light of suggested innovations and advice on constructability issues, and remains part of a single team to deliver the project. At all times the asset is owned by the project's commissioning client.

We can see from this perspective that alliancing lies at the opposite pole of a continuum to PPP procurement approaches. Between these poles there is a range of procurement choices in which collaboration between the client and project delivery teams occurs but that asset life ownership levels vary.

In the turnkey approach, for example, the appointed turnkey contractor takes a client specification brief (similar to the case of a PPP or Build-Own-Operate-Transfer (BOOT) arrangement), develops a total project design and construction package and hands over a commissioned functional asset. Literally the key is turned to unlock the asset's use and ownership by the client. The turnkey contractor is usually selected for specialised and rare expertise in engineering design and construction, access to full interim financing during the period up to asset hand over and other reasons. These reasons are often related to its capacity to deliver an integrated design and delivery solution that has high levels of novelty and intellectual property that the client and other potential competing turnkey consortia may lack. The rationale for choosing this pseudo-monopolistic approach is often based on negotiations driven by the desire to gain expertise and/or due to the high cost (or speed) of gaining the necessary know-how about systems integration to deliver the facility. It becomes a cost-benefit decision about the cost and/or time for testing the market (Brady *et al.*, 2005; Ahola *et al.*, 2008).

Design and construct (D&C) project delivery options involve competing syndicates that take the client brief and develop an integrated solution. In more traditional D&C projects the client will invite several syndicates to compete, through a selected filtered process based on competence and capacity. This is undertaken without much, if any, hands-on collaborative involvement beyond defining the brief and providing guidance on what may be judged an acceptable outcome (Productivity Commission, 2014). This guidance may be improved further by the client collaborating with a design team to produce a partial design solution that will be subsequently refined, leading to a novated D&C solution (Doloi, 2008; Walker and Rowlinson, 2008). This still involves the client being the asset owner. The novated design takes the essence of the original partial design and infuses practical and commercial input to develop a solution that provides the expected project outcome performance but is guided by the partial design. In one sense the conceptual ownership of the asset's final physical form is created from the client and original design team while idea ownership is transformed by the novated D&C entity.

There are other project procurement forms along the continuum of the client commissioning an asset that is owned by the client upon delivery of the asset. What remains constant is that the client owns the asset at the project handover stage. Figure 2.5 illustrates the project phases, the degree and nature of client involvement and the locus of asset ownership. Client hands-on involvement is illustrated by a dashed line and client hands-off by a solid line. Asset ownership is illustrated by the key symbol.

Incubation of the project idea begins at the project initiation phase where the idea for the scope and scale of the project and its benefits are conceptualised. A business case is prepared to justify proceeding and committing the necessary resources to complete the project. The client uses in-house experts and advisors and accesses advice from external sources. If the project is mandated to proceed, then the full brief is developed that specifies expected outcomes and performance levels in the form of KRAs.

In the case of PPP projects, KRAs become the performance guidelines along with a client-developed series of specific KPIs that inform the bidding

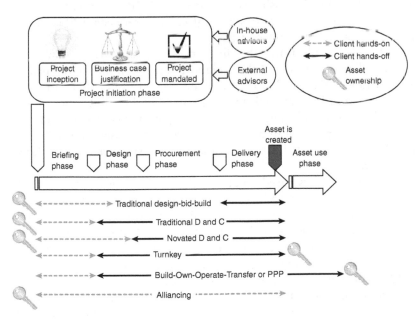

Figure 2.5 Project phases and client hands-on involvement

PPP consortia design. Concession fees are charged for the service provided through use of that asset. KRAs and KPIs inform the design phase for both turnkey and PPP approaches. The asset is owned by the consortia developing the asset until the project asset is completed and functioning. The asset is handed over to the client for turnkey projects upon commissioning for use but this does not happen for PPPs until the end of the concession period. The key image in Figure 2.5 represents the point at which the client owns the project's assets.

The brief informs the design phase for traditional design-bid-build, D&C forms and alliancing. The client owns the project concept and developed asset for all these delivery forms and so the key symbol appears at the left-hand side of Figure 2.5 to illustrate asset ownership.

To summarise, at the start of a PPP the client has a hands-off relationship with the project tender and delivery. With an alliance the client is highly involved with both the tender phase and delivery phase and so has a 'hands-on' relationship with the team delivering the project. Thus from a hands-on/off perspective PPPs and alliancing do rest at opposite ends of the spectrum for that dimension.

Opposites or a continuum? Asset owner project development involvement

From a *relationship-based* perspective *within* the project delivery teams, the two project procurement forms (PPP and alliancing) share characteristics and attributes but the relationship is complex. The RBP Taxonomy provides a useful tool to explore similarities and differences between a PPP and an alliance team relationships (refer to Table 2.1).

Figure 2.6 clearly illustrates the differences and similarities between a PPP and an alliance in terms of the collaboration taxonomy developed by Walker and Lloyd-Walker (2015). Alliancing, by nature of its contractual forms with the three contractual limbs discussed earlier (but more importantly the alliance principles and behavioural requirements) demands greater focus on relationships between parties that develop the asset than is the case for PPPs. While Figure 2.6 illustrates points of close similarity it also shows significant divergence.

Naturally Figure 2.6 is based on an idealised archetypical form of a PPP or alliance, so individual PPP and alliance projects may show variation from the patterns illustrated. However, this visualisation, when combined with an understanding of the concept of client hands-on and hands-off involvement and engagement, provides a useful tool to understand better the distinction between these two project delivery forms. Walker and Jacobsson's case study (2014) provides an example of an alliance operating within a PPP. This may show the start of a trend towards introducing and incorporating many of the positives found in alliancing arrangements into other non-alliance forms.

Table 2.1 Comparison of PPP with alliancing

Taxonomy sub-element	PPP – rating and notes	Alliancing – rating and notes
1 – Collaboration environment motivation and context	Rating 3.5 – Contractor, designer and facilities management operations team collaborate at the bid phase but the client has no input. The asset delivery often uses a D&C approach.	Rating 4.5 – Contractor, designer and project owner team collaborate at the bid phase but often the facilities management operations team have to rely on the project owner representative to voice any potential issues.
2 – Joint project governance	Rating 3.5 – Some SPV-level joint governance and high use of KRAs and KPIs but the D&C delivery approach tends towards a business-as-usual (BAU) culture.	Rating 4.5 – High use of KRAs and KPIs and alliance leadership team and alliance management team governance structures and processes to guide a common best-for-project mindset.
3 – Integrated risk management strategy	Rating 4.0 – High levels of risk management/risk sharing and mitigation conversations. The SPV assumes all/most risk.	Rating 4 – High levels of risk management/sharing and mitigation conversations. The client is actively involved in risk sharing strategy.
4 – Joint communication strategy	Rating 4.0 – High levels of inter-operability of systems and communications but each party has its separate systems.	Rating 4.5 – Very high levels of effort on ensuring common platforms and system integration rather than relying upon system inter-operability.
5 – Substantial co-location	Rating 4.0 – High levels of co-location of design and delivery teams from tender through to delivery	Rating 4.5 – Very high levels of co-location of design and delivery teams from tender through to delivery with close involvement of client/owner.
6 – Degree of authentic leadership	Rating 4.0 – High levels of genuine appreciation of each team's strengths and weaknesses but D&C demands may privilege commercial outcomes.	Rating 5.0 – Very high levels of integrity, respect and alliance principles and values shaping behaviours and the leadership style.

(continued)

Table 2.1 (continued)

Taxonomy sub-element	PPP – rating and notes	Alliancing – rating and notes
7 – Trust-control balance	Rating 3.5 – Moderate levels of trust, autonomy with high levels of structural focus on accountability and low/moderate mutual adjustment provisions. Higher emphasis on control.	Rating 4.5 – High accountability levels of jointly agreed and actioned decision making that leads to mutual adjustment provisions to help each party to meet commitments. Higher emphasis on trust.
8 – Commitment to be innovative.	Rating 4.5 – High levels of achieving synergistic team effort and focus on smart solutions to emerging problems.	Rating 4.5 – High levels of achieving synergistic team effort and focus on smart solutions to emerging problems.
9 – Common best-for-project mindset and culture	Rating 4.0 – High levels of design and contractor commitment to achieve a successful outcome but often the D&C contract mentality inhibits this mindset extending to major sub-contractors.	Rating 5.0 – The alliance contract form enshrines a best-for-project norm into the culture and this is often extended to included major subcontractors (see for example Walker and Jacobsson, 2014).
10 – No-blame culture	Rating 3.5 – Moderate focus on removing a 'claims mentality' due to the D&C delivery approach. High potential for dispute resolution but moderate level of dispute avoidance.	Rating 5.0 – High levels of focus on dispute avoidance, on trying to perceive other participant's perspective and jointly work through issues and problems.
11 – Consensus decision making	Rating 3.0 – Moderate focus on gaining a consensus but each party to the SPV guards their interest and balances that with the common interest.	Rating 4.5 – High levels of consensus in decision making is a core alliance norm. All parties accept its logic in reducing the need for defensive routines and supporting a no-blame culture.
12 – Focus on learning and continuous improvement	Rating 4.0 – High levels of focus on opportunities to develop and apply innovation to achieve better outcomes. No structural requirements for organisational learning.	Rating 4.5 – Very high levels of focus on opportunities to develop and apply innovation to achieve better outcomes. KRAs and KPIs are used to reinforce this organisational learning focus with innovation registers kept and maintained.

13 – Incentive arrangements	Rating 3.0 – Moderate focus on incentives provided by the SPV dependent on the D&C contract with a reliance on benefiting from SPV success as an incentive.	Rating 5.0 – High levels of focus on incentives through pain-sharing and gainsharing formulae based on KRAs and KPI performance. This is one defining feature of an alliance.
14 – Pragmatic learning-in-action	Rating 4.0 – High levels of focus on opportunities to learn from experimentation and innovation though this is not explicitly part of the SPV contract.	Rating 4.5 – Very high levels of focus on learning from experimentation and innovation and this is explicitly part of the alliance agreement and part of the workplace culture.
15 – Transparency and open-book processes	Rating 3.5 – Moderate to high levels of focus providing SPV partners with access to verification about the capacity to be part of the SPV and lower levels of ability to audit and access cost information.	Rating 5.0 – Very high levels of focus on transparency and accountability with all alliance partners having the ability to see what each other is doing. This culture extends beyond cost openness to general knowledge and information sharing.
16 – Mutual dependence and accountability	Rating 4.0 – High levels of focus on the SPV parties making the PPP work, though the D&C contract places the asset delivery end dependent on each party but not through the concession stage. Often the contractor SPV parties sell their interest after a few years.	Rating 5.0 – Very high levels of focus on mutual dependency and accountability. The alliance agreement stresses this as a core value and the 'sink or swim together' mindset is reinforced by the incentive arrangements that are based on project and not participant KPI performance.

Rating 1 = low to 5 = high; refer to Walker and Lloyd-Walker (2015) for detailed rating description

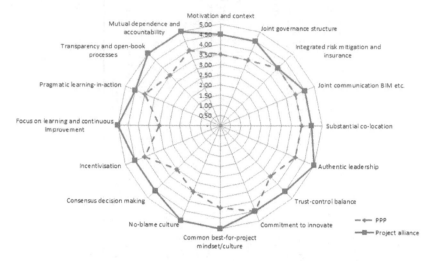

Figure 2.6 Differences and similarities between PPPs and alliances

Chapter summary

Forms, or types, of project relationships are constantly changing and adapting. This offers project parties the opportunity to consider carefully the type of relationship that will best support achievement of their desired outcome – what the project is expected to deliver. PPPs and alliances are only two of these possible forms; however, they provide an excellent basis for comparing and contrasting attributes of relationships and their appropriateness for a specific project type. This chapter compared and contrasted PPPs and project alliances using the RBP Taxonomy developed by Walker and Lloyd-Walker (2015). In answering the question posed in the chapter abstract we have taken two general perspectives.

The client has a 'hands-off' involvement in the collaboration process in PPPs. Instead they are intensely involved in preparing a project brief to deliver the asset so that a service can be delivered (toll for road/bridge/tunnel access or fee for patient/prisoner/consumer access to a social infrastructure service), but have little or no involvement thereafter other than a contracts administration role with the concessionaire. The commissioning client only gains ownership of the asset upon completion of the agreed concession period. However, in an alliance, the client adopts a 'hands-on' role and is intensely engaged in collaboration with the asset development team and owns from its inception the asset that is being developed. Thus, from the client's asset *ownership perspective*, it is clear that PPPs and alliances are poles apart.

When we compared and contrasted PPPs with alliances from a *relationship perspective* we saw a very different picture emerge. The degree of collaboration with the design and development team during the design and delivery stage is starkly different for each of these two approaches. This is because the client adopts a hands-off stance for PPPs and a hands-on stance for alliances. Table 2.1 provided a summary based on the Walker and Lloyd-Walker RBP Taxonomy (2015) that allowed us to develop a radar diagram, presented in Figure 2.6, to visualise the relationship intensities of teams involved in the project design and delivery phases. Figure 2.6 is based on two idealised forms of PPPs and alliances. When considered from this perspective, PPPs and alliances share some similar positions along the 16 dimensions illustrated in Figure 2.6.

Similarities and differences are clearly evident and are linked to the hands-on or hands-off nature of the client's role as determined by the procurement form. PPPs ostensibly embrace high degrees of cross-team collaboration. When PPP syndicates are observed during the proposal and bid phases we see intense collaboration between the design and construction teams with perhaps some lesser collaboration intensity with financiers and operational facilities management teams. However, for PPPs reward and performance is highly team-individualised, and the design and construction teams often on-sell their stake in the SPV not long after project completion. This can be contrasted with alliancing in which all teams that form the alliance remain locked into joint pain- or gain-sharing based on the total project delivery performance. Hence the different mindsets drive different behaviours. Additionally, the client is a full and active participant in alliancing throughout the design and delivery phases.

Figure 2.6 provides a useful tool in helping us understand the position that PPPs and alliances maintain along the RBP Taxonomy's 16 dimensions. It provides a means by which a specific PPP may be compared to a specific alliance.

Reflections

1 Define both the PPP and alliance procurement approaches and identify the key differences between them.
2 How has procurement choice evolved since the work of Walker and Rowlinson (2008)?
3 The Australian Government's Department of Infrastructure and Transport published a 'Guide to Alliance Contracting' in 2011. In can be viewed at: https://www.infrastructure.gov.au/infrastructure/nacg/files/National_Guide_to_Alliance_Contracting04July.pdf.
4 Project alliancing and project partnering are often confused. Definitions and differences of the two are discussed, and still relevant today, in Walker *et al.* (2002).
5 Given the continued support of alliance contracting by various public sector clients, particularly by Australian state governments, identify recent case study projects and the key issues that have made this procurement model successful.

References

Ahola, T., Laitinen, E., Kujala, J. and Wikström, K. (2008). Purchasing strategies and value creation in industrial turnkey projects. *International Journal of Project Management.* **26** (1): 87–94.

Akintoye, A., Hardcastle, C., Beck, M., Chinyio, E. and Asenova, D. (2003). Achieving best value in private finance initiative project procurement. *Construction Management and Economics.* **21** (5): 461–470.

Brady, T., Davies, A. and Gann, D. M. (2005). Creating value by delivering integrated solutions. *International Journal of Project Management.* **23** (5): 360–365.

Cheung, F. Y. K., Rowlinson, S., Jefferies, M. and Lau, E. (2005). Relationship contracting in Australia. *Journal of Construction Procurement.* **11** (2): 123–135.

Clifton, C. and Duffield, C. F. (2006). Improved PFI/PPP service outcomes through the integration of alliance principles. *International Journal of Project Management.* **24** (7): 573–586.

Department of Finance and Treasury Victoria (2010). *The Practitioners' Guide to Alliance Contracting.* Melbourne, Department of Treasury and Finance, Victoria.

Department of Infrastructure and Transport (2011). *National Alliance Contracting Policy Principles.* Department of Infrastructure and Transport A. C. G. Canberra, Commonwealth of Australia.

Doloi, H. (2008). Analysing the novated design and construct contract from the client's, design team's and contractor's perspectives. *Construction Management & Economics.* **26** (11): 1181–1196.

Grimsey, D. and Lewis, M. K. (2005). Are Public Private Partnerships value for money?: Evaluating alternative approaches and comparing academic and practitioner views. *Accounting Forum.* **29** (4): 345–378.

Hauck, A. J., Walker, D. H. T., Hampson, K. D. and Peters, R. J. (2004). Project alliancing at National Museum of Australia: collaborative process. *Journal of Construction Engineering & Management.* **130** (1): 143–153.

Hodge, G. A. (2004). The risky business of Public-Private Partnerships. *Australian Journal of Public Administration.* **63** (4): 37–49.

Hutchinson, A. and Gallagher, J. (2003). *Project Alliances: An Overview.* Melbourne, Alchimie Pty Ltd, Phillips Fox Lawyers.

Jacobsson, M. and Roth, P. (2014). Towards a shift in mindset: partnering projects as engagement platforms. *Construction Management and Economics.* **32** (5): 419–432.

Jooste, S. F., Levitt, R. and Scott, D. (2011). Beyond 'one size fits all': how local conditions shape PPP-enabling field development. *Engineering Project Organization Journal.* **1** (1): 11–25.

Lahdenperä, P. (2009). *Project Alliance: The Competitive Single Target-Cost Approach.* Espoo, VTT.

Lahdenperä, P. (2010). Conceptualizing a two-stage target-cost arrangement for competitive cooperation. *Construction Management and Economics.* **28** (7): 783–796.

Lau, E. and Rowlinson, S. (2009). Interpersonal trust and inter-firm trust in construction projects. *Construction Management and Economics.* **27** (6): 539–554.

Lloyd-Walker, B. and Walker, D. (2011). Authentic leadership for 21st century project delivery. *International Journal of Project Management.* **29** (4): 383–395.

Love, P. E. D., Davis, P. R., Chevis, R. and Edwards, D. J. (2011). A risk/reward compensation model for civil engineering infrastructure alliance projects. *Journal of Construction Engineering and Management.* **137** (2): 127–136.

Nyström, J. (2005). The definition of partnering as a Wittgenstein family-resemblance concept. *Construction Management and Economics.* **23** (5): 473–481.

Productivity Commission (2014). *Public Infrastructure Productivity Commission Inquiry Report Volume 1*, Canberra, Commonwealth of Australia.

Raisbeck, P., Duffield, C. and Xu, M. (2010). Comparative performance of PPPs and traditional procurement in Australia. *Construction Management and Economics.* **28** (4): 345–359.

Reed, H. and Loosemore, M. (2012). Culture shock of alliance projects. In *Twenty-Eighth ARCOM Annual Conference*, Edinburgh, 5–7 September, ed. Smith, S., Association of Researchers in Construction Management: 543–552.

Regan, M., Love, P. and Smith, J. (2013). Public-Private Partnerships: capital market conditions and alternative finance mechanisms for Australian infrastructure projects. *Journal of Infrastructure Systems.* **19** (3): 335–342.

Ross, J. (2003). Introduction to project alliancing. Alliance Contracting Conference, Sydney, 30 April 2003, Project Control International Pty Ltd.

Ross, J. (2008). Price competition in the alliance selection process. *PCI Alliance Services, Infrastructure Delivery Forum.* Perth, WA, Main Roads Department of Western Australia.

Rowlinson, S., Cheung, F. Y. K., Simons, R. and Rafferty, A. (2006). Alliancing in Australia: no litigation contracts; a tautology? *ASCE Journal of Professional Issues in Engineering Education and Practice.* **132** (1, Special Issue on Legal Aspects of Relational Contracting): 77–81.

Rowlinson, S., Walker, D. H. T. and Cheung, F. Y. K. (2008). Culture and its impact upon project procurement. In *Procurement Systems: A Cross Industry Project Management Perspective*, ed. Walker, D. H. T. and S. Rowlinson. Abingdon, Oxon, Taylor & Francis: 277–310.

Walker, D. H. T. and Hampson, K. D. (2003a). Enterprise networks, partnering and alliancing. In *Procurement Strategies: A Relationship Based Approach*, ed. Walker, D. H. T. and K. D. Hampson. Oxford, Blackwell Publishing: 30–73.

Walker, D. H. T. and Hampson, K. D. (2003b). Procurement choices. In *Procurement Strategies: A Relationship Based Approach*, ed. Walker, D. H. T. and K. D. Hampson. Oxford, Blackwell Publishing: 13–29.

Walker, D. H. T. and Jacobsson, M. (2014). A rationale for alliancing within a Public-Private Partnership. *Engineering, Construction and Architectural Management.* **21** (6): 648–673.

Walker, D. H. T. and Lloyd-Walker, B. M. (2014a). The ambience of a project alliance in Australia. *Engineering Project Organization Journal.* **4** (1): 2–16.

Walker, D. H. T. and Lloyd-Walker, B. M. (2014b). Collaborative project performance. *Megaprojects: Theory Meets Practice.* Love, P. Perth, Australia, Curtin University.

Walker, D. H. T. and Lloyd-Walker, B. M. (2014c). Understanding relationship based procurement in the construction sector: summary of research findings. *PMI Research Conference.* Messikomer, C. Portland, Oregon, 27–29 July, Project Management Institute.

Walker, D. H. T. and Lloyd-Walker, B. M. (2015). *Collaborative Project Procurement Arrangements.* Newtown Square, PA, Project Management Institute.

Walker, D. H. T. and Rowlinson, S. (2008). Project types and their procurement needs. In *Procurement Systems: A Cross Industry Project Management Perspective*, ed. Walker, D. H. T. and S. Rowlinson. Abingdon, Oxon, Taylor & Francis: 32–69.

Walker, D. H. T., Hampson, K. and Peters, R. (2002). Project alliancing vs project partnering: a case study of the Australian National Museum Project. *Supply Chain Management: An International Journal.* 7 (2): 83–91.

Walker, D. H. T., Harley, J. and Mills, A. (2013). *Longitudinal Study of Performance in Large Australasian Public Sector Infrastructure Alliances 2008–2013.* Melbourne, RMIT University, Centre for Integrated Project Solutions.

Walker, D. H. T., Lloyd-Walker, B. M. and Mills, A. (2014). Enabling construction innovation: the role of a no-blame culture as a collaboration behavioural driver in project alliances. *Construction Management and Economics.* 32 (3): 229–245.

Williams, T. (2010). Analysis of the London Underground PPP failure. *Engineering Project Organizations Conference*, Taylor J. E. and P. Chinowsky, South Lake Tahoe, CA, Engineering Project Organization Society (EPOS).

Wilson, D. I., Pelham, N. and Duffield, C. F. (2010). A review of Australian PPP governance structures. *Journal of Financial Management of Property and Construction.* 15 (3): 198–215.

Wood, P. and Duffield, C. (2009). *In Pursuit of Additional Value: A Benchmarking Study into Alliancing in the Australian Public Sector.* Melbourne, Department of Treasury and Finance, Victoria.

Yeung, J. F. Y., Chan, A. P. C. and Chan, D. W. M. (2007). The definition of alliancing in construction as a Wittgenstein family-resemblance concept. *International Journal of Project Management.* 25 (3): 219–231.

3 Public-Private Partnerships

A review of the UK Private Finance Initiative

Graham M. Winch and Sandra Schmidt

Chapter introduction

Public-Private Partnerships (PPPs) have been widely advocated for the provision of public infrastructure as governments have found themselves caught between constraints on their ability to borrow from the private sector due to disciplines of international financial markets on government debt, and the unwillingness of voters to accept increases in government revenues through taxation. Globally, the Organisation for Economic Cooperation and Development (e.g. 2008) has long been promoting their use, while they form an important element of current European Union policy for infrastructure (Commission of the European Communities 2009). This chapter will focus on a particularly innovative type of PPP – the Private Finance Initiative (PFI). PPPs are typically used to provide 'user-pays' infrastructure such as turnpikes for which private users pay a fee directly to the concessionaire. The innovation of PFI was to develop a fee-based form of PPP where concessionaires are reimbursed directly by the sponsoring government body which allows them to be used for a much wider range of public projects than 'classic' infrastructure.

After briefly defining terms, the first contribution of this chapter is to provide both a narrative and a thematic review of the complete cycle of experience of the UK government with PFI over 20 years from the launch of the policy in 1992 to its near cessation in 2012. In this period 732 projects were financed at a total nominal value at financial close of £60.5bn. (These data include a relatively small number of non-PFI toll-financed concessions such as the North Birmingham Relief Road in 1992 and the Mersey Gateway Bridge in 2010.) The narrative review will update Pollitt (2005) and Hellowell (2010) to cover the entire cycle of the PFI policy experiment. The empirical basis of the thematic review is largely the work of the UK National Audit Office's (NAO) Value for Money Reports which constitute a unique database on this topic. The second contribution is to provide an overall evaluation of the UK Private Finance Initiative with the aim of contributing evidence-based practical experience (Raadschelders 2008) in this contentious area of public policy where hard evaluation has been lacking (Hodge and Greve 2007).

Defining the Private Finance Initiative

There is very much a language game surrounding PPPs (Hodge and Greve 2007; see also Delmon 2010; OECD 2008), so we need to define our terms carefully. Virtually all provision of social and economic infrastructure involves both the public and private sectors; PPPs involve government agencies promoting projects and contracting with special purpose vehicles (SPVs) which will finance, construct and operate the infrastructure asset as concessionaire for a specified period which is typically around 25 years. In the 'classic' model, the returns on this investment are recouped by charging users of the facility directly through tolls – the prototype arrangement is the turnpike. Regan *et al.* (2011) provide a review of Australian experience with this type of arrangement.

The originality of PFI was to replace direct user charges with a government-paid fee for facility availability rather than actual use through either a *unitary charge* or, in some cases, a *shadow toll*. Thus the return on investment for the private sector comes directly from the sponsoring government body rather than from real users. This innovation allowed the use of private finance for social infrastructure such as hospitals and schools which in the UK are free at the point of use as a matter of cross-party political agreement, and for roads where tolls would have been politically unacceptable. This arrangement stimulated major national programmes of rebuilding hospitals, schools, prisons and the like and is the focus of the present chapter. The UK has also made limited use of the 'classic' infrastructure concessions which will not be covered here; nor will we discuss sale and lease back of property assets or urban regeneration partnerships (Edelenbos and Teisman 2008; Kort and Klijn 2011) which tend to be more loosely coupled (Steijn *et al.* 2011) and have also been used in the UK. There is also an important debate around accountability (Shaoul 2005; Forrer *et al.* 2010) to which this chapter does not contribute.

The UK experience with the Private Finance Initiative: a narrative

The Conservative administration launched PFI in 1992 following a decade of privatisations and experimentation with tolled infrastructure concessions. The level of PFI activity over the subsequent years is presented in Figure 3.1, which shows the HM Treasury data adjusted for inflation by the new construction output price index to show the real value of deals signed by year of financial close. (These data exclude the massive London Underground PPP.) As can be seen, progress was initially slow as both government and potential concessionaires learned how to put together PFI deals – only two were signed in the period 1992 to 1995. Support for the initiative came from 1993 on from the Private Finance Panel (PFP); in 1994, the Design Build Finance and Operate (DBFO) method of procuring roads was launched, which relied

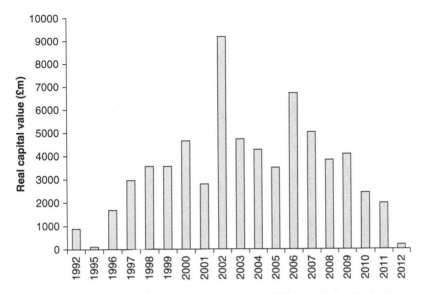

Figure 3.1 Real capital value of private finance projects by year of financial close
(Source: HM Treasury (www.hm-treasury.gov.uk/ppp_pfi_stats.htm;
accessed 16 October 2014) and Office for National Statistics 'Output
Price Indices for New Construction' (www.statistics.gov.uk; accessed
16 October 2014). Data exclude the London Transport PPP)

upon shadow tolls. The Public Private Partnership Programme (4ps; now
Local Partnerships) was set up in 1996 to extend the initiative to local government. In areas such as health and schools, legal problems regarding the
status of National Health Service (NHS) trusts and local authorities delayed
deals, while more generally, the bidding procedures were widely criticised as
costly and time-consuming. Despite some high-profile successes in achieving
deals in the transport and prison sectors, PFI was in some difficulty by the
time of the election in May 1997 (Winch 2000).

The incoming Labour administration acted swiftly by abolishing the
PFP and launching a full review of PFI. As a result, bidding procedures
were overhauled and legislation was introduced to clarify the status of
NHS trusts. A Treasury Taskforce was established to promote PFI and the
rate of deal closure began to pick up. A second review led to the establishment of Partnerships UK in 2000 as a promoter of private finance at
national level to complement 4ps. These two reviews formed the basis of
the subsequent strong development of PFI from 2000 on, particularly as
it aligned well with the 'third way' rhetoric of the day (Connolly *et al.*
2008). As Figure 3.1 shows, the *real value* of deals peaked in 2002 and
was sustained at a high level above £3 billion per annum in real terms
from 1998 until 2009 (the drops in 2001 and 2005 appear to have been

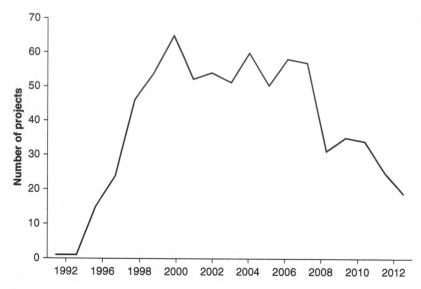

Figure 3.2 Number of private finance projects by year of financial close (Source: HM Treasury (www.hm-treasury.gov.uk/ppp_pfi_stats.htm; accessed 16 October 2014). Data exclude the London Transport PPP).

fully compensated in the following years). Through this peak period, private finance accounted for 15–37 per cent of capital investment in UK public services (Hellowell 2010), with its importance in some sectors such as health being predominant. The value of deals then fell away rapidly to the lowest level since 1995 in 2012. As Figure 3.2 shows, the *number* of deals also fell away significantly after 2007 to the lowest level since 1996 by 2012.

A review conducted by HM Treasury (2003) pronounced general satisfaction with PFI policy but identified two problem areas:

- Information technology projects had a much poorer record of delivery, and most of the successful IT projects renegotiated their contracts and moved away from the PFI model.
- Smaller projects – defined as less than £20m – tended to suffer from disproportionately high transaction costs.

The report therefore recommended that PFI deals be abandoned for IT facilities, and that smaller deals such as schools and primary health care facilities be bundled into programmes called 'strategic partnerships' serving an administrative area. A number of other factors started to work against PFI. These include:

- Growing awareness of the limits to private finance, particularly in the provision of so-called soft facilities management (FM) such as 'hotel' services in hospitals.
- Changes in HM Treasury policy in the context of a more benign environment for public finances meant that public finance became more readily available. This was reinforced by revised HM Treasury guidelines on value-for-money in 2005.
- Some spectacular failures such as the termination of the contract during construction for the National Physical Laboratory PFI in 2004 and the abandonment of the Paddington Health Campus PFI in 2005 after millions had been spent on development costs.
- Preparations for the shift to International Financial Reporting Standards (IFRS), implemented in 2009, which mandated a much stiffer test for assets, and hence the borrowing associated with them, to be placed off the public sector balance sheet (McQuaid and Scherrer 2010).

Developments in the financial markets during 2007 and 2008 posed a greater threat to the use of PFI than internal policy changes. Private finance in the UK is usually provided in ratios of around 90 per cent debt finance and 10 per cent equity finance. Debt financiers charge interest at the sum of the reference rate which represents general market risks plus loan margin which represents project-specific risks. Debt on larger projects is typically sold as bonds which are then rated by credit rating agencies. The purchase of credit insurance then made these bonds acceptable to long-term investors such as pension funds. In late 2007 confidence in the providers of credit insurance collapsed and made it difficult to sell bonds. The non-bond source of debt finance was syndicated bank loans, with the banks often drawing the sums required from the wholesale money markets, but the collapse of Lehman Brothers in late 2008 led to a very difficult period for debt capital from banks which the continuing Eurozone crisis reinforced.

These crises in the financial markets dried up the sources of debt that had enabled the UK boom in private finance as loan margins doubled rendering many projects unaffordable. The NAO (2010b) calculated that these increases in financing cost raised annual charges by around 7 per cent – this means around £200k extra each year for a typical school project for instance. Some 'shovel ready' projects were still allowed to go ahead despite increased financing costs during 2008 and 2009 due to their economic stimulus effect. In July 2010 the new Conservative/Liberal Democrat administration cancelled all BSF (Building Schools for the Future) projects that were not near financial close on value-for-money grounds. Similarly, 7 of the 18 municipal waste PFI projects were cancelled in October 2010 leaving the real value of deals signed in 2010 below the 1997 level. By 2011, PFI was in serious difficulty in the UK. In response to these challenges HM Treasury (2012) provided a further review which introduced the concept of Private Finance 2 (PF2). Perhaps most radically, this proposed that the public sector could take an equity interest in privately financed projects.

Issues in private finance: a thematic review

The UK has built up considerable experience in PFI which is unique in its depth and comprehensiveness. It has seen a cycle from the depths of the recession in the early 1990s through the subsequent long boom to the present return to recession. Many lessons have been learned (NAO 2011), and the UK has actively promoted its expertise across the world. The aim of this section is to provide a thematic review of some of the principal lessons.

The cost of capital

Despite the protestations of finance theorists, private finance is more expensive than public finance in practice. In theory, the public sector should be adding the same uplifts to its cost of capital as the private sector because they are related to the risks of the project, not the final user (Jenkinson 2003). In practice, however, the government tends to act as a self-insurer and spreads its project risks thereby reducing its effective cost of capital. For instance the percentage point uplift for private finance over the public finance baseline for the weighted average cost of capital (WACC) was between +2.05 and +7 per cent in 2010 (Infrastructure UK 2010; see also Ball *et al.* 2002). These higher effective rates for loan capital can be compounded by higher than normal market returns on the equity in the SPV (NAO 2012).

Project performance

The NAO has reported twice on PFI asset delivery performance in the construction sector (2003, 2009a) and has shown that PFI does improve project execution performance on schedule and budget compared to conventional procurement. The NAO cautions against reading too much into these figures due to the differences in the project mix between surveys. Furthermore there are weaknesses in the method of the earlier studies (Pollock *et al.* 2007), but the trend is clear and has also been reported for Australia (Raisbeck *et al.* 2010).

This improvement in delivery performance has not been at the expense of conformance to specification – the NAO reports general satisfaction in this respect. However, questions were raised regarding design specification quality in some early PFI schools (Audit Commission 2003; Northern Ireland Audit Office 2004). As a result, the Commission for Architecture and the Built Environment developed guidance in 2002, and in 2007 was funded to provide designs reviews for SPV proposals under the BSF programme. NAO data (2009a) suggest that satisfaction with the design quality of facilities procured increased between 2002 and 2008.

Innovative facilities

During the early phases of the advocacy of PFI it was suggested that this procurement route would stimulate innovation, particularly with respect to through-life costs of the facility but there is little evidence that this is actually happening. A set of case studies of innovation in PFI hospitals (Barlow and Köberle-Gaizer 2008) supports this conclusion. Reasons include (NAO 2009b):

- lack of available data on whole-life costing that would allow informed decisions to be made on specifications to achieve better whole-life costs of the facility;
- risk-averse financiers (banks and bond holders) who perceive that innovation increases project risk on all three dimensions of budget, schedule and specification;
- lack of experience on the commissioning side within government and difficulties in writing the output specification;
- separation between designers and users mediated poorly through the SPV;
- poor arrangements for learning from project to project;
- de-scoping of the project to achieve affordability criteria.

However, innovation rates do appear to be higher where the whole service is outsourced such as with DBFO roads and detention facilities (Construction Industry Council 2000), rather than the final service delivery remaining in the public sector as in hospitals and schools.

Operational performance and flexibility

One of the principal differences between procurement of infrastructure through conventional public finance and through private finance is that the SPV retains operational responsibility for ensuring the delivery of the services provided by the asset for the life-cycle of the contract. Operational performance complemented by the possibility of flexibility in the required performance to meet changing operational needs is therefore essential.

For the local authority sector, 4ps examined the operational performance of 30 privately financed facilities (2005) and found general satisfaction with operational performance. Research by KPMG (2010) suggests that the patient environment and cleanliness of PFI hospitals on a standard self-assessed protocol are better compared to conventionally procured ones of a similar age. A review of the operational performance of PFI hospital contracts by the NAO (2010a) concluded that operational performance was satisfactory – service levels were generally as agreed in the contract. However, the NAO also found that while performance was perceived to be a little better than in non-PFI hospitals, operational costs were no lower. Lower catering costs tended to be counterbalanced by higher maintenance costs due to the higher standards written into the contract within the unitary charge. Market testing of 'hotel' or 'soft FM' costs (catering, portering and cleaning) every few years or so is typically allowed under the contract, but this rarely achieves savings. The NAO also noted that mechanisms to enable continuous improvement in maintenance costs were rare in the contracts.

The difficulties here are reflected in the decision to remove soft FM from the scope of PF2 (HM Treasury 2012).

Turning now to flexibility, one-third of the respondents to the 4ps (2005) research expressed concern regarding the future flexibility of the contract although they had generally found it possible to negotiate changes in the contract as requirements evolved. NAO research (2008b) found that while there was general satisfaction with flexibility to make changes as requirements evolved, the public sector is typically paying over the market rate for those changes. In some cases it is not feasible to invite competitive tenders for the changes, but even where this is feasible, this was not often done. In addition to the cost of the actual works associated with the changes, SPVs were charging fees to manage the changes and amend the life-cycle cost model of the facility. The NAO was moved to describe some of these additional costs as 'not always justifiable'. The NAO analysis suggests that flexibility is more expensive under PFI than conventional procurement, and so is an additional cost which needs to be counterbalanced by benefits elsewhere.

Transaction costs

Cost of capital is not the only additional cost incurred by private finance; transaction costs are also significantly increased. HM Treasury (2006) accepts that transaction costs will be higher on privately financed projects due to the much greater rigour required in writing the output specification compared to an input one and the much greater costs of tendering and associated contract negotiation. Indeed, this is the principal reason why it moved to exclude projects valued at under £20m from PFI due to their inability to support the transaction cost overhead.

Project promoter and bidding costs tend to disappear into overheads, but the extended length of the procurement period for privately financed projects suggests that these will be higher compared to conventional procurement. NAO (2007) found that the average private finance contract took 34 months to close, roughly double the time for large conventional projects. It may be inferred that transaction costs were around double as well. Many public authorities underestimated the resources required for negotiations and overran on their budgets for advisors; opportunity costs for internal staff were not costed in the research. So far as identifiable transaction costs are concerned, the NAO (2007) put the figure for third-party advisors at an average of 2.6 per cent of project costs for projects closed between 2004 and 2006; for the first batch of 12 Scottish schools closed in the late 1990s (Audit Scotland 2002) this figure averaged 7 per cent.

In principle, transaction costs are counterbalanced by the production cost savings which suppliers are incentivised to share with clients through 'competitive tension' in the supply market, although growing difficulties were experienced in maintaining competitive tension. NAO (2007) found that many projects only received two serious bids, and extensive negotiations often took place after the selection of the preferred bidder leading to repricing in the absence of competitive tension. The problem here is that competitive tension is the principal mechanism for reducing 'opportunism' by suppliers (Lonsdale 2005); where this is absent it is more difficult for the public sector to achieve value for money. The evidence on flexibility discussed above suggests opportunism by some SPVs when making changes during operation; it is likely that final negotiations with preferred bidders also experience such opportunism.

The working conditions of staff

Continuing concern has been expressed by the UK's principal trade union for public sector employees (UNISON 2002) because staff previously employed by the public sector – most frequently in soft FM services such as cleaning and catering – are transferred to the SPV in many PFI projects. Particularly for lower-paid workers, employment terms and conditions are often superior to those in the private sector. However, such transfers are covered by European Union employment legislation - the Transfer of Undertaking (Protection of Employment) (TUPE) regulations – which provides protection for such employees. Of more concern are those people hired following the transfer who are not protected, and pensions which are not covered by TUPE. Newly hired employees with some of the early SPVs did appear to have been disadvantaged so, in a series of policy steps between 2001 and 2005, the government moved to the position that newly hired SPV employees should be offered broadly comparable terms and conditions including pensions to those who were transferred (NAO 2008a).

Value for money

The public sector comparator (PSC) lies at the heart of the value-for-money calculus. It works by calculating a notional cost of using public finance (the PSC) and comparing it to the private finance alternative (PFA). Given the higher cost of private capital as discussed above, the default case will always be that the public finance option is better value for money for a project that is viable in terms of its base case. In order to achieve greater value for money for private finance additional factors need to be taken into account in the calculation. The principal way in which this has been done is to calculate the value of the risk transferred to the private sector supplier by using private finance in comparison to using traditional procurement. However, this is a complex process that compounds the inherent issues in investment appraisal (Coulson 2008), and is based on little empirical evidence (NAO 2013).

The underlying problem here is the closeness of many PFAs to PSCs – sometimes less than 1 per cent (e.g. Northern Ireland Audit Office 2004) – which suggests that there is a good chance that the PSC will actually be lower than the PFA on a given project. In the case of the MoD

Main Building refurbishment for instance (NAO 2009b, figure 6), the difference was 0.0001 per cent rendering the chances of the PSC being less than the PFA at around 50 per cent. For one PFI hospital in procurement during 2011 this figure was 0.03 per cent and largely accounted for by an assumption of additional third-party income under the PFA alone (House of Commons Treasury Committee 2011). Despite regular statements to the contrary by HM Treasury, the impression still lingers in the public sector that PFI finance is the only option available for projects. As late as 2011, a local political leader defending the choice of PFI for a hospital redevelopment stated 'I know that it doesn't provide Value for Money now or in the future, but it's the only game in town' (House of Commons Treasury Committee 2011: 33).

In 2013, the NAO returned to this vexed question and examined the financial modelling for a number of recently closed deals in more detail. They found that in all but one of the cases, the purported benefit of the PFA was marginal over the PSC, and highly sensitive to input assumptions. The report also clarified that, according to HM Treasury, the point of this comparison was not to compare the value for money of private versus public finance, but to assess the value of using private finance compared to the project sponsor using their existing publicly financed budgetary allocation. These budgetary allocations are set by HM Treasury for government as a whole and hence not in scope for the decision-making of the project sponsor. This position implies that even if the PSC is cheaper, it does not necessarily release public funds. Additionally, the NAO (2013) argues that this approach favours private finance because only the PFA and not the PSC is discounted over time. Reworking the analysis to a true value-for-money comparison shows that the PFA was consistently more expensive than the PSC.

Additionality

Additionality is the provision of private capital that provides investment funds for public infrastructure additional to those available from public sources because they lie outside the calculus of the Public Sector Net Debt (PSND) by defining the asset as being off the public sector's balance sheet. In effect, this allows governments to raise more investment capital than the prudential constraints of public finance allow. Where such techniques simply shift the capital cost from the capital account to the revenue account through the payment of unitary charges as in the UK's PFI, then the beneficial effect on the level of investment in public sector assets is only short term (Ball *et al.* 2002). Audit Scotland (2002: 20) put the issue succinctly: 'ultimately the additional funding [to repay the unitary charge] must come through reductions in other areas of expenditure, increased council tax, or a combination of both'. Only if the investment raises additional income, as with tolled infrastructure, is this problem avoided.

Review and discussion

Table 3.1 attempts a balance sheet for the UK experience with PFI, drawing on the data from the various reviews of PFI performance by the NAO. The 'conventional' column provides a baseline for each of the performance criteria, while the sign in the 'PFI' column indicates whether, on the balance of the evidence, PFI provides a greater, equivalent or lesser benefit to the public sector authority in comparison. It should be noted that the PFI column is a relative assessment compared to conventional, and the analysis may not apply to other forms of PPP.

It can be seen that the benefits of short-term additionality and of a reduced risk of budget and schedule overruns associated with project execution indicated by the plus signs in Table 3.1 need to outweigh the greater costs of PFI arising from the cost of capital, transaction costs and flexibility indicated by the minus signs. So far as we are aware, no data are available which would allow a calculation of this balance sheet to be made (NAO 2011). However, we tentatively conclude from this analysis that the general case for PFI is not proven; the benefits gained are too nebulous for us to be sure that they are outweighing the known additional costs.

In sum, the UK experience with PFI over 30 years shows that:

- Making PFI the only option for economic and social infrastructure procurement can lead to rampant strategic misrepresentation (Flyvbjerg 2007) and hence the funding of inappropriate projects with inevitable consequences later on. PFI can only ever be one of a number of procurement options for public infrastructure.
- Accountability is essential for the improvement in policy. The role of the NAO's careful research has been central to the development of the policy. Mediated through the House of Commons Public Accounts Committee, these data have continually pushed HM Treasury to nuance its policy regarding private finance. Freedom of Information requests can also yield important insights into the decision-making process (Cuthbert and Cuthbert 2010).

Table 3.1 The Private Finance Initiative balance sheet

Performance criterion	*Conventional*	*PFI*
Additionality	0	+
Value for money	0	0
Cost of capital	0	−
Project execution risk	0	+
Level of innovation	0	0
Whole-life costs	0	0
Flexibility	0	−
Transaction costs	0	−
Working conditions	0	0

- The effective promotion of a national PFI policy requires specialist government agencies to continuously refine contractual standards, provide project promoters with appropriate advice and 'matchmake' promoters and potential concessionaires (c.f. OECD 2010).
- The public sector is very short of the commercial skills that are required to negotiate these complex deals. Despite initiatives such as the Office of Government Commerce, which was founded in 2000 and absorbed into the Efficiency and Reform Group of the Cabinet Office in 2010, and experience of negotiating hundreds of PFI contracts, this remains a serious problem (NAO 2011) which actively hampers the pursuit of value for money by the public sector on behalf of UK taxpayers.
- More generally, we can suggest that private finance can provide benefits for the provision of public infrastructure in appropriate conditions but it can be difficult to confidently establish value for money in the investment appraisal. Moreover, net benefits are typically only achieved where the investment stimulates economic growth – that is the private finance is used for economic rather than social infrastructure – otherwise it can lead to overconsumption of social infrastructure.

These experiences are not confined to the UK (Bougrain *et al.* 2005; Hodge and Greve 2005; Reeves 2008; Winch *et al.* 2011). In Australia, which has also been energetic in the pursuit of PPPs, the balance sheet also appears mixed with improvements in project delivery but significant issues remain over the value for money of the facilities delivered which raise intergenerational equity issues (Regan *et al.* 2011). We suggest, therefore, that the UK experience with PFI is important in the broader international arena of public policy and administration because it is one of the most sustained experiments over 20 years in this distinctive new form of organisation of the supply of social and economic infrastructure. It has thereby developed an extensive evidence base on its strengths and weaknesses. Other countries presently considering similar initiatives can thereby learn much from this experience.

Chapter summary

PFI in the UK has been a remarkable policy experiment over the last 20 years. From 1992 onwards, the apparent benefits of additionality during a period of tight restraint on PSND encouraged government agencies to promote a wide variety of infrastructure projects, some of which were more appropriate for private finance than others, under the flag of the Private Finance Initiative. The early period saw some significant successes, many projects with mixed results, and some spectacular failures. The incoming Labour government of 1997 quickly assuaged the doubts about PFI it had expressed in opposition and moved to tighten up on a number areas of practice ranging from the standardisation of contracts to clarifying the employment terms and conditions of transferred workers. The period from 1998 to 2007 saw a remarkable expansion of PFI as many of the problems that challenged the early projects were addressed. While value for money was the official policy, project promoters continually expressed concerns that they were being pushed into PFI because there was no alternative means of financing the required facility.

After 2007 the flow of deals signed slowed and then fell away rapidly from 2010. As more and more projects moved into the operational phases it became clear that a high proportion of facilities were delivering on the expectations of their users, but that few additional benefits were being achieved. Difficulties in raising finance started to become apparent due to changes in international financial markets, and the private sector's appetite for new PFI deals apparently became sated. The number of privately financeable projects dropped as value-for-money criteria became tighter – a trend that was reinforced with the adoption of the new IFRS in 2009.

Thus by 2012, PFI was in serious decline in the UK. If capital market conditions improve, we may see the revival of some projects – the current UK government certainly has this aspiration – but the issues summarised in Table 3.1 have not been resolved (House of Commons Treasury Committee 2011, 2014) and little actual progress has been made with PF2. The use of additionality in a very tight period for government spending might be attractive, but is now further constrained by the adoption of IFRS. The most viable candidates for private finance remain tolled rather than fee-based infrastructure, but a country such as the UK with mature infrastructure offers relatively few opportunities for that kind of investment. Where investment is urgently needed in areas such as urban transit, private finance is difficult to justify under current HM Treasury policies, and where it is urgently needed in areas such as education and health, affordability criteria will set a low cap on activity.

Acknowledgements

We are extremely grateful for the support given for the research reported here by the Association for Chartered and Certified Accounts of London.

Reflections

1   What issues led to the development of PFI as an alternative procurement approach to traditional methods for infrastructure provision?
2   PFI was first launched by the UK government in 1992 and became high-profile public policy for over 20 years. Summarise the successful aspects of this innovative form of procurement?
3   PFI is not a UK-specific approach as it has had an enormous influence on the international procurement scene. How is the PPP approach used by governments globally?

4   Concerns with value for money led to the development of a revised PFI approach in 2012, namely PF2 (www.gov.uk/government/publications/private-finance-2-pf2), has this led to widespread improvement?

5   The UK government recently created the Infrastructure Act (2015) (www.legislation.gov.uk/ukpga/2015/7). How will the Act have an impact on procurement methods such as PPP?

References

4ps (2005). *4ps Review of Operational PFI and PPP Projects* (4ps, London).

Audit Commission (2003). *PFI in Schools* (Audit Commission, London).

Audit Scotland (2002). *Taking the Initiative: Using PFI Contracts to Renew Council Schools* (Accounts Commission, Edinburgh).

Ball R, Heafey M, and King D (2002). "The Private Finance Initiative and Public Sector Finance" *Environment and Planning C: Government and Policy* 20 57–74.

Barlow J, and Köberle-Gaizer M (2008). "The Private Finance Initiative, Project Form and Design Innovation: The UK's Hospitals Programme" *Research Policy* 37 1392–1402.

Bougrain F, Carassus J and Colombard-Prout M (2005). *Partenariat public-privé et bâtiment en Europe: quels enseignements pour la France?* (Presses Ponts et Chaussées, Paris).

Commission of the European Communities (2009). *Mobilising Private and Public Investment for Recovery and Long Term Structural Change: Developing Public Private Partnerships* (CEC COM 615, Brussels).

Connolly C, Martin G, and Wall A (2008). "Education, Education, Education: The Third Way and PFI" *Public Administration* 86 951–968.

Construction Industry Council (2000). *The Role of Cost Saving and Innovation in PFI Projects* (CIC, London).

Coulson A (2008). "Value for Money in PFI Proposals: A Commentary on the UK Treasury Guidelines for Public Sector Comparators" *Public Administration* 86 483–498.

Cuthbert M, and Cuthbert J (2010). "The Royal Infirmary of Edinburgh: A Case Study on the Workings of the Private Finance Initiative" *Public Policy and Management* 30 371–378.

Delmon J (2010). *Understanding Options for Public-Private Partnerships in Infrastructure: Sorting Out the Forest from the Trees: BOT, DBFO, DCMF, Concession, Lease . . .* (World Bank, Washington DC).

Edelenbos J, and Teisman G R (2008). "Public-Private Partnership: On the Edge of Project and Process Management. Insights from Dutch Practice: The Sijtwende Spatial Development Project" *Environment and Planning C: Government and Policy* 26 614–626.

Flyvbjerg B (2007). "Policy and Planning for Large-infrastructure Projects: Problems, Causes, Cures" *Environment and Planning B: Planning and Design* 34 578–597.

Forrer J, Kee J A, Newcomer K E, and Boyer E (2010). "Public-Private Partnerships and the Public Accountability Question" *Public Administration Review* 70 475–484.

Hellowell, M (2010). "The UK's Private Finance Initiative: History, Evaluation, Prospects". In G A Hodge, C Greve and A D Boardman (eds) *International Handbook on Public-Private Partnerships*. 307–332 (Edward Elgar, Cheltenham, UK).

HM Treasury (2003). *PFI: Meeting the Investment Challenge* (HMSO, Norwich).

HM Treasury (2006). *PFI: Strengthening Long-term Partnerships* (HMSO, Norwich).

HM Treasury (2012). *A New Approach to Public Private Partnerships* (HM Treasury, London).

Hodge G A, and Greve C (eds) (2005). *The Challenge of Public-Private Partnerships: Learning from International Experience* (Edward Elgar, Cheltenham, UK).

Hodge G A, and Greve C (2007). "Public-Private Partnerships: An International Performance Review" *Public Administration Review* 67 545–558.

House of Commons Treasury Committee (2011). *Private Finance Initiative: HC 1146* (The Stationery Office, London).

House of Commons Treasury Committee (2014). *Private Finance 2: HC 97* (The Stationery Office, London).

Infrastructure UK (2010). *National Infrastructure Plan 2010* (HM Treasury, London).

Jenkinson T (2003). "Private Finance" *Oxford Review of Economic Policy* 19 323–334.

Kort M, and Klijn E-H (2011). "Public-Private Partnerships in Urban Regeneration Projects: Organizational Form or Managerial Capacity?" *Public Administration Review* 71 618–626.

KPMG International (2010). *Operating Infrastructure Healthcare Infrastructure: Analysing the Evidence* (KPMG International, London).

Lonsdale C (2005). "Post-contractual Lock-in and UK Private Finance Initiative (PFI): The Cases of National Savings and the Lord Chancellor's Department" *Public Administration* 83 67–88.

McQuaid R W, and Scherrer W (2010). "Changing Reasons for Public-Private Partnerships (PPPs)" *Public Money and Management* 30 27–34.

NAO (National Audit Office) (2003). *PFI: Construction Performance: HC 371* (The Stationery Office, London).

NAO (National Audit Office) (2007). *Improving the PFI Tendering Process: HC 149* (The Stationery Office, London).

NAO (National Audit Office) (2008a). *Protecting Staff in PPP/PFI Deals* (NAO, London).

NAO (National Audit Office) (2008b). *Making Changes in Operational PFI Projects: HC205* (The Stationery Office, London).

NAO (National Audit Office) (2009a). *Performance of PFI Construction* (NAO, London).

NAO (National Audit Office) (2009b). *Private Finance Projects* (NAO, London).

NAO (National Audit Office) (2010a). *The Performance and Management of Hospital PFI Contracts: HC68* (The Stationery Office, London).

NAO (National Audit Office) (2010b). *Financing PFI Projects in the Credit Crisis and the Treasury's Response: HC287* (The Stationery Office, London).

NAO (National Audit Office) (2011). *Lessons from PFI and Other Projects: HC920* (The Stationery Office, London).

NAO (National Audit Office) (2012). *Equity Investment in Privately Financed Projects: HC 1792* (The Stationery Office, London).

NAO (National Audit Office) (2013). *Review of the VFM Assessment Process for PFI* (The Stationery Office, London).

NIAO (Northern Ireland Audit Office) (2004). *Building for the Future: A Review of the PFI Education Pathfinder Projects* (TSO, Norwich).

OECD (Organisation for Economic Cooperation and Development) (2008). *Public-Private Partnerships: In Pursuit of Risk Sharing and Value for Money* (OECD, Paris).

OECD (Organisation for Economic Cooperation and Development) (2010). *Dedicated Public-Private Partnership Units: A Survey of Institutional and Governance Structure* (OECD, Paris).

Pollitt M (2005). "Learning from the UK Private Finance Initiative". In G Hodge and C Greve (eds) *The Challenge of Public-Private Partnerships: Learning from International Experience*. 207–230 (Edward Elgar, Cheltenham, UK).

Pollock A M, Price D, and Player S (2007). "An Examination of the UK Treasury's Evidence Base for Cost and Time Overrun Data in the UK Value-for-Money Policy and Appraisal" *Public Money and Management* 27 127–134.

Raadschelders J C N (2008). "Understanding Government: Four Intellectual Traditions in the Study of Public Administration" *Public Administration* 86 925–949.

Raisbeck P, Duffield C, and Xu M (2010). "Comparative Performance of PPPs and Traditional Procurement in Australia" *Construction Management and Economics* 28 345–359.

Reeves E (2008). "The Practice of Contracting in Public Private Partnerships: Transaction Costs and Relational Contracting in the Irish Schools Sector" *Public Administration* 86 969–986.

Regan M, Smith J, and Love P (2011). "Infrastructure Procurement: Learning from Public-Private Partnerships 'Down Under'" *Environment and Planning C: Government and Policy* 29 363–378.

Shaoul J (2005). "A Critical Financial Analysis of the Private Finance Initiative: Selecting Financing Methods or Allocating Economic Wealth?" *Critical Perspectives on Accounting* 16 441–471.

Steijn B, Klijn E-H, and Delenbos J (2011). "Public Private Partnerships: Added Value by Organizational Form or Management?" *Public Administration* 89 1235–1252.

UNISON (2002). *PFI: Failing our Future* (UNISON, London).

Winch G M (2000). "Institutional Reform in British Construction: Partnering and Private Finance" *Building Research and Information* 28, 2, 141–155.

Winch G M, Onishi M, and Schmidt S E (eds) (2011). *Taking Stock of PPP and PFI Around the World*. London, ACCA Research Report 126.

4 London Underground's Public-Private Partnership

Lessons learned by the public sector

Mark Gannon

Chapter introduction

London Underground's (LU) £15bn Public-Private-Partnership (PPP) was the most significant project in terms of capital value, stakeholder controversy, innovation and complexity that has been undertaken in the United Kingdom (UK). Whilst this major transport project was abandoned after seven and a half years, despite an intended 30-year contract, there are vital lessons to be learned for the public sector when moving from a PPP policy to implementation. The public sector, in this case, covers a range of stakeholders: the contracting authority LU, central government and civil servants who were integral to LU's PPP development and implementation.

The research for this chapter was based on secondary research that utilised content analysis to obtain key themes for lessons learned from a public sector's perspective. Documents were sourced from the UK parliament, the National Audit Office, Transport for London, LU's PPP project, trade journals, quality newspaper articles and the author's own publications. It should also be noted the author worked for the LU PPP Implementation team during bid preparation and for the Metronet consortium during their Best and Final Offer (BAFO) stage. This has brought a further insight to the context of secondary sources.

Background

Under-investment backlog

If the UK is to exploit London's competitive position as a world-class city an efficient, reliable and safe underground system is essential. At the heart of London is London Underground (LU), which has been running since 1890 and merits both public and private sector investment (London Transport, 1996). Since the early 1990s investment in LU's system by central government had been declining. In 1997, London Transport (LU's parent company now Transport for London) estimated the under-investment backlog to be £1.2 billion (Gannon, 2011; NAO, 2004a). Uncertain annual capital funding levels from central government combined with the legal inability to borrow funds, as governed by the London Regional Transport Act 1984, forced LU in to an inefficient annual planning cycle (HC, 1998), the consequence of which led to the degradation of system assets, rising service costs and lack of customer satisfaction due to poor performance, particularly poignant at a time when passenger numbers were increasing.

Major project delivery

The Jubilee Line Extension (JLE) and Central Line Upgrade were two major projects funded by central government managed by LU during the 1990s that experienced major problems. The planned construction cost for JLE, at the Parliamentary Bill stage 1989, was £1.26bn and outturn cost was £3.50bn in 2000; additionally the project was delivered two years late (Mitchel, 2003). According to Mitchel (2003) the construction programme was ambitious with a budget that failed to consider various requirements of legislation, regulation and fitness for purpose. The Central Line Upgrade failed to deliver the required journey time improvements and was also delivered late. The outcomes of these projects provided central government with a lack of confidence in LU's management capabilities to manage large-scale project investment in an economic and efficient manner using traditional procurement (NAO, 2004a).

Private Finance Initiative

The Private Finance Initiative (PFI) launched by the Conservative government in the 1992 Autumn statement, was perceived by central government as another source of funds along with grants and revenues that could alleviate LU's under-investment backlog (Gannon, 2002). During LU's PFI period, 1992 to 1999, LU grasped this initiative and signed five major contracts (Gannon, 2006) – one line-based contract: the Northern line trains service contract, a 20-year contract to design, build operate and maintain trains for the Northern line, and four network-wide contracts: Power – a 17-year contract to provide and maintain a new power system with pick up from the National Grid; Oyster (formerly LT Prestige) – a 17-year contract to provide and maintain a new ticketing system for LU and London buses; Connect – a 20-year contract to provide and maintain a new integrated radio and communications network for LU and its interfaces; and British Transport Police – a 23-year contract to construct and maintain police station facilities for British Transport Police. At 1998 prices, LU's total PFI contract value was estimated to be £6bn over the contract term (HC, 1998). However, despite LU making full use of PFI and developing a strong knowledge base, this initiative was not an efficient and economic way to solve LU's under-investment backlog. Each PFI had a long lead time, high setup cost, complex contractual interfaces and severely impacted on LU's gross margin leading to less retained revenue to fund future investment (HC, 1998). A solution that considered the whole investment programme in a more economic manner was urgently required. During LU's PFI period the incumbent government, the Conservatives, were considering privatisation to solve LU's investment problems; however, the Labour party were politically opposed to this approach and proposed a PPP model in their 1997 manifesto. Labour won the May 1997 election whereupon a PPP funding model was utilised to solve LU's under-investment backlog (Gannon, 2011).

Central government's PPP funding policy

Policy objectives and options

Shortly after being elected Labour formed a working group mid 1997, supported by financial advisors, to identify business structures to meet three policy objectives. These included obtaining private sector investment and expertise to modernise LU, guaranteeing value for money for the taxpayer and passenger and safeguarding the public interest and delivering a safe operation (NAO, 2004a). LU were not part of this working group on the basis LU's 'presence could inhibit a frank exchange' of views (NAO, 2004a). In parallel LU conducted their analysis on four options. LU's preferred option was to keep LU in the public sector with a consistent level of borrowing through bonds or hypothecated revenues to fund infrastructure improvements. However, the government's working party concluded their preferred option against their criteria was a private sector, horizontal structure with one or more contracts to maintain and renew infrastructure with one public sector operator. Consequently it was concluded by LT that LU's PPP funding model was a more-or-less done deal politically (Gannon, 2011).

PPP announcement

On 20 March 1998 the Deputy Prime Minister announced in the House of Commons the future plans for LU in outline. The company was to be dismantled into an Operating company (Opsco) and Infrastructure companies (Infracos). Opsco would remain a public sector entity responsible for the trains, stations and network control; and Infracos would be private sector entities responsible for maintaining and renewing the system. The PPP was expected to deliver £7bn of investment and to be signed by April 2000 with £110m development costs to cover re-organisation and procurement, as estimated by the external financial advisors in October 1997 to be (NAO, 2004a). At this stage it was unclear whether one, two or three contracts would be awarded and whether the contract term would be 15 or 30 years.

Planning and implementation

Re-organisation and implementation

The Department for Transport (DfT) along with HM Treasury (HMT) took the lead in deciding the form of PPP and 'relied largely on LU to develop and procure the deals' (NAO, 2004a). However, the entire planning and implementation process was under the stewardship of HMT and DfT civil servants and their respective ministers. LU and LT (now TfL) was re-organised under the instructions of the secretary of state with a number of commercially experienced senior directors and managers departing the organisation (HC, 1998; Gannon, 2008). In the summer of 1998 LU set up a PPP implementation team headed by the PPP Implementation Director, a policy group supported by financial advisors, legal advisors and technical advisors assigned to provide external advice to the PPP and support LU's commercial and PFI skills shortage.

The PPP Implementation Director and the government were keen to communicate internally and externally the uniqueness of LU's PPP compared to the other five PFI contracts undertaken by LU. PFI contracts were considered more rigid with a fixed scope, with a set period of time to design and build, whereas with LU's PPP design and build would be continuous. The PPP architects were also keen to learn from the documented problems of PFI (Hutton, 2000). However, despite these differences a greater understanding of how LU's PPP might work in practice was required to be established by LU to enable central government's PPP funding policy to be implemented. Hence a market soundings exercise was essential in establishing the appetite of potential bidders.

Market soundings

The government's proposals for LU were further developed by the PPP implementation team utilising feedback from the market soundings exercise conducted between September and October 1998. Twenty potential suppliers were asked a range of face-to-face questions on the proposed structure, funding and risk allocation for LU's PPP. Responses were analysed and resulted in a revised timetable with contract signature now moved from April 2000 to December 2000. The market soundings provided sufficient information for LU to now commence procurement. From the market soundings analysis it was established by LU that bidders would be bidding to finance, maintain and renew three Infracos: two deep tube Infracos known as BCV (Bakerloo, Central and Victoria lines) and JNP (Jubilee, Northern and Piccadilly lines) and SSL a sub-surface Infraco (Circle, Hammersmith and City, District and Metropolitan lines).

Procurement

In March 1999 a periodic indicative notice (PIN) was published to inform the market of LU's PPP intentions, whereupon 114 companies registered interest. Surprisingly to potential bidders, late May 1999, central government withdrew SSL Infraco from the procurement process. This action by the secretary of state for transport facilitated discussions and negotiations with Railtrack (now Network Rail the main UK railway infrastructure provider) for the SSL Infraco on a single bidder privileged status (Gannon, 2011).

An OJEC was published on 15 July 1999 calling a competition for two deep tube Infracos (BCV and JNP Infracos). This was followed by a supplier conference prior to pre-qualification in the British Library on 23 July 1999. Although six consortiums submitted pre-qualification submissions, five pre-qualified for each of the two Infracos; three consortiums were invited to bid for both Infracos and two consortiums were invited to bid for one Infraco each. One consortium, Team London, submitted a pre-qualification response that wanted to retain LU under public ownership and did not entertain the spirit of the government's PPP; it hence performed poorly against the pre-qualification criteria. After a period of six months consortiums submitted priced bids, which were evaluated for compliance and against the bid evaluation criteria. Two consortiums for each Infraco were invited to submit best and final offers (BAFOs). Bids were submitted twice thereafter, firstly due to being unaffordable and requiring the contract to be de-scoped and secondly due to additional power required for new trains thus resulting in a five-month delay (NAO, 2004a).

In September 1999, LU was organised in to four public sector units to shadow run operations and maintenance and ensure the PPP contracts and performance regimes between Opsco and the three Infracos were fully tested prior to contract award. Shadow running eventually lasted for a period of three years. In essence the private sector bidders would be acquiring existing entities that had full accounts and operations.

Discussions between LU and Railtrack collapsed and SSL Infraco returned to the PPP procurement process in December 1999 albeit five months behind the shortlisting of bidders for the two deep tube Infracos (BCV and JNP). Four consortiums pre-qualified for SSL Infraco. Again the project was de-scoped due to unaffordable bids. The Tube Lines consortium won the JNP Infraco and the contract was awarded December 2002. Metronet won the BCV and SSL Infracos; however, there was a delay with Metronet raising the required finance and this delayed contract close until April 2003. Table 4.1 provides a chronology of LU's PPP procurement stages and the progression of consortiums.

Formation of Greater London Authority and mayoral legal challenges

One of Labour's main pledges in its May 1997 manifesto was to hold a referendum to setup a democratically elected strategic authority for London headed with an elected mayor and assembly. In October 1999 the Greater London Authority (GLA) Act received its royal assent. The most high-profile candidate running for London Mayor was Ken Livingstone, not the Labour party's preferred candidate, as he was diametrically opposed to the partial privatisation of LU.

Table 4.1 Chronology of procurement and progression of consortiums

PPP Procurement Stage	Deep Tube Lines			Sub-surface Lines	
	Date	BCV Infraco	JNP Infraco	Date	SSL Infraco
OJEC Published	15 July 1998	In excess of 100 responses received to be involved in consultation and 20 potential candidates invited to discuss proposals and views			Railtrack exclusive negotiations collapsed
Market soundings process	Sept–Oct1998				
Periodic Indicative Notice (PIN)	March 1999	114 companies expressed interest and responses to 40 questions		30 May – Nov 1999	
Official Journal European Communities (OJEC) Published	15 July 1999			Dec 1999	Expressions of interest from five consortiums
Supplier conference	23 July 1999	Supplier conference held in the British Library.			
Pre-qualification Issue and response	22 July 1999/ 7 Oct 1999	Six consortiums responded (Team London not shortlisted) LINC, Metronet, Tube Lines and New Metro	LINC, Metronet, TubeRail and Tube Lines Group	Feb 2000 April –Sept 2000	LINC, Metronet, TubeRail, New Metro and Subsurface Lines Group
Invitation to Tender (ITT) Issue and response	Oct 1999/ March 2000				LINC, Metronet and Surface Lines Group
Shortlisting for best and final offer (BAFO)	July 2000	LINC and Metronet	LINC, Metronet, Tube Lines and New Metro Tube Lines	Feb-July 2001	Metronet and LINC
Best and final offer (BAFO)	Nov 2000				
De-scoping and re-bidding	Jan 2001				
Power change and re-bidding	April 2001				
Preferred bidder	May 2001	Metronet		Sept 2001 Feb 2002	Metronet
Committed finance offers	Dec 2001	–			
Financial close	Dec 2002 April 2003	Metronet	–	April 2003	

Source: NAO (2004a); Gannon (2002)

Greater London's Mayor, Ken Livingstone, was elected on the 4 May 2000 for a five-year term along with the 25 assembly members accountable for the strategic government of Greater London. A large part of the Mayor's strategic remit was transport in London of which the tube is a critical component. The Mayor, along with his Transport Commissioner, was vehemently opposed to LU's PPP.

The Mayor and TfL applied for two judicial reviews, heard in April 2001 and July 2002 (NAO, 2004b), at a cost of £4m (Evening Standard, 2002). In the first judicial review TfL claimed that being bound in a 30-year contract would not facilitate the implementation of the Mayor's transport strategy under the GLA Act 1999. This review was unsuccessful. In the second judicial review TfL were claiming LU's PPP breached EC procurement rules, the contracts could not satisfy the value-for-money (VfM) test and it was unaffordable. The second challenge was withdrawn as the mayor and TfL were advised they had no complaint.

Stakeholder reviews and concerns

The House of Commons Transport Select Committee, industry think tanks and leading academics expressed their concerns regarding central government's decision to partially privatise LU. The concerns from stakeholders initially centred upon using PPP rather than a bond issue to finance LU's investment programme. However, as the government ignored these concerns and opted for a PPP funding model rather than a bond issue stakeholders concerns then focused upon achieving value for money, separation of infrastructure from operations, contractual complexity and safety.

In the summer of 2000, prior to contract signature, the Transport Select Committee instructed the NAO to review the PPP's financial analysis. The NAO (2000) concluded that the financial analysis provided 'useful but incomplete insight in to the value for money alterative; and it is essential that the decision makers understand what lies behind the figures before reaching a conclusion'. Furthermore whilst the NAO commended LU for undertaking a thorough assessment of the VfM options attention was drawn to the limitations of the analysis and its ability to take into consideration other factors that influence the VfM for LU's PPP. These factors highlighted by the NAO (2000) included: strategic issues that allowed flexibility to incorporate future changes within the contract; having a PPP contractual framework to ensure future changes, ensuring bidders had sufficient financial capacity so as not to request increases in the Infrastructure Service Charge (ISC) for future requirements; a performance specification that had the ability to incentivise bidders to perform avoiding perverse behaviour; the need to build long-term partnerships to eliminate strategic and contractual risks especially as the contract would be handed over to TfL, who were not involved or supportive of LU's PPP; and the project having a comprehensive risk analysis and management to show clearly which parties bear the risk.

Deloitte and Touche were instructed by the Mayor (TfL) to review the VfM. LU and LT temporarily banned the Mayor from publishing this report; however, after a High Court hearing the ban was lifted to allow an edited version to be published in August 2001 (Butcher, 2012). The report indicated that 'neither a 30 year for 7.5 year comparison provides a satisfactory basis for establishing VfM' and the adjustments made to the Public Sector Comparator (PSC) were 'judgemental, volatile or statistically simplistic' (Deloitte and Touche, 2001). They reported that LU had added £2.5bn to the PSC despite the figure having been approved by external consultants and the NAO. Central government appointed Ernst and Young, October 2001, to conduct an independent review of the PSC. The report concluded the methodology was robust; however, the analysis that the PPP would achieve value for money was subjective with the analysis undertaken (Ernst and Young, 2002).

Contract overview

Contractual structure

An overview of LU's PPP contractual structure is shown in Figure 4.1. Two contracts were awarded to the Metronet consortium and one contract to the Tube Lines consortium. In July 2003 ownership of LU was transferred from central government to TfL (PAC, 2005). Metronet predominantly utilised a tied supply chain model whereby consortium members competed for the majority of the work, whereas Tube Lines selected a project-management-based model and competitively tendered work externally and were able to take advantage of the market discipline expected from LU's PPP (Gannon et al., 2013).

The contracts were predominantly output-based, leaving the Infracos to determine how they would procure and deliver their contractual obligations. Station refurbishment and modernisation improvements were input-based contracts. LU's PPP contract was expected to attract £15.7bn of investment over a 30-year contract term with £9.7bn (2002–3 prices) of investment within the first review period (NAO, 2004a). An annual capital grant of £1bn per annum was paid from the DfT to LU via TfL and the GLA, to subsidise the PPP; thus totalling nearly £9bn over the first review period (Gannon, 2006).

LU was responsible for operating the trains and stations and would collect the fares and secondary revenue. Consortiums were responsible for financing, maintaining, renewing and upgrading assets on the tube's system for a 30-year period. Infracos had scarce resource contracts between each other covering supply of LU's scarce assets. Existing system-wide PFI contracts, discussed earlier, were retained by LU whereas line-based PFIs Northern line trains service contract was novated to Tube Lines' PPP contract.

The arbiter's role

The arbiter's role was set out under the GLA Act 1999. In essence the role was to provide advice, guidance and directions to relevant contractual parties to meet the PPP's objectives. Critical to the arbiter's role was establishing effective dialogue with PPP parties and other stakeholders (NAO, 2004a). If there were any difference between the price paid to the Infracos by LU or about being economic and efficient the arbiter would resolve such matters independently (PPP Arbiter, 2005). The arbiter's role was also to determine what constitutes good industry practice and the level of cost and performance that would be economic and efficient.

Payment mechanism

A bonus/penalty payment mechanism linked to the service specification was utilised. LU customer service requirements were represented within the Infrastructure Service Charge (ISC) paid by LU to consortiums. The ISC

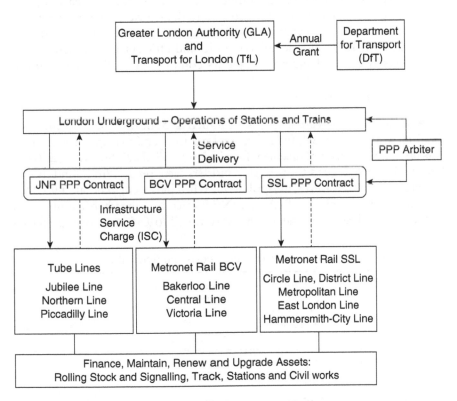

Figure 4.1 London Underground's PPP contractual structure (Source: Gannon, 2002)

comprised three components: capability (reduced journey times), availability (of assets) and ambience (condition and cleanliness) that incentivised consortiums to deliver improvements to the systems.

Periodic reviews

The 30-year contract comprised four periodic review periods occurring every seven and a half years. As LU was unable to predict its service requirements over the 30-year period the PPP contract allowed LU to restate its requirements every seven and a half years and the ISC would be reviewed to reflect changes in costs for an efficient 'Notional Infraco' (Bolt, 2003). Consortiums bid fixed prices for the first review period (seven and a half years) and for periods thereafter the arbiter would oversee the tender process along with LU. The notional Infraco, dubbed 'fantasy Infraco' (RMT, 2009), assumed that the Infraco delivered its activities in an efficient and economic manner in accordance with 'good' rather than best industry practice under the same contractual obligations as a PPP Infraco. The notional Infraco was used to compare how much the Infraco should be earning and performing compared to an Infraco's actual historic and future performance. Extraordinary reviews could be triggered between the arbiter and an Infraco where despite being economic and efficient the Infraco's costs had adversely changed and it had used its contingency (PPP Arbiter, 2005).

Timescales and costs

Development and procurement costs totalled £450m of which LU incurred £180m covering technical, financial, legal advisors and management consultants (see Table 4.2); reimbursement for bidders totalled £270m (NAO, 2004a). Bid reimbursement costs for preferred bidders were set at £8m (£4m/Infraco) in October 1999. However, in April 2000, when bids were to be submitted for SSL Infraco (BAFO bids for BCV and JNP Infraco had already been submitted by bidders), bid reimbursement costs were raised to £45m to cover preferred

Table 4.2 Planned and outturn development costs and timescales

Contract award	PPP policy Feb 1999 (a)	Contract close (2002–3) (b)	Difference between cols (a) and (b)
	April 2000	Dec 2002 and April 2003	2.75 to 3 years late
LUL development	£150 m	£180 m	120%
Bidders costs reimbursed	£57 m	£270 m	increase in development
Total Costs	£207 m	£450 m	and procurement cost

Source: NAO (2004a)

bidders. Bid reimbursement costs were again raised in December 2002 to £270m, after JNP Infraco's financial close (NAO, 2004a). The rationale for the increase with the bid reimbursement costs was to maintain competition and cover the opportunity costs of consortiums bidding for other projects.

Consortium's operational performance

Review of the deal

In 2004, the NAO investigated 'whether LU's PPP were good deals?' and 'whether LU's PPP was likely to work successfully?' The first report concluded that LU had broadly delivered the government's PPP policy objectives; however, there was 'limited assurance' the price paid to bidders was reasonable due to the complexity of the PPP. Furthermore the periodic review every seven and a half years left uncertainty over the eventual price, transaction costs were high as a result of two rounds of bidding and finally, while PPP offered an 'improved prospect compared to the pre-1997 investment regime', there was no certainty that the upgrades would be delivered (NAO, 2004a). The second report concluded the performance against benchmarks was mixed, good partnerships were being built between LU and consortiums, however, there were many tests ahead in a 30-year contract (2004b).

After two and three years

Two years into the PPP, the London Assembly Transport Committee (part of GLA) reported overall performance on the network was improving; however, Metronet's station refurbishments programme was behind schedule by four months and there were an increasing number of engineering overruns for both Infracos (London Assembly, 2005). Notably Metronet's shareholders replaced their Chief Executive with an aim of bringing about change.

Three years into the PPP contract the London Assembly Transport Committee reported LU's PPP had developed an interesting paradox in that Tube Lines demonstrated that LU's PPP worked, however there were concerns over Metronet's performance (London Assembly, 2007). Metronet's station refurbishment programme was significantly behind schedule after three years, having completed 14 of the 35 stations for refurbishment, each of these being late (London Assembly, 2007). Furthermore the report focused on Metronet's tied supply whereby the majority of its work was tendered to its own consortium members, reported to be 60 per cent of project capital costs (HC, 2010). This arrangement was in not in the spirit with the intended market discipline of the PPP (Gannon et al., 2013).

In May 2007 Metronet reported that it anticipated overspending by more than £1bn, significantly higher than the arbiter's £750m anticipated in November 2006 (Guardian, 2007). Metronet blamed its position on LU's over specification and TfL blamed this position on Metronet's failure to manage the contract. In June Metronet was refused access to loan facilities by banks and therefore approached the arbiter to commence an Extraordinary Review for BCV Infraco and increase the ISC to recover cost increases over the first review period prior to completing the review (PPP Arbiter, 2007). Metronet requested an increase in the ISC of £551.1m. However, the arbiter would only allow £121m to cover a period June 2007 to July 2008, as Metronet had not acted in an economic and efficient manner (NAO, 2009a). In July 2007 Metronet was forced into administration leaving TfL with debts of £1.7bn to repay the consortium's banks, with an overall direct loss to the taxpayer estimated to be between £170m and £400m (NAO, 2009a). The BCV and SSL Infracos returned to the public sector.

The NAO (2009a) identified Metronet's failure primarily as poor corporate governance, leadership and its tied supply chain. Many decisions had to be agreed unanimously by five shareholders, who were also Metronet's suppliers, each with different motivations. Executive management frequently changed and was unable to manage its tied supply chains (a shareholder dominated supply chain) (Gannon *et al.*, 2013). Suppliers within the supply chain were in control of the scope of work, being paid for extra work and had access to better cost information than management (NAO, 2009a).

First periodic review

Tube Lines were in negotiations with LU at the start of 2010 in preparation for the second review period (2010–17). The failings of Tube Lines dominated proceedings rather than the earlier success as reported by the London Assembly in 2007. Tube Lines estimated the upgrade of the Jubilee Line signalling had cost an additional £327m due to LU's scope changes and also contested LU had not allowed sufficient weekend closures (HC, 2010). The upgrade was scheduled to be complete at the end of December 2009; however, it was not to be finished until October 2010. The HC (2010) also reported that Tube Lines were in danger of repeating Metronet's tied supply chain where work would be seconded to parent companies without the required expertise and this had contributed towards upgrade delays. Tube Lines responded that its parent companies were used only after a competitive tendering exercise had taken place. The poor relationship between LU and Tube Lines further continued in to the periodic review (HC, 2010). During 2009 planning for the next review period (2010–17) LU and Tube Lines were unable to agree a cost figure for the future programme of work. LU forecast £4.00bn, Tube Lines £5.75bn and the PPP arbiter £4.40bn (PPP Arbiter, 2009). All parties were unable to agree a forecast cost for the second review period so TfL bought the Tube Lines shares for £310m in May 2010 and acquired £1.3bn of Tube Lines £1.6bn debt between August and October 2011 (TfL, 2011). Thus, JNP Infraco returned to public sector ownership.

Outcome versus policy

In 2005 the House of Commons Committee of Public Accounts published their report on LU's PPP having heard evidence from DfT, LU and external advisors. They were critical of the PPP's scope of works, the use of VfM as conclusive evidence, use of PPP rather than a bond issue, lenders' political risk perception leading to costs increasing by £450m, and bidders' significant success fees being reimbursed. In this report the committee showed how well PPP had delivered on its policy objectives set in 1997/8, contract close and the first review period, see Table 4.3.

With the demise of the PPP before the first review period in 2010 it is not possible to assess whether VfM was achieved as the contract was to be re-priced at each review period (every seven and a half years). However, the PPP did provide a stable source of funds for investment for nearly a period of seven and a half years that was not matched by central government. Whilst there was some evidence of private sector innovation and expertise the transfer of risk was limited.

Lessons learned

Optimistic timetables and costs

The NAO (2004a) focused their attention on three main areas to explain why the costs and timescales overran on the original DfT estimates (see Table 4.2). These included not knowing the final price until firm bids were submitted requiring contracts to be de-scoped to ensure the project was affordable, LU's constraints of not providing power for new trains required another round of re-bidding, and the political and legal interventions made by TfL. Furthermore, central government's intervention with the procurement process delayed the timetable by five months adding further costs. It is often the case that politicians work to a 'political' timetable for a project set to meet their own political agenda – in this case, to have contracts in place within a two-year period, before the Mayor was elected in May 2000. Additionally the DfT's time frame for project completion and costs were optimistic considering the project's complexity and the average time LU took to complete each of the five PFI's was three to four years.

Stakeholder support

Omitting LU, a key stakeholder, from initial discussions during the early stage of the project and not incorporating their analysis and informed opinions of independent experts and politicians meant that central government's preferred solution would be met with strong resistance. The House of Common's Transport Select Committee continuously criticised and questioned whether it was the right solution and ultimately labelled LU's PPP model fatally flawed (PAC, 2005). Due to the declining support for the PPP, central government

Table 4.3 Achievement of LUL's PPP policy objectives

PPP policy objectives	PPP policy May 1997/8	Contract award (2002/3)	First review period (2002/3–10)
Obtaining private sector investment and expertise to modernise the Tube			
Efficiency and innovation	Significant savings to be achieved.		Project efficiency savings of 17% – not achieved; some evidence of innovation however limited
Price competition with Infracos	Separate consortiums for each Infraco	Metronet awarded two Infracos, Tube Lines one infraco	Comparison until 2007 when Metronet went in to administration
Risk transfer to private sector	Cost and time overruns and performance to be largely transferred to private sector	Limitations on risk for known assets and exclusion of risk transfer to private sector for unknown assets. Risk remains with taxpayer – as 95% of senior debt guaranteed in event of termination	DfT paid £1.7bn repaying 95% of Metronet's debt obligation; and £1.3bn of Tube Lines debt
Guaranteeing value for money for taxpayers and passengers			
Stable long term funding	Remove inadequate and uncertain annual funding		
Minimal or no capital grant subsidy	Reduction of subsidy over time	Agreed for a seven-and-a-half-year period. £1.1bn per annum capital grant subsidy Infracos	
Better value than public investment programme	Financial analysis showed better than PSC	Not possible to determine due to review periods	Collapse of Metronet in 2007 and Tube Lines shares bought in 2010. Unable to assess VfM.
Safeguarding the public interest, which included the safe operation of the Tube			
Passenger operation, safety, marketing and ticketing remain in public sector	LUL manages operations – stations and trains; also network-wide PFI contracts (Prestige, Power and Connect, BT Police).		

Source: adapted from PAC (2005)

undertook a major offensive campaign through LU in defence of the partial privatisation (Lovelace, 2001). The appropriate involvement of key stakeholders within the PPP process is critical if the benefits of partnerships are to be realised (Fischbacher and Beaumont, 2005). However, with LU's PPP political support from key elected and unelected stakeholders was polarised and this continued from the project's start through to its eventual demise.

Central government's role

The Transport Committee expressed their concern over HMT's role during the PPP process, in particular the bidding process. The committee concluded 'HMT has recently begun to exert too much influence over the policy areas which are properly the business of other Departments and that is not necessary in the best interest of the Treasury or Government as a whole' (HC, 2002). The committee also concluded that the Chancellor of the Exchequer had imposed the PPP despite opposition from experts. The committee requested that a minister from HMT gave evidence to the committee; however, this was refused by HMT. Central government's dominance was evident in May 1999 when prior to the call for competition the government intervened and withdrew one of the three Infraco contracts. The government was possibly motivated to offer Railtrack exclusive single bidder status on the basis that national rail could be integrated with sub-surface lines and provide a pseudo version of Crossrail (Gannon, 2011). Delaying the procurement for SSL by five months offered consortiums a reduced opportunity to each win an Infraco, preventing bidders from achieving economies of scale bidding for three contracts. However, perversely this did improve bidding competition, for LU, restricting four consortiums to compete for two Infracos. The interference by central government did, however, come at a cost, in order to make the position attractive to bidders, reimbursement costs to maintain competition needed to be increased.

Limited risk transfer

The HC (2008) reported that Metronet's anticipated return was disproportionate with the level of risk associated with the contract; and the limited risk transfer to Metronet was ineffective. Metronet was posed to make nominal returns of up to 20 per cent (NAO, 2004a). Although considered a higher rate than other PFI/PPP contracts LU's PPP was comparable to road projects at 15 per cent (HC, 2008). The parent company's risk liability was £70m; having invested in Metronet they would only bear their own cost of inefficiency of up to £50m for cost overruns. Surprisingly, 95 per cent of senior debt was underwritten by central government given a BBB rating by agencies. Bidders perceived risks between the funding from central government and possible events at the end of each review period. This led to a cost increase of around £450m more in interest on debt payments (PAC, 2005).

Bid cost reimbursement

Bid reimbursement for successful and unsuccessful bidders totalled £270m according the NAO (2004a) and £275m according to PAC (2005), see Table 4.2. However, PAC (2005) reported unsuccessful bidders (LINC, Surface Lines Group) were reimbursed £25m, successful bidder Tube Lines was reimbursed £134m (£95m for winning JNP Infraco of which £39m covered success fees and £95m for resource costs of which £7m covered submission for unsuccessful BCV Infraco) and Metronet was reimbursed £116m (£50m for success fees and £65m for resource costs). The success fees were intended to compensate bidders for the opportunity cost of bid resources that could be used on other projects. The success fees are typically recovered through the ISC; however, pressure on affordability probably meant HMT wanted to treat this outside the ISC. However, if the success fees are excluded then Metronet's bid reimbursement costs (£65m) for two Infracos were disproportionate to Tube Lines' bid reimbursement costs (£88m) for one Infraco and one unsuccessful bid. Whilst Tube Lines provided their rationale for the substantial bid costs, PAC (2005) did highlight that some of Tube Lines' design work should have been covered in the ISC and there was a risk of double counting.

Scope uncertainty

The 'partnership' between LU and consortiums came under significant pressure particularly around the station modernisation programme specified within the contract, resulting in the scope for each station having to be agreed for each station (NAO, 2009a). The minimal specifications for stations led to different expectations between LU and consortiums; LU had more 'extensive interpretations' and consortiums 'less costly interpretations' (Gannon, 2010). The London Assembly (2005) reported LU and Metronet had major disagreements over scope of works that delayed work and added costs. However, Tube Lines took an alternative approach of carrying out the work and then negotiating with LU over scope at the end of the works rather than at the start (London Assembly, 2007). Scope and expectations need to be clear prior to contract award. This major uncertainty led to serious delays and costs for both the Metronet and the Tube Lines consortiums.

Value for money

The VfM analysis came under severe scrutiny and criticism by the Mayor's advisors and leading academics regarding its subjectivity and validity of assumptions for greater efficiency within the private sector, risks assessments, assumptions regarding bond financing and the 30-year period of comparison selected by LU (Shaoul, 2002). A fundamental question arises whether the assumptions used to justify the PPP model were entirely realistic? Whilst possible risks were identified as part of the risk analysis and adjustments made to LU's cost base to reflect possible risks and efficiencies of the private sector, these were later shown to be insufficient considering the scope changes and cost increases resulting from station modernisation. The NAO (2009b) reported that financial modelling can contain errors and too much decision-making emphasis can be placed on subjective judgements of risks that results in showing the PPP options being cheaper and leading to VfM. Using a 30-year period to assess VfM was also inappropriate especially as bids were only fixed up to the first review period. Furthermore, the wider factors highlighted by the NAO that influence the VfM analysis needed to be taken into consideration with bid evaluation. It was also apparent that expectations for risk transfer and efficiency gains were optimistic as it was evident the private sector faced the same problems of major cost overruns and time delays for the same reason as the public sector did when managing large-scale investments. The Industrial Society (Hutton, 2000) noted the average 17 per cent efficiency gains for PFI schemes was derived from redundancies and changes in work practices that would be difficult to replicate in an unionised LU and remarked the efficiency gains are 'probable rather than certain expectations'.

Metronet's tied supply chain

Metronet utilised a tied supply chain model that came under major scrutiny when contractual obligations were not being delivered. Whilst Metronet's tied supply chain used the traditional PFI/PPP model it failed to separate shareholder's interests sufficiently from the supply chain and to have in place the necessary corporate governance arrangements (HC, 2008). This arrangement is illustrated in the early years of LU's PPP whereby station modernisations work was awarded to Trans4m, a subsidiary of Metronet dominated by four of Metronet's five shareholders (Gannon *et al.*, 2013), without competitively tendering the work. Metronet's weak corporate governance arrangements prevented its Board capturing many of the issues that arose within the supply chain within the first review period. The Industrial Society (Hutton, 2000) highlighted this weakness in the PPP's supply chain, 'Infracos' members have an intense interest in wining business from the company they are supporting, so there is an ever present danger of sweetheart deals or contracts awarded in less than competitive circumstances' (Hutton, 2000). The HC (2008) reports 'Government must not allow this blurring between the roles of shareholder and supplier in future bids to carry out work by the private sector'. Where the shareholder and supplier role is blurred then a full investigation is required during bid evaluation by the contracting authority (HC, 2008), as frequently subcontracted work competitively tendered is likely to lead to the best price and bring the market discipline expected of PPP.

Chapter summary

Central government had a significant dilemma in attempting to solve LU's under-investment backlog: surrender to the popular voice of bond finance and keep LU in the public sector running the risks of continued poor public sector project delivery or opt for a PPP model and assume the private sector will deliver efficiencies and stable funding and remove LU as a financial liability from HMT. Key stakeholders were not convinced a PPP model was preferable to retaining LU in the public sector supported by stable funding conditional on LU delivering efficiency savings and performance improvements. HMT were fully aware of the principal-agent conflicts prior to the options analysis and therefore had no choice except to dominate the project direction and intervene where necessary, frequently moving outside of their sphere of operation and causing a major disruption to bidders and procurement costs. Despite the PPP architect's dogma, practice demonstrated the private sector could not deliver the expected efficiencies, risk transfer expected and VfM developed by advisors to justify the PPP option. The private sector experienced many of the same issues experienced by the public sector under traditional central government grant funding. Furthermore the House of Commons Transport Committee stated LU's PPP model was flawed; it could not provide the capacity enhancement required as the scope was reduced due to unaffordability, provide value for money and adequately transfer risk to the private sector. These flaws were further compounded by the poor evaluation of consortiums' supply chains at the bid evaluation stage. The, mistakenly, preferred PPP option justified by central government's perception of the public sector's poor record in major project delivery consequently failed. This perversely exposed the public sector to major risks and costs under an initiative intended to transfer risk to the private sector. Since the failure of LU's PPP and revocation of three Light Rail Transit schemes in 2006, there have been no further major metro PFI/PPP's undertaken in the UK (Gannon and Smith, 2009). Since the failure of LU's PPP and revocation of three light rail transit schemes in 2006, there has been no further PFI/PPP scheme of this type undertaken in the UK (Gannon and Smith, 2009) other than an extension to the tram system in Nottingham that was signed in 2011. Despite the costly mistakes to the public sector detailed within this chapter, lessons have been learned by the public sector that have supported the development of PFI2 launched by the UK government in 2012.

Reflections

1   What are the key lessons learned from the London Underground PPP and how have they influenced the procurement of subsequent transport-related PPPs in the UK?
2   What aspects of the LU PPP model could have been different in order to avoid the project being abandoned?
3   Since the 'failure' of the LU PPP, this model of procurement has been used globally to deliver transport projects. Can you provide any successful international examples of rail PPPs?
4   The report published by UK's National Audit Office (NAO), referred to in this chapter, provides further perspective on the LU PPP with a

particular focus on VfM. The report can be accessed at http://www.nao. org.uk/report/london-underground-ppp-were-they-good-deals.
5   The World Bank provides excellent resources on PPPs and a good place to start is http://ppp.worldbank.org/public-private-partnership.

References

Bolt C (2003). *Regulating London Underground,* The Annual Beesley Lecture, 13 November 2003. http://chrisbolt.me.uk/resources/Regulating+a+state+owned+co mpany.pdf (accessed 21 July 2014).

Butcher L (2012). London Underground PPP: Background. SN1307. House of Commons Library. http://www.parliament.uk/business/publications/research/briefing-papers/ SN01746/london-underground-after-the-ppp-2007 (accessed 19 July 2014).

Deloitte and Touche (2001). *London Underground Public-Private-Partnership Emerging Findings.* Deloitte and Touche Corporate Finance, 17 July.

Ernst and Young (2002). *London Underground PPP's: Value for Money Review.* Ernst and Young, 5 February.

Evening Standard (2002). £4m Bill as Ken Ends PPP Battle, *London Evening Standard,* 26 July. www.thisislondon.co.uk (accessed 10 August 2011).

Fischbacher M and Beaumont P (2005). PFI, Public-Private-Partnerships and the Neglected Importance of Process: Stakeholders and the Employment Dimension. *Public Money and Management,* **23** (3): 171–176, July.

Gannon M J (2002) Funding London's Underground Using PPP? European Transport Conference, PTRC, Hommerton College, Cambridge University, 9–11 September 2002. PTRC VR Ref: ETC2001.

Gannon M J (2006). Funding London Underground's Investment Programme Using Public-Private-Partnerships. *Focus: The Journal of the Chartered Institute of Logistics and Transport Focus,* **8** (3): 21–26, April.

Gannon M J (2008). Tube Failure Was Signalled, *Evening Standard,* 21 May, p49. Invited contribution by Evening Standard letters editor in response to Professor Tony Travers' (LSE) article on 'Boris Johnson: The London Mayor'.

Gannon M J (2010). Has Labour's Public-Private-Partnership Funding Policy for London Underground Finally Come to a Halt? *Focus: The Journal of the Chartered Institute of Logistics and Transport,* **12** (2): 24–29, February.

Gannon M J (2011). A Re-examination of the Public-Private-Partnership Discourse: Was PPP the Way to Upgrade London Underground's Infrastructure? European Transport Conference, Glasgow, 10–12 October.

Gannon M J and Smith N J (2009). *The Rise and Fall of Public-Private-Partnerships: How Should Public Sector Rail Transport Infrastructure Be Funded in the United Kingdom?* Association of European Transport, Netherlands, 9–11 October.

Gannon M, Male S and Aitken J (2013). Tied Supply Chains in Construction Projects: Lessons from London Underground's Public-Private-Partnership. In: Smith S D and Ahiaga-Dagbui D D (eds) *Procs 29th Annual ARCOM Conference,* Reading, UK, 2–4 September 2013, Association of Researchers in Construction Management, pp819–826.

Guardian (2007). Tube Contractor Pins Hopes on Review as Overshoot Hits £1bn, *The Guardian,* 23 May.

HC (1998). London Underground. HC 715. Select Committee on Environment, Transport and the Regional Affairs Committee, Seventh Report of Session 1997–8, 15 July.

HC (2002). London Underground. HC 387. Transport, Local Government and the Regions Committee, Second Report of Session 2001–2, 5 February.

HC (2008). The London Underground and the Public-Private-Partnership Agreements: Government Response to the Committee's Second Report of Session 2007–8. HC 461. 26 March 2008.

HC (2010). Update on the London Underground and the Public-Private-Partnerships (PPP) Agreements. House of Commons Transport Committee, Seventh Report of Session 2009–10, 17 March.

Hutton W (2000). *The London Underground Public Private Partnership: An Independent Review by The Industrial Society, Executive Summary.* Industrial Society, September.

London Assembly (2005). *The PPP: Two Years In, The Transport Committee's Scrutiny into the Progress of the PPP.* London Assembly Transport Committee, June.

London Assembly (2007). *A Tale of Two Infracos: The Transport Committee's Review of the PPP.* London Assembly Transport Committee, January.

London Transport (1996). *Planning London's Transport to Win as a World Class City.* London Transport, London.

Lovelace N (2001). LUL Offensive Promotes Private Tube Upgrade, *New Civil Engineer*, 23 August.

Mitchel B (2003). *Jubilee Line Extension from Concept to Completion.* Thomas Telford, London.

NAO (2000). The Financial Analysis for the London Underground Public Private Partnerships. HC 54. Session 2000–1, 15 December 2000.

NAO (2004a). London Underground PPP: Were They Good Deals? HC 645. National Audit Office, Report for the Comptroller and Auditor General, Session 2003–4, 17 June 2004.

NAO (2004b). London Underground PPP: Are They Likely to Work Successfully? HC 644. National Audit Office, Report for the Comptroller and Auditor General, Session 2003–4, 17 June 2004.

NAO (2009a). The Failure of Metronet. HC 512. Department of Transport, Session 2008–9, 5 June 2009.

NAO (2009b). *Private Finance Projects.* A Paper for the Lords Economic Affairs Committee, October.

PAC (2005). *London Underground Public Private Partnerships.* HC 446. House of Commons Committee of Public Accounts, Public Seventeenth Report of Session 2004–5, 9 March 2005.

PPP Arbiter (2005). Role of the PPP Arbiter, Office of the PPP Arbiter. http://www.ppparbiter.org.uk/role_of_arbiter_html (accessed 29 October 2005).

PPP Arbiter (2007). Arbiter Confirms Receipt of Extraordinary Review Reference from Metronet. Press Notice 02/07, 29 June 2007.

PPP Arbiter (2009). PPP Arbiter Announces Draft Decisions on Tube Lines Periodic Review. Press Notice 02/09, 17 December 2009.

RMT (2009). Understanding the PPP Arbiter's Draft Directions, 20 December 2009. http://www.rmtlondoncalling.org.uk/node/1145 (accessed 21 July 2014).

Shaoul J (2002). A Financial Appraisal of the London Underground Public-Private Partnership. *Public Money and Management,* **22** (2): 53–60, April–June.

TfL (2011). Transport for London Acquires Tube Lines' Bonds. Press Notice, 19 October.

5 Risk management in PPPs

Emerging issues in the provision of social infrastructure

Jennifer Firmenich and Marcus Jefferies

Chapter introduction

Life-cycle orientation is a driver of economical, ecological and social sustainability in three main ways. First, the optimization of life-cycle cost instead of investment cost usually leads to cost savings in the long run and thus economical sustainability. Second, the economical sustainability is linked to ecological sustainability via the likes of energy costs. For example, the level of current and future energy costs justifies higher initial investments for energy and thus cost saving. Third, the social sustainability of the infrastructure's users is influenced by the design of the operation, maintenance and replacement processes over the infrastructure's life time.

Life-cycle orientation is a key element of Public-Private Partnership (PPP) projects. PPP procurement is usually an alternative to traditional procurement. In theory, PPP is pursued if the Public Sector Comparator indicates that PPP generates efficiency gains leading to cost savings over the life cycle in comparison with traditional procurement. In practice, PPP is often used by the public client as a mean to finance the initial investment. In this case, the initial investment will be provided by the private consortium using loans and private equity. The funding is provided either by users or by the public authorities with an availability payment during the contractual operation period.

Therefore risk management of such life-cycle-oriented projects is an essential success factor to achieve the intended life-cycle optimization and/or alternative financing. The findings of this chapter provide a thorough theoretical grounding for the specific case study findings discussed by Jefferies and McGeorge in Chapter 9.

Risk management and allocation

PPP procures unique, long-term, multi-player projects over their life-cycle with private financing of the initial investment. The resulting complexity can only be managed with an appropriate project risk management. The risk management provides two main services: first, the pre-contractual project risk management aims for an initial project set-up that optimizes the risk allocation among the contracting parties on the one hand and ensures risk-bearing capacity for the project as a whole and the project players on the other hand. Second, the contractual risk management controls the project progress and allows for project adaption to unforeseen events in good time. In terms of a cost-benefit approach, the efforts made before contract signing should compensate for the problems avoided consequently in the contract period. The proactive pre-contractual risk management implies the potential for project optimization and is therefore the focus of this chapter. In practice, however, the pre-contractual risk management is often neglected due to time and cost pressure.

One major aspect of PPP risk management and an important precondition for a long-term stable and successful PPP project is the optimal risk allocation. The PPP's risk allocation has significance for each of the multiple PPP project players:

- The *public client* uses PPP to aim for efficiency, life-cycle orientation, alternative financing and risk transfer. If the risk can be quantified it should be taken by the player that minimizes the sum of net risk cost and risk mitigation cost. Therefore, the project's risk allocation should minimize the risk cost.
- The *special purpose vehicle* (SPV) represents the joint venture (JV) consisting of private companies towards the public client. On a first risk transfer level risks are transferred from the public client to the SPV. On a second risk transfer level the SPV transfers the risks to the other JV. The SPV will transfer as many risks as possible received from the public client to the operative players. The SPV collects user fees or availability payments and uses the cash for amortization and interest, operational cost, replacement reserves and dividends. The providers of financing want the SPV to take as little risk as possible.
- The *lenders* usually provide the majority of the initial investment's financing as loans to the SPV. It is their priority that amortization and interest are paid back as agreed. Therefore, they have a strong interest that the project is sound and set-up with the ability to deal with unforeseen events. They will ask for a due diligence to prove the risk-bearing capacity. The better the risk-bearing capacity, the lower the interest will be.
- The *sponsors* provide private equity to the financing of the initial investment. The equity is a crucial element of risk coverage and thus risk-bearing capacity. The share of equity required by the lenders depends on the macroeconomic circumstances and the project. The higher the equity share, the more expensive the financing cost of the project becomes though. The sponsors have an interest as well to have their equity repaid with an appropriate interest. Consequently, they share an interest with the lenders in appropriate risk allocation and risk-bearing capacity.

- The *contractor* is responsible for the construction, and often the design, of the infrastructure procured. The *operator* is responsible for the operation and maintenance, and potentially the replacements of the infrastructure procured. These operative players are at the end of the risk transfer chain and are therefore the ultimate risk recipients. They have an interest to only take those risks that they can actually manage and those risks for which they receive an appropriate risk premium from the SPV.

At least two risk transfer levels have to be considered for a PPP project assuming that the public client is the default risk taker. The first risk transfer level relates to the decision of what risks should be transferred from the public client to the private PPP JV in general. The JV's representative and main contact of the client is the special purpose vehicle (SPV). The more risks the public client transfers, the higher the total risk premium the risk sender has to pay to the risk recipient because the resulting risk load of the recipient needs to be covered. Those risks transferred to the SPV can be transferred in a next step to the players within the private PPP JV, which is the second risk transfer level. This concept of several risk transfer levels is visualized in Figure 5.1. This overview indicates the importance of appropriate project risk management and the optimal risk allocation for all players involved in a PPP.

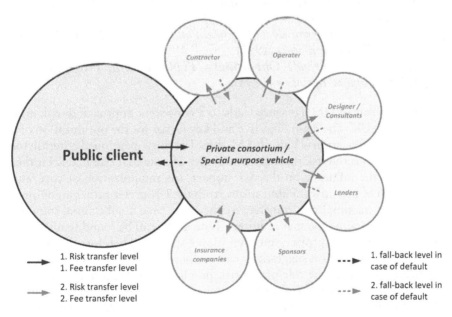

Figure 5.1 Risk transfer levels of PPP projects (Source: Firmenich, 2013)

State of practice and research regarding PPP risk management

To better understand the subject, a literature review has been conducted to determine the state of practice and the state of research regarding PPP risk management. It is assumed that the state of practice can be approached by analyzing reports, guidelines and other documents from public or other PPP-focused institutions. The state of research has been analyzed with a focus on peer-reviewed papers, PhD theses or other comparable scientific output. Primary geographical focus is laid on the United Kingdom, Germany, Switzerland and Australia for both analyses.

State of practice

The most mature PPP market in Europe is the United Kingdom (UK). Due to the implementation of the Private Finance Initiative (PFI) in 1992 more than 700 projects worth more than £55bn have been implemented (HM Treasury, 2012: p. 5ff.). The most relevant institutions for PPP in the UK are HM Treasury, Treasury Task Force, House of Commons, and National Audit Office. A selection of publications relevant for the PPP risk management comprises:

- HM Treasury: *A New Approach to Public Private Partnerships* (2012)
- The Treasury Committee: *Private Finance Initiative* (2011)
- National Audit Office: *Private Finance Projects* (2009)
- HM Treasury: *Value for Money Assessment Guidance* (2004)
- Treasury Task Force: *Private Finance: Technical Note No. 5 – How to Construct a PSC* (1999)
- Treasury Task Force: *Private Finance: Partnerships for Prosperity – The Private Finance Initiative* (1997)
- HM Treasury: *Private Opportunity, Public Benefit: Progressing the Private Finance Initiative* (1995).

The central role of HM Treasury leads to a consistent approach to risk allocation in the UK. The main objective and key factor for the optimization of a project's risk allocation is Value for Money. The most mentioned criterion for risk allocation is the risk management ability of potential risk taker. Further criteria are the influence on the risk impact, the minimization of cost, and the creation of incentives. Publications from the UK prefer rather an optimal instead of a maximal risk transfer. At the same time a substantial minimal risk transfer is pointed out as required. Tendencies can be found that the risk allocation should be project and even player specific. Therefore, hardly any standard risk allocation suggestions can be found. Further aspects in the UK publications include the role of the risk premium which should be minimal and the awareness that the ultimate responsibility always lies with the public client. In particular, it had been criticized that excessive risk transfer leads to an unnecessary increase of project cost. Additional dissatisfaction with slow and expensive tender processes, inflexible contracts over long contract terms, and insufficient transparency led to the launch of the PF2 initiative in the UK. The consequences of this initiative for risk allocation are that more risks are to be retained by the public to save cost and to preserve more flexibility in the face of unforeseeable risks. In Chapter 3, Winch and Schmidt provide a *review* of the UK's PFI approach and in Chapter 4, Gannon focuses on the *lessons learnt* from a case study of the London Underground PPP.

By contrast to the UK market, the German PPP market is advanced but less mature than the UK PPP market. An analysis of relevant German PPP institutions and according reports, guidelines, etc. reflects on the one hand a model role of the UK PPP market and on the other hand a different perception of optimal risk allocation. German publications on PPP risk management agree that optimal risk allocation goes hand in hand with maximal efficiency which can be understood as a synonym for Value for Money. The risk management ability and the influence of risk impact are mentioned as criteria for risk allocation as well. Furthermore, German PPP publications pick up the idea of optimal instead of maximal risk transfer too. They state that excessive risk transfer would lead to high-risk premiums and therefore reduce project efficiency. Additionally to the UK, the influence of the risk cause, the importance of risk-bearing capacity, and the willingness to take risks are brought up as relevant factors for PPP risk allocation. German publications mention further fresh thoughts in the context of PPP risk allocation:

- If the PPP risk allocation is not optimal, this might lead to opportunistic behavior.
- If a risk is not quantifiable, it should not be transferred to the private player.
- Risks should be considered over time

German publications suggest, more often than UK publications, a standard risk allocation. These suggestions are not clarified with traceable criteria for the determination of the risk allocation presented. This represents a contradiction to the assumption that a risk allocation needs to be project specific in accordance with the project's uniqueness.

The state of practice of very immature PPP markets, for example Switzerland, draws from the experience of more mature PPP markets and is then adapted to country-specific circumstances. Apart from country-specific publications, a couple of international institutions on the European or worldwide level have picked up the subject of PPP risk allocation. Relevant publications include:

- Organisation for Economic Cooperation and Development (OECD): *Recommendations of the Council on Principles for Public Governance of Public-Private Partnerships* (2012)
- IMF / De Palma, Leruth, Prunier: *Towards a Principle-Agent Based Typology of Risks in Public-Private Partnerships* (2009)

- World Bank / Delmon: *Private Sector Investment in Infrastructure Project Finance: PPP Projects and Risks* (2009)
- United Nations Economic Commission for Europe: *Guidebook on Promoting Good Governance in Public-Private Partnerships* (2008)
- OECD: *Public-Private Partnerships: In Pursuit of Risk Sharing and Value for Money* (2008)
- World Bank / Irwin: *Government Guarantees: Allocating and Valuing Risk in Privately Financed Infrastructure Projects* (2007)
- International Monetary Fund: *Public Private Partnerships* (2004)
- European Investment Bank: *The EIB's role in Public-Private Partnerships* (2004)
- EU Commission: *Green Paper on Public-Private Partnerships and Community Law on Public Contracts and Concessions* (2004)
- European Commission: *Guidelines for Successful Public-Private Partnerships* (2003)
- World Bank / Irwin, Klein, Perry, Thobani: *Dealing with Public Risk in Private Infrastructure* (1997)
- United Nations Industrial Development Organization (UNIDO): *BOT Guidelines* (1996).

These publications mostly pick-up the insights from the country-specific publications regarding PPP risk management and PPP risk allocation and put them in a different context. In particular, Delmon (2009) observes that in practice risks are allocated rather according to bargaining strength than according to maximal efficiency. This leads to risk takers that are not able bear the risk and therefore projects that are potentially instable over the long contract period.

In Australia, PPPs have become deeply embedded as an integral part of Government procurement strategies. As discussed by Hodge and Duffield (2010), Australian Governments have made very large commitments with the private sector under the aegis of PPPs. Quiggin (2004) subscribes to the view that the 2001 "Partnerships Victoria" policy document was a watershed in the development in Australia PPPs, and that it was this document that is representative of the approach adopted by the other Australian states and territories. Selected Australian Government publications include:

- Victorian Government: *Partnerships Victoria: Guidance Material* (2001)
- New South Wales Government: *Working with Government: Guidelines for Privately Financed Projects* (2001)
- Australian Commonwealth Government: *National Public Private Partnership Policy and Guidelines* (2008).

State of research

The state of research picks up the geographical differentiation of the state of practice and is then structured according to various *think tanks* to get an overview of the publication activities. In the next step, the identified publications are structured and analyzed according to their content.

The main PPP research think tanks in the mature European PPP market of the UK appear to be Glasgow Caledonian University, University of Central Lancashire, and University of Manchester. The contributing authors from these institutions include Akin Akintoye, Anthony Merna and Graham Winch. Key publications from these authors include Merna and Adams (1996); Merna and Owen (1998); Akintoye, *et al.* (2001); Merna and Njiru (2002); Akintoye, *et al.* (2003a); Akintoye, *et al.* (2003b); Merna and Lamb (2003); Li, *et al.* (2005); Akintoye and Beck (2008); and Winch, *et al.* (2011). Further relevant research from the UK consists of published work by Construction Industry Council (2000); Dixon, *et al.* (2003); Boussabaine and Kirkham (2004); Boussabaine (2007); Shaoul (2005); Medda (2007); Gannon (2010); and Gannon, *et al.* (2013).

Distinguished PPP think tanks in one of the advanced European PPP markets, Germany, are the Bauhaus University Weimar, the Technical University Berlin, and the Technical University Darmstadt. Switzerland, as an example for an immature European PPP market regarding practical implementation, does, however, have a PPP research think tank, the Federal Institute of Technology (ETH) Zurich.

In Australia it is the University of Melbourne, University of Newcastle, and Queensland University of Technology (QUT) that need to be added to the PPP research think tanks. The most relevant contributions are being made by Collin Duffield, Xioa Hua Jin, Pauline Teo, Adrian Bridge, and Marcus Jefferies. Publications from these authors include Jefferies, *et al.* (2002); Duffield (2005); Jefferies (2006); Jefferies and McGeorge (2008); Jin and Doloi (2008); Jefferies and McGeorge (2009); Jin and Doloi (2009); Hodge and Duffield (2010); Jin (2010a); Jin (2010b); Teo, *et al.* (2011); Jin and Zhang (2011); and Teo, *et al.* (2013).

Other relevant scientific publications regarding PPP risk management include: Yamaguchi, *et al.* (2001); Grimsey and Lewis (2002); Spiegl (2002); Abednego and Ogunlana (2006); Sun and Yang (2006); Zou, *et al.* (2008); Yun and Wei (2008); Wiggert (2009); Li and Ren (2009); and Brewer, *et al.* (2013).

The main *research content*, and indeed research questions, investigated in these publications tended to focus on risk management in general, and often and more specifically on optimal risk allocation. Due to the complex nature of the problem the question can be divided into several sub-problems on the one hand and can be approached from different point of views on the other hand. To understand the state of research differentiations can be made regarding what kind of research approach, what kind of theoretical background, and what kind of research methodology have been used.

The *empirical research* approach that aims to understand the state of practice is mostly used in the UK and Australia. Empirical work is also used in German publications (Girmscheid and Pohle, 2010; and Alfen, *et al.*, 2010) but less often. The building of descriptive models was conducted amongst others by the publications from the University of Manchester and the Bauhaus University Weimar. Generally, this includes a broader context like PPP risk management, PPP project phases, PPP players, etc. The building of decision models was pursued in particular by publications from ETH Zurich (Girmscheid, 2013; Firmenich, 2014).

The publications from German-speaking countries tend to use New Institutional Economics as a theoretical background, if any explicit theoretical background is used at all. Alternative theoretical backgrounds would be Neoclassical Theory, Behavioral Economics, etc. Usually, New Institutional Economics consists of three sub-theories that have been used exclusively by different relevant publications:

- Property Rights Theory: Pfnür (2009); Pfnür, *et al.* (2010); Frank-Jungbecker (2011)
- Principal Agent Theory: Kowalski (2004); Miksch (2007); De Palma, *et al.* (2009); Schetter (2010); Frank-Jungbecker (2011)
- Transaction Cost Theory: Construction Industry Council (2000); Beckers and Miksch (2002); Kowalski (2004); Miksch (2007); Frank-Jungbecker (2011).

It can be observed that various methodologies are applied to solve the problem of optimal risk allocation, amongst others the following:

- Monte Carlo Simulation
- real options
- artificial neuronal networks
- fuzzy logic
- stochastic processes
- game theory.

Simulations, in particular *Monte Carlo Simulation* and its variations, are a popular methodology to deal with the problem of optimal PPP risk allocation. As every project is unique it is not possible to work probabilistically on an empirical data basis. Instead a fictive set of projects is simulated. Monte Carlo Simulation as a tool for PPP risk management has either been recommended or used by the following publications: Merna and Owen (1998); Boussabaine and Kirkham (2004); Smith, *et al.* (2006); Cadez and Streuer (2006); Pfnür and Eberhardt (2006); Gürtler (2007); Alfen, *et al.* (2010); and Stichnoth (2010).

Real options consider and quantify future decision options. Publications that mention real options in the context of PPP are amongst others Miksch (2007); Chiara and Garvin (2007); Cheah and Garvin (2009); and Bi and Wang (2009).

Artificial neuronal networks help to model causalities as a pre-condition for decision models. The set-up network needs to be trained with empirical data to be used on real world problems. Publications that mention or use artificial neuronal networks in the context of PPP are amongst others Boussabaine and Kirkham (2004); Jin (2010b); and Jin and Zhang (2011).

Fuzzy logic is used to model imprecise and colloquial values. It is often used to translate qualitative subjective expressions into precise quantitative values using fuzzy sets and special functions. Publications that use fuzzy logic in the context of PPP are amongst others Luu, *et al.* (2006); Jin and Doloi (2007); Yun and Wei (2008); Bi and Wang (2009); Jin and Doloi (2009); and Jin (2010b). Sachs (2007) shows in a broader context how fuzzy logic can be used to quantify subjective expert estimations.

Stochastic processes can be used to simulate a probabilistic value over time. This has been done in the context of PPP amongst others by Schöbener, *et al.* (2007) and Schetter (2010).

Game theory is a quantitative approach to understand the behavior of players in certain situations under given constraints. Game theory helps to understand incentives, information asymmetries, and opportunistic behavior. Furthermore, it enables deriving and testing actions, for example to improve the outcome of the contract development for PPP projects. Publications that use game theory include amongst others Roggencamp (1999) and Medda (2007).

Other promising and emerging approaches are *portfolio theory* or *Bayesian networks*.

The major scientific PPP *collaboration* in Europe is the COST Action TU1001 "Public Private Partnerships in Transport: Trends and Theory." One of the work groups focuses on risk management. The whole collaboration is supposed to be finished successfully in 2014 with more published results expected.

Findings

The literature review shows that the relevant publications are numerous and multi-sided. This requires structuring for better understanding and analysis. In this literature review the publications have been structured according to the following criteria:

- the geography of their origin (the UK as representation of a mature European PPP market, Germany as representation for an advanced European PPP market, Switzerland as representation for an immature PPP market, and other geographical origins, such as Australia, as an emerging PPP market);
- the nature of their origin (reports and guidelines from public or other PPP institutions vs those from scientific think tanks);
- the nature of the content in case of scientific publications.

The publications that determine the state of practice tend to ensure practicability and put the problems in an appropriate, bigger context. The problem solving suggestions, however, are often too simple or superficial to address the complexity of the underlying problem properly. The publications that determine the state of research, on the contrary, tend to focus on specific sub-problems that are solved on a very detailed level using complex methodologies. These solutions are often not embedded in a practical context and therefore the knowledge transfer to the industry cannot take place.

A meta-analysis (Firmenich, 2011) using systematic review confirmed that the research activity regarding PPP risk allocation is intense. At the same time the study shows that no consensus could yet be achieved in the research community regarding what the main objectives, decision criteria and influence factors for an optimal risk allocation should be. This seems to be reinforced by language barriers. The publications from German-speaking countries are still often not published in English. This would explain why German publications adapt the findings from the UK but not vice versa (for example the concept of risk-bearing capacity).

Project risk management in a nutshell

Based on the findings from the literature review and research conducted by the author, the following sections present the most important aspects of PPP project risk management.

Understanding risk

Depending on the context, risk is perceived and defined in many different ways. According to Schnorrenberg, *et al.* (1997), risk is seen as a variation of uncertainty. If the probability of occurrence (PO) of an event is PO = 0 or PO = 1, this event is not a risk but a certainty. If the event was identified and the probability of occurrence as well as the impact can be quantified, this event is called a risk. If the event was identified but the probability of occurrence could not be quantified, this event counts as uncertainty of first degree. All unidentified events count as uncertainty of second degree.

In the context of life-cycle-oriented infrastructure projects, risks are related to uncertain events that have a cause and lead to an impact. This impact usually represents a deviation from the reference scenario. Potential positive deviations are chances and potential negative deviations are threats. Many risks have this two-sided character. This reflects the potential for entrepreneurial action and success. To understand a risk it is not only important to differentiate between cause and impact. In particular for complex projects it should be clear what phases, players, and tasks may cause a risk or may be affected by a risk.

Once identified, risks can be assessed qualitatively (i.e. low, middle, high) or quantitatively. A quantitative assessment facilitates the relation between a risk analysis and the reference scenario in the given case. If the risk assessment is done quantitatively, there is still the alternative of deterministic and probabilistic assessment. For quantitative deterministic risk assessment a single risk is assessed regarding its probability of occurrence $(0 < PO < 1)$ and the impact (I as monetary unit). The deterministic risk costs R are the result of the multiplication of these two values $(R = PO * I)$. The most complex and powerful risk assessment is the quantitative probabilistic approach. It can be processed relatively easily by first determining the probability of occurrence independently as before. Secondly, the impact is assessed as a three-point assessment where experts estimate the minimal, the modal and the maximal value of the risk's potential impact. Thirdly, this information can be transferred into a probability and/or distribution function with the help of according software. These functions can be used to derive the total aggregated project risk with Monte Carlo Simulation (Hertz, 1964; Metropolis and Ulam, 1949; Vose, 1996; and Firmenich and Girmscheid, 2013a).

When trying to understand risks, the human brain is usually confronted with cognitive limitations that often lead to misconceptions (Vose, 1996). As countermeasure it should be considered to work in preferably interdisciplinary teams and, most of all, to follow a systematic and well-documented approach to project risk management.

Project risk management process

The typical risk management process begins with the risk identification. Only identified risks can be managed. The risk identification is followed by the risk assessment. The risk assessment should be as detailed as necessary and as simple as possible. Complex projects, however, certainly benefit from a refined assessment of identified risks, for example in a quantitative probabilistic way. An optional step is the risk classification. A risk classification is useful if only part of the identified risk should or can be managed further. This should be done only if limited resources require it. The identified and assessed risks will then be analyzed in the light of risk mitigation alternatives (see Table 5.1). If possible, the risks expected should be mitigated proactively by reducing either the probability of occurrence or the impact of the risk or both before signing the contract. The cost of a risk mitigation action should be less than the difference between gross and net risk cost. In the end, the risk has to be accepted by one of the contracting parties or has to be avoided. The risk management steps take place in particular for the pre-contractual project risk management to determine the contractual risk allocation and a financially sound project (see Figure 5.2). The steps are repeated with an increasing level of detail over the typical project phases before contract signing (adequacy test, preliminary efficiency test, tender phase, and Public Sector Comparator). During the contract period risk controlling and reactive risk mitigation become more important.

Every risk management module has another objective, needs different inputs, and produces different outputs. There is a set of methodologies that helps to succeed with every step. An overview is given in Figure 5.3. These risk management tasks are mainly qualitative information acquisition. The information needed for pre-contractual project risk management cannot be acquired by statistical or experimental analysis. The information needs to be estimated subjectively by experts.

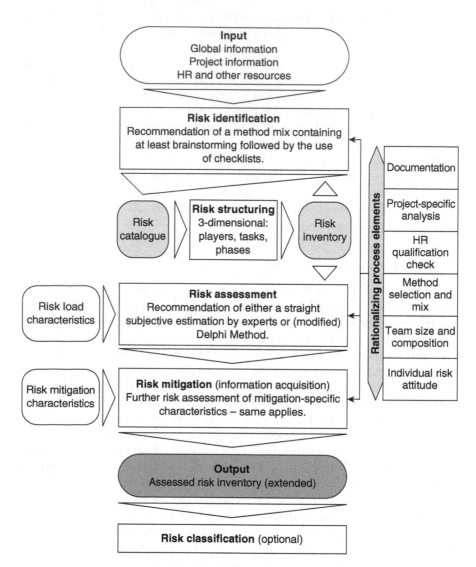

Figure 5.2 Rational information acquisition for pre-contractual risk management
(Source: Firmenich and Girmscheid, 2013b)

Risk transfer principles

The risk allocation for PPP contracts needs to be seen in the context of a
risk mitigation concept. The transfer of risk is one of several risk mitigation
alternatives as demonstrated in Table 5.1. The underlying assumption is
that all risks are taken by the client by default (potential risk sender). The
PPP contracts are a tool to transfer the risks to the other contract parties

	Module: risk identification	Module: risk assessment	Module: risk classification	Module: risk mitigation
Objective	Identification of as many risks as possible related to the PPP project under consideration of the cost-benefit ratio	Assessment of identified risks (impact, probability and further characteristics)	Classification and structuring of identified and assessed risks regarding their importance relative to risk costs	Identification and assessment of risk mitigation measures for each risk and player under consideration of measure cost and residual / net risk
Input	• **Specific project information** • **Global environmental information** • Human resources • Other resources	• **Risk catalogue or inventory** • Specific project information • Global environmental information • Human resources • Other resources	• **Risk inventory** • Human resources • Other resources	• **Risk inventory** • Specific project information • Global environmental information • Human resources • Other resources
Output	Risk catalogue (unstructured) or risk inventory (structured)	Assessed risk inventory	Assessed and classified risk inventory	Assessed and classified risk inventory (extended)
Process elements	Determination of risk structuring	Identification of relevant risk load characteristics	Method application	Identification of relevant risk mitigation characteristics
	Documentation, HR qualification check-up, selection and mix of process specific methods			
	• Project-specific risk analysis • Group size and composition	• Project-specific risk analysis • Group size and composition • Risk attitude and risk strategy		• Project-specific risk analysis • Group size and composition • Risk attitude and risk strategy
Methodical re-commendation	• Contract analysis • **Brainstorming** • Risk check lists	Human estimation: • Expert estimation or • Classic round-based Delphi Method or • Modified real-time **Delphi Method**	**Portfolio analysis** • ABC analysis • Equi-Risk-Contour method • Impact analysis • Sensitivity analysis	• Decision table • Decision tree method • **Utility analysis** • Simulation
Threats	• Important risks not identified • Inappropriate resource use	• Inappropriate assessment of risk load characteristics because of bounded rationality, opportunism, subjectivity or group dynamics • Inappropriate resource use	• Non-consideration of important risks because of risk classification results • Interpretation or conclusion mistakes because of bounded rationality, opportunism, subjectivity or group dynamics	• Inappropriate assessment of risk handling because of bounded rationality, opportunism, subjectivity or group dynamics • Inappropriate resource use

Figure 5.3 Overview pre-contractual risk management modules (Source: Firmenich and Girmscheid, 2013b)

Table 5.1 Overview risk mitigation alternatives

Risk mitigation alternative	Risk consequence	Net risk	Financial consequences for sender	Differentiation	Risk type
Avoidance	PO = 0	Gross risk eliminated	Risk mitigation cost	Cause oriented	Active
Reduction, via HR, technical, or organisational measures	PO↓ Impact↓	Net risk < gross risk (reduced)	Risk mitigation cost + net risk cost		
Transfer (insurance) Transfer (contract parties): 1. level, 2. level, etc.	(Residual) risk eliminated for sender and premium paid from risk sender to risk recipient	Net risk = gross risk (un-changed)	Insurance premium Transfer premium	Impact oriented	Passive
Acceptance	Risk coverage necessary		Gross risk cost		

PO = Probability of occurrence

Source: Firmenich (2013)

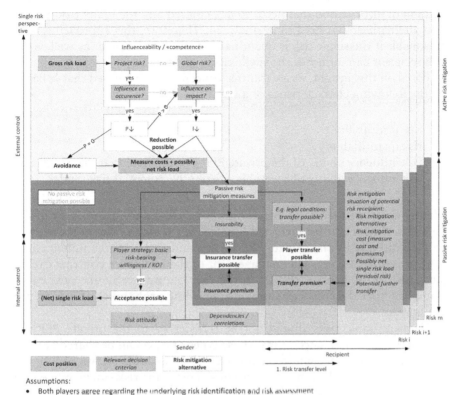

Figure 5.4 Elements and causalities of risk mitigation (Source: Firmenich, 2013)

(potential risk recipient). If the transfer takes place, the price for the con-tracted product and/or service rises.

Girmscheid (2013) proposes an approach to allocate risks quantitatively according to minimal cost under consideration of the player's risk mitiga-tion options. On the one hand, the inherent logic of a quantitative approach serves the aimed-for rationality. On the other hand, the cost-oriented optimi-zation represents an acceptable criterion to achieve Value for Money.

In theory, the risk mitigation alternatives have to be analyzed for every single risk and a risk mitigation concept for all relevant players needs to be developed. If this is done for the potential risk sender and the potential risk recipient, the player's risk mitigation situation and the according cost can be compared. In the end, the risk should be taken by the player that can handle the risk cheapest under consideration of potential residual risk. The elements of a risk mitigation analysis and the causalities are displayed in Figure 5.4.

In practice, this approach requires a lot of resources because of the amount of information that needs to be acquired and the many risk mitigation alternatives that need to be analyzed. Therefore, this approach is only reasonable if the single risk is quantifiable and the risk sender as well as the risk recipient can influence the single risk.

Based on this insight, a risk transfer concept can be derived that is based on the following characteristics of a risk:

- risk insurability
- risk quantifiability
- risk influence ability of the private PPP JV / the potential risk recipient
- risk influence ability of the public client / the potential risk sender.

Mainly due to reduction of uncertainty it is advised to insure every insurable risk. This might be neglected for strategic reasons. The premium will be paid directly by the private PPP JV and indirectly by the public client via the availability fee. The management of the insurance contract can be delegated to the private PPP JV. If the risk is uninsurable the next question is whether the risk is quantifiable regarding probability of occurrence and impact.

Unquantifiable risks should not be transferred to the private PPP JV. Due to the according uncertainty, those types of risks can be shared at most between the public client and the private PPP JV including a risk cap for the private player. Finally, it needs to be considered if the risk can be influenced by the potential risk sender and/or the potential risk recipient. The most important transfer principles are proposed:

- If the risk is uninsurable, quantifiable, and can only be influenced by one player, it should be taken by this player.
- If the risk is uninsurable, quantifiable, and can be influenced by both players, this requires further analysis and leads to a quantitative decision-making approach based on comparison of the players' risk mitigation situation.
- If the risk is uninsurable, quantifiable, and no player can influence the risk, it should be shared (without cap).
- If the risk is uninsurable, unquantifiable, and the private PPP JV cannot influence it, the risk should be taken by the public client.
- If the risk is uninsurable, unquantifiable, and the private PPP JV can influence it, the risk should be shared with a risk cap for the private PPP JV.

The described causalities are visualized in Figure 5.5. The qualitative approach proposed includes rational and traceable criteria. Furthermore, it is resource friendly and thus practicable for industry application.

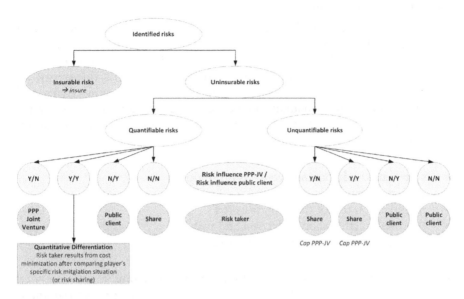

Figure 5.5 Risk transfer decision tree (Source: Firmenich, 2013)

Risk-bearing capacity testing

The risk allocation should be determined as rationally as possible for the benefit of all players involved. In any case, the risk transfer proposed should be tested against a side condition to ensure long-term project success, the risk-bearing capacity. The risk-bearing capacity depends on the risk allocation and the risk premiums paid. The risk load resulting from the risk transferred with a certain probability should be covered with according financial reserves resulting from the fee paid by the risk sender to the risk recipient. If the risk coverage is too low, risk transfer needs to be reduced or fees need to be increased. If the risk coverage is too high, more risks can be transferred or the fees can be reduced. Under consideration of these causalities, the risk-bearing capacity testing not only ensures a sound risk allocation but creates a traceable relation between risk transfer and risk premium.

Even after risk transfer, the risk sender bears the ultimate responsibility. If the risk recipient defaults the risk falls back to the risk sender (see Figure 5.1). Therefore, the risk sender has an interest only to transfer those risks that can be managed better by the risk recipient and pay the appropriate risk premium for the transfer. It the project's risk allocation is optimal and the risk transfer is accompanied by appropriate risk premiums, the whole project is sound. Furthermore, all players share a common destiny. Due to the project's dependencies, this ideal risk allocation and risk premium situation creates the right incentives: every player has an interest in the other players not failing.

Rational information acquisition

The weakest chain in the link of processes described above is the rational acquisition of information needed for decision making. It is inevitable that the rational information acquisition will be based on subjective expert opinions. Therefore, the quality of the outcome is threatened by bounded rationality (in the sense of cognitive limitations), subjectivity, and the opportunism of individuals as well as social dynamics within teams. Typical actions to avoid those threats of rational information acquisition are (see Figure 5.2):

- systematic and documented decision making
- project-specific risk analysis
- qualification test of the experts asked
- thorough methodology selection and mix
- appropriate team size and composition.

Considering the complexity of a PPP project it is strongly recommended to use interdisciplinary and diverse expert teams for the project risk management to enhance the output quality and usability. For practicability it is crucial though to optimize the cost-benefit ratio at the same time.

Chapter summary

Due to the diversity and multi-sidedness of publications so far, it would be desirable that further publications on PPP risk management would consolidate the status quo. Research dealing with PPP risk management is part of the applied sciences. It should be ensured that the knowledge developed by the scientific community can and will be transferred to the industry. This can be supported by not only solving a singular sub-problem. Research should be put in a bigger context and should take place in cooperation with the industry to ensure that their requirements are considered. The interdisciplinary nature of the subject asks for systematic knowledge transfer for actual scientific problem solving from different disciplines like mathematics, engineering, operations research, business, economics, finance, law, etc. It is assumed that the project risk management in the industry could benefit from a more interdisciplinary approach as well. The cost and time pressure in the pre-contractual project phase is a fact that should not be ignored by the academic research efforts, if aiming for practicability of the results. At the same time, the priorities of the industry players need to adapt. It is desirable that the industry changes from a "reactive short-term cost minimization" approach to a "proactive long-term profit maximization" approach. This becomes particularly evident in the handling of the project risk management.

Reflections

1   To date, what have been the main aspects of risk management investigated in PPPs and is a new research gap emerging?
2   How do the varying PPP markets of UK, Europe, and Australia compare?
3   The European Commission (EC) and the UK and Australian Governments have produced various *Guidelines for PPPs* that include significant *risk* components. Are these risk profiles still relevant? The recent UK example from 2012 is referred to by Winch and Schmidt in Chapter 2, the Australian example, published in 2008, can be viewed at: http://www.infrastructureaustralia.gov.au/public_private/files/National_PPP_Guidelines_Overview_Dec_08.pdf. The EC version, published in 2003, can be viewed at: http://ec.europa.eu/regional_policy/sources/docgener/guides/ppp_en.pdf.
4   Has the recent global economic crisis influenced the way in which Governments procure public infrastructure?
5   What research methods are appropriate for investigating current issues associated with risk management in PPP projects?

References

Abednego, M. P. and Ogunlana, S. O. (2006). Good project governance for proper risk allocation in public-private partnerships in Indonesia. *International Journal of Project Management*, 24: 622–634.
Akintoye, A. and Beck, M. (eds) (2008). *Policy, Management and Finance for Public Private Partnerships*. Wiley-Blackwell, Oxford.
Akintoye, A.; Beck, M.; Hardcastle, C.; Chinyio, E.; Asenova, D. (2001). Risk mitigation practices under the Private Finance Initiative environment. *COBRA*, Glasgow Caledonian University, pp. 1–12.
Akintoye, A.; Beck, M.; Hardcastle, C. (2003a). *Public-Private Partnerships: Managing Risks and Opportunities*. Blackwell Science, Oxford.
Akintoye, A.; Hardcastle, C.; Beck, M.; Chinyio, E.; Asenova, D. (2003b). Achieving best value in private finance initiative project procurement. *Construction Management and Economics*, 21 (5): 461–470.
Alfen, H. W.; Rieman, A.; Leidel, K.; Fischer, K.; Daube, D.; Frank-Jungbecker, A.; Gleissner, W.; Wolfrum, M. (2010). *Lebenszyklusorientiertes Risikomanagement für PPP-Projekte im öffentlichen Hochbau - Abschlussbericht zum Forschungsprojekt*. Bauhaus-Universität Weimar, Weimar.
Beckers, T. and Miksch, J. (2002). *Die Allokation des Verkehrsmengenrisikos bei Betreibermodellen für Strasseninfrastruktur*. Technische Universität Berlin, Berlin.
Bi, X. and Wang, X. F. (2009). The application of fuzzy-real option theory in BOT project investment decision-making. *IEEE 16th International Conference on Industrial Engineering and Engineering Management*, Beijing, 21–23 October, pp. 289–293.
Boussabaine, A. and Kirkham, R. J. (2004). *Whole Life-cycle Costing Risk and Risk Responses*. Blackwell Publishing, Oxford.
Boussabaine, H. A. (2007). *Cost Planning of PFI and PPP Building Projects*. Taylor & Francis, New York.

Brewer, G.; Gajendran, T.; Jefferies, M.; McGeorge, D.; Rowlinson, S.; Dainty, A. (2013) Value through innovation in long-term service delivery: facility management in an Australian PPP. *Built Environment Project and Asset Management*, 3 (1): 74–88.

Cadez, I. and Streuer, U. (2006). Stochastische Risikoanalyse bei Public Private Partnership-Infrastrukturprojekten - Monte-Carlo-Simulation als Hilfsmittel zur Durchführung der Risikoanalyse für Sponsoren, Kreditgeber, Monoliner und Finanzberater. In Kapellmann, K. D. and Vygen, K. (eds) *Jahrbuch Baurecht*, Werner, Neuwied, pp. 287–324.

Cheah, C. Y. J. and Garvin, M. J. (2009). Application of real options in PPP infrastructure projects: opportunities and challenges. In Akintoye, A. and Beck, M. (eds) *Policy, Finance and Management for Public-Private Partnerships*, Wiley-Blackwell, Chichester, UK, pp. 229–249.

Chiara, N. and Garvin, M. J. (2007). Using real options for revenue risk mitigation in transportation project financing. *Transportation Research Record*, 1993: 1–8.

Construction Industry Council (2000). *The Role of Cost Saving and Innovation in PFI Projects*. London.

De Palma, A.; Leruth, L.; Prunier, G. (2009). *Towards a Principal-Agent Based Typology of Risks in Public-Private Partnerships*. IMF Working Paper WP/09/177.

Delmon, J. (2009). *Private Sector Investment in Infrastructure Project Finance: PPP Projects and Risks*. Wolters Kluwer, Austin, TX.

Dixon, T.; Jordan, A.; Marston, A.; Pinder, J.; Pottinger, G. (2003). *Lessons from UK PFI and Real Estate Partnerships*. BRE, London.

Duffield, C. F. (2005). PPPs in Australia. In Ng, Thomas S. (ed.) *Public Private Partnerships: Opportunities and Challenges*, Centre for Infrastructure and Construction Industry Development, University of Hong Kong, pp. 5–14.

European Commission (2003). *Guidelines for Successful Public-Private-Partnerships*. Brussels.

European Commission (2004). *Green Paper on Public-Private Partnerships and Community Law on Public Contracts and Concessions*. Brussels.

European Investment Bank (2004). *The EIB's Role in Public-Private Partnerships (PPPs)*.

Firmenich, J. (2011). Identification of parameters for a quantitative risk allocation model with systematic review. *Sixth International Structural Engineering and Construction Conference: Modern Methods and Advances in Structural Engineering and Construction*, Zurich, pp. 267–272.

Firmenich, J. (2013). Risk allocation decision-making concept for PPP projects. *Global Challenges in PPP: Cross-sectoral and Cross-disciplinary Solutions?* Antwerp, pp. 229–238.

Firmenich, J. (2014). *Rationale Risikoallokation und Sicherstellung der Risikotragfähigkeit für PPP-Projekte im Hochbau*. Unpublished PhD Thesis, ETH Zurich.

Firmenich, J. and Girmscheid, G. (2013a). Probabilistic time-specific risk load of a PPP building project. In Kajewski, S., Manley, K. and Hampson, K. (eds) *Construction and Society: Proceedings of the 19th International CIB World Building Congress*, paper no. 223, Queensland University of Technology, Brisbane.

Firmenich, J. and Girmscheid, G. (2013b). Rational information acquisition for quantitative holistic risk allocation of PPP projects in an entrepreneurial context. *PPP Conference* Preston, UK, pp. 229–238.

Frank-Jungbecker, A. (2011). *Verkehrsmengenrisiko bei PPP-Projekten im Strassensektor - Determinanten effizienter Risikoallokation.* Unpublished PhD Thesis, Bauhaus-Universität Weimar.

Gannon, M. J. (2010). Has Labour's public-private-partnership funding policy for London Underground finally come to a halt? *Focus: The Journal of the Chartered Institute of Logistics and Transport*, 12 (2): 24–29.

Gannon, M.; Male, S.; Aitken, J. (2013). Tied supply chains in construction projects: lessons from London Underground's public-private-partnership. In Smith, S. D and Ahiaga-Dagbui, D. D (eds) *Procs 29th Annual ARCOM Conference*, Reading, UK, 2–4 September 2013, Association of Researchers in Construction Management, pp. 819–826.

Girmscheid, G. (2013). Risk allocation model (RA model): the critical success factor for public-private partnerships. In de Vries, Piet and Yehoue, Etienne B. (eds) *The Routledge Companion to Public-Private Partnerships*, Routledge, Abingdon, UK, pp. 249–300.

Girmscheid, G. and Pohle, T. (2010). *PPP - Stand der Praxis: Risikoidentifizierung im Strassenunterhalt und Risikoverteilungskonzept.* ETH Zürich, Zürich.

Grimsey, D. and Lewis, M. K. (2002). Accounting for public private partnerships. *Accounting Forum*, 26 (3–4): 245–270.

Gürtler, V. (2007). *Stochastische Risikobetrachtung bei PPP-Projekten.* Unpublished PhD Thesis, Technische Universität Dresden, Renningen.

Hertz, D. B. (1964). Risk analysis in capital-investment. *Harvard Business Review*, 42: 95–106.

HM Treasury (1995). *Private Opportunity, Public Benefit: Progressing the Private Finance Initiative.* London.

HM Treasury (2004). *Value for Money Assessment Guidance.* London.

HM Treasury (2012). *A New Approach to Public Private Partnerships.* London.

Hodge, G. A. and Duffield, C. F. (2010). The Australian PPP experience: observations and reflections. In Hodge, G. A., Greve, C. and Boardman, A.F. (eds) *International Handbook on Public-Private Partnerships.* Edward Elgar, Cheltenham, UK, pp. 399–438.

International Monetary Fund (2004). *Public-Private Partnerships.*

Irwin, T. C. (2007). *Government Guarantees: Allocating and Valuing Risk in Privately Financed Infrastructure Projects.* World Bank, Washington, DC.

Irwin, T. C.; Klein, M.; Perry, G. E.; Thobani, M. (1997). *Dealing with Public Risk in Private Infrastructure.* World Bank, Washington, DC.

Jefferies, M. C. (2006). Critical success factors of public private sector partnerships: a case study of the Sydney SuperDome. *Engineering, Construction and Architectural Management*, 13 (5): 451–462.

Jefferies, M. and McGeorge, D. (2008). Public-private partnerships: a critical review of risk management in Australian social infrastructure projects. *Journal of Construction Procurement, Special Edition: Building Across Borders*, 14 (1): 66–80.

Jefferies, M. and McGeorge, W.D. (2009). Using public-private partnerships (PPPs) to procure social infrastructure in Australia. *Engineering, Construction and Architectural Management*, 16 (5): 415–437.

Jefferies, M. C.; Gameson, R.; Rowlinson, S. (2002). Critical success factors of the PPP procurement system: reflections from the Stadium Australia case study. *Engineering, Construction and Architectural Management*, 9 (4): 352–361.

Jin, X. H. (2010a). Determinants of efficient risk allocation in privately financed public infrastructure projects in Australia. *Journal of Construction Engineering and Management (ASCE)*, **136** (2): 138–150.

Jin, X. H. (2010b). Neurofuzzy decision support system for efficient risk allocation in public-private partnership infrastructure projects. *Journal of Computing in Civil Engineering*, **24** (6): 525–538.

Jin, X. H. and Doloi, H. (2007). Risk allocation in public-private partnership projects: an innovative model with an intelligent approach. *The Construction and Building Research Conference of the Royal Institution of Chartered Surveyors*, Atlanta, GA, pp. 1–13.

Jin, X. H. and Doloi, H. (2008). Interpreting risk allocation mechanism in public-private partnership projects: an empirical study in transaction cost economics perspective. *Construction Management and Economics*, **26** (7): 707–721.

Jin, X. H. and Doloi, H. (2009). Modeling risk allocation in privately financed infrastructure projects using fuzzy logic. *Computer-Aided Civil and Infrastructure Engineering*, **24** (7): 509–524.

Jin, X. H. and Zhang, G. M. (2011). Modelling optimal risk allocation in PPP projects using artificial neural networks. *International Journal of Project Management*, **29** (5): 591–603.

Kowalski, T. (2004). *Darstellung vertraglicher Risikotransfer-Mechanismen bei PPP-Projekten unter besonderer Beachtung der Anreiz-Beitrags-Problematik.* Braunschweig.

Li, B. and Ren, Z. M. (2009). Bayesian technique framework for allocating demand risk between the public and private sector in PPP projects. *6th International Conference on Service Systems and Service Management*, Xiamen, China, pp. 624–628.

Li, B.; Akintoye, A.; Edwards, P. J.; Hardcastle, C. (2005). The allocation of risk in PPP/PFI construction projects in the UK. *International Journal of Project Management*, **23** (1): 25–35.

Luu, D. T.; Ng, S. T.; Chen, S. E.; Jefferies, M. C. (2006). A strategy for evaluating a fuzzy case-based construction procurement selection system. Advances in *Engineering Software,* 37 (3): 159–171.

Medda, F. (2007). A game theory approach for the allocation of risks in transport public private partnerships. *International Journal of Project Management*, **25** (3): 213–218.

Merna, A. and Adams, C. (1996). Appraising risk in concession projects. In Merna, A. and Smith, N. J. (eds) *Projects Procured by Privately Financed Concession Contracts*, Asia Law & Practice Ltd, Hong Kong.

Merna, A. and Lamb, D. (2003). *Project Finance: The Guide to Value and Risk Management in PPP Projects.* Euromoney Books, Oxford.

Merna, A. and Owen, G. (1998). *Understanding the Project Finance Initiative.* Asia Law & Practice Ltd., Hong Kong.

Merna, T. and Njiru, C. (2002). *Financing Infrastructure Projects.* Thomas Telford, London.

Metropolis, N. and Ulam, S. (1949). The Monte Carlo Method. *Journal of the American Statistical Association*, **44** (247): 335–341.

Miksch, J. (2007). *Sicherungsstrukturen bei PPP-Modellen aus Sicht der öffentlichen Hand, dargestellt am Beispiel des Schulbaus.* Unpublished PhD Thesis, Technische Universität Berlin.

National Audit Office (2009). *Private Finance Projects*. London.

Organisation for Economic Cooperation and Development (2008). *Public-Private Partnerships: In Pursuit of Risk Sharing and Value for Money*. OECD Publications, Paris.

Organisation for Economic Cooperation and Development (2012). *Recommendations of the Council on Principles for Public Governance of Public-Private Partnerships*. OECD Publications, Paris.

Pfnür, A. (2009). Möglichkeiten und Grenzen der Risikoallokation zur Effizienzsteigerung von PPP-Projekten. In *Unternehmertum und Public Private Partnership - wissenschaftliche Konzepte und praktische Erfahrungen*. Gabler, Wiesbaden, pp. 27–52.

Pfnür, A. and Eberhardt, T. (2006). Allokation und Bewertung von Risiken in immobilienwirtschaftlichen Public Private Partnerships. In Budäus, D. (ed.) *Kooperationsformen zwischen Staat und Markt*, Nomos Verlagsgesellschaft, Baden-Baden, pp. 159–188.

Pfnür, A.; Schetter, C.; Schöbener, H. (2010). *Risikomanagement bei Public Private Partnerships*. Springer, Berlin.

Quiggin, J. (2004). Risks, PPPs and the public sector comparator. *Australian Accounting Review*, 14: 51–62.

Roggencamp, S. (1999). *Public Private Partnership - Entstehung und Funktionsweise kooperativer Arrangements zwischen öffentlichem Sektor und Privatwirtschaft*. Peter Lang, Frankfurt am Main.

Sachs, T. (2007). *Quantifying Qualitative Information on Risks (QQIR) in Structured Finance Transactions*. Unpublished PhD Thesis, Nanyang Technical University, Singapore.

Schetter, C. (2010). *Finanzierung öffentlicher Infrastrukturmassnahmen im Rahmen von Public Private Partnership*. Unpublished PhD Thesis, Technische Universität Darmstadt.

Schnorrenberg, U.; Goebels, G.; Rassenberg, S. (1997). *Risikomanagement in Projekten Methoden und ihre praktische Anwendung*. Vieweg, Braunschweig.

Schöbener, H.; Schetter, C.; Pfnür, A. (2007). *Reliability of Public Private Partnership Projects under Assumption of Cash Flow Volatility*. Technische Universität Darmstadt, Darmstadt.

Shaoul, J. (2005). A critical financial analysis of the Private Finance Initiative: selecting a financing method or allocating economic wealth? *Critical Perspectives on Accounting*, 16: 441–471.

Smith, N. J.; Merna, T.; Jobling, P. (2006). *Managing Risk in Construction Projects*. Blackwell Publishing, Oxford.

Spiegl, M. (2002). *Ein alternatives Konzept für Risikoverteilung und Vergütungsregelung bei der Realisierung von Infrastruktur mittels Public Private Partnership unter International Competitive Bidding*. Unpublished PhD Thesis, Universität Innsbruck.

Stichnoth, P. (2010). *Entwicklung von Handlungsempfehlungen und Arbeitsmitteln für die Kalkulation betriebsphasenspezifischer Leistungen im Rahmen von PPP-Projekten im Schulbau*. Unpublished PhD Thesis. Universität Kassel.

Sun, D. X. and Yang, L. M. (2006). Risk allocation of BOT project based on the optimal risk responding strategies. *CRIOCM 2006 International Research Symposium on Advancement of Construction Management and Real Estate*, Beijing, pp. 401–407.

Teo, P. L.; Bridge, A. J.; Gray, J.; Jefferies, M. C. (2011). Developing a research method to test a new first-order decision making model for the procurement of public sector major infrastructure. *Proceedings of the 27th Annual Association of Researchers in Construction Management (ARCOM) Conference*, Bristol, UK.

Teo, P. L.; Bridge, A. J.; Gray, J. (2013). A new first order decision making model for the procurement of public sector infrastructure: procedures and testing. In Kajewski, S., Manley, K. and Hampson, K. (eds) *Construction and Society: Proceedings of the 19th International CIB World Building Congress*, paper no. 1, Queensland University of Technology, Brisbane.

The Treasury Committee (2011). *Private Finance Initiative*. London.

Treasury Task Force (1997). *Partnerships for Prosperity: The Private Finance Initiative*. London.

Treasury Task Force (1999). *How to Construct a Public Sector Comparator*. London.

United Nations Economic Commission for Europe (2008). *Guidebook on Promoting Good Governance in Public-Private Partnerships*. Geneva.

United Nations Industrial Development Organization (1996). *Guidelines for Infrastructure Development through Build-Operate-Transfer (BOT) Projects*. UNIDO, Vienna.

Vose, D. (1996). *Quantitative Risk Analysis: A Guide to Monte Carlo Simulation Modelling*. Wiley, Chichester, UK.

Wiggert, M. (2009). *Risikomanagement von Betreiber- und Konzessionsmodellen*. Unpublished PhD Thesis, Technische Universität Graz.

Winch, G. M.; Onishi, M.; Schmidt, S. E (2011). *Taking Stock of PPP and PFI around the World*. ACCA Research Report 126, London.

Yamaguchi, H.; Uher, T. E.; Runeson, G. (2001). Risk allocation in PFI projects. In Akintoye, A. (ed.) *17th Annual ARCOM Conference*, University of Salford, 5–7 September. Association of Researchers in Construction Management, **1**: 885–894.

Yun, C. and Wei, T. (2008). The model of risk allocation in BOT Expressway project. *International Conference on Information Management, Innovation Management and Industrial Engineering*, Taiwan, pp. 283–286.

Zou, P. X. W.; Wang, S. Q.; Fang, D. (2008). A life-cycle risk management framework for PPP infrastructure projects. *Journal of Financial Management of Property and Construction*, **13** (2): 123–142.

6 Design management of infrastructure projects

A comparative case study analysis of Design-Build (DB) and Construction Manager/General Contractor (CMGC) projects in the U.S.

Giovanni Migliaccio and Edward Minchin

Chapter introduction

Relationship contracting (RC) is a fluid concept that can manifest itself differently in construction practice. The pervasive nature of RC is seen in the way construction entities cluster at the project level, through the practices of joint venturing and alliancing independently from the project delivery system and contract, or strategically over time across projects that are delivered through different delivery systems. Conversely to this fluid nature, there is a widespread understanding that RC strongly relies on the establishment of alignment among all parties involved, including the client, the designer, and the builder.

A recent study has analyzed the content of 11 standard contracts for different project delivery systems to identify their adherence to eight behavior norms underlying relational contract theory: Role Integrity, Reciprocity, Flexibility, Contractual Solidarity, Reliance and Expectations, Restraint of Power, Propriety of Means, Harmonization of Conflict (Harper and Molenaar, 2014, p. 1335). Harper and Molenaar found that RC norms often permeate at various levels from any standard construction contract "[making] construction contracts, regardless of the contracting method, relational exchange agreements in theory" (2014, p. 1337). Still, they found Integrated Project Delivery contracts to be more relational than Design-Build (DB), and Construction Management/General Contracting (CMGC) contracts with Design-Bid-Build (DBB) being the less relational as expected by the adversarial nature of these contracts. Findings of this previous study rely on the analysis of contracts for construction services—with the exception of DB contracts that include design and construction services in the scope of work. However, the well-known Influence and Expenditures Curve for the Project Life Cycle (CII, 1995, p. 6) has highlighted the importance of early lifecycle phases on overall project success, including planning and design.

Whereas achieving alignment among all parties may be more effective during the design phase, very little is known in terms of how RC may manifest itself during this phase. In the United States, the three main methods for delivery transportation projects are DBB, DB, and CMGC with DBB maintaining a *de facto* separation between design and construction functions. Relying on the results of a study funded by the National Academies through their National Cooperative Highway Research Program (NCHRP) (Minchin *et al.*, 2014), this chapter analyzes design management practices for DB and CMGC infrastructure projects. The narration will focus on how project outcomes have been successfully achieved through effective design management and will provide a critical comparison of various approaches. The conclusions will also relate the findings on contract administration to the same RC norms that were used by Harper and Molenaar in their study.

Background

The traditional approach for separately procuring and managing design and construction services for public works has served North American public transportation agencies well for most of the past century. The foundation of this approach, often called Design-Bid-Build (DBB), is the principle of selecting designers based on qualifications (Brooks Act – Public Law 92–582) and selecting builders based on competitive sealed bids with award of a lump sum or unit price contract to the lowest responsive and responsible bidder. Bid prices are almost always based on complete or almost complete design (see Figure 6.3 part a). Over the decades, DBB has provided US taxpayers with a large portfolio of functional, safe, and efficient transportation facilities at the lowest initial price that responsible, competitive bidders can offer. The Interstate system as well as almost all state and county roads have been delivered through this traditional approach. DBB has in most cases effectively prevented favoritism in spending public funds and has provided checks and balances through separate contracts with the designer and the builder while stimulating competition in the private sector. Under DBB, the agency has retained full control over design management, often seen as an advantage of this approach. In this chapter, the authors have defined design management as the approach used by agencies to organize and oversee the process of designing the transportation infrastructure.

While adequate for most construction projects, DBB has also demonstrated various drawbacks, including fostering adversarial relationships among the project parties, limiting innovation, and often resulting in serious growth in project cost and duration. In addition, it may not necessarily provide the best value to the owner for all project circumstances or types. In recent years, this issue has become more pressing for public works agencies in the United States, as deteriorating infrastructure and increasing population create tremendous pressure to move critical projects quickly from the planning stage through design and into construction without a commensurate increase in available funding. Underlying these external budget and time pressures is the basic requirement of maintaining quality in all phases of the highway program. Thus, there is a continuing need for highway agencies to review and evaluate alternative procurement and contracting procedures that promote improved efficiency and quality. As a result, other approaches to procuring and delivering transportation projects have been introduced over the last 20 years that better promote RC.

The wide range of options for project delivery methods available today is a relatively recent development for publicly funded infrastructure projects in the United States. DBB was the only way to deliver transportation construction projects until the introduction of Design-Build (DB) in the Intermodal Surface Transportation Efficiency Act of 1991. Another step was taken in 1996 when the Federal Acquisition Reform Act explicitly authorized the use of DB for federal projects. After that, the TEA-21, Public Law 105–178, allowed state departments of transportation (DOTs) to receive federal funding for DB contracts if the enabling state-level legislation was in force (TEA-21 1998). Figure 6.1 shows a detailed timeline of events that led to widespread adoption of DB for public works. Subsequent to the successful experience of using DB on several projects, many states passed new legislation and codes to allow alternative project delivery methods such as DB and Construction Manager/General Contractor (CMGC),

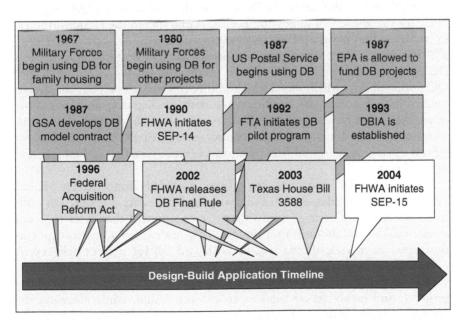

Figure 6.1 Design-Build timeline (Source: Molenaar *et al.*, 1999; Rein *et al.*, 2004)

which was derived from Construction-Management-at-Risk (CMR) and remains quite similar.

Many public agencies have implemented DB to accelerate project delivery with positive outcomes demonstrated by consistent research results (Hale *et al.*, 2009; Minchin *et al.*, 2013). DB has advantages, including single point of responsibility (combining the designer and builder under a single contract), accelerated delivery, collaboration, and incentivization for innovation. However, its currently prevalent form, which is based on a contract awarded competitively mostly on price, also presents certain disadvantages to agencies, including less agency control over design and a preference on the part of most designers to work closely with the owner instead of acting as consultants to a builder (Knight *et al.*, 2002). In fact, under DB, design and construction services are provided by a single contractual entity, the Design-Builder that often contracts out design services. Whether design is self-performed by the Design-Builder or a service provided by a design consultant to the Design-Builder, the management of design services is substantially different from what agencies use under DBB. Whereas the agency is highly involved in design activities under DBB, involvement is limited under DB to contractually-allocated responsibilities for quality control and assurance (QC/QA) except on occasions where the owner might produce a preliminary design. Any further involvement results in potential change orders and affects the initial DB contract price, which is usually set at a 10–30 percent design complete range as shown in Figure 6.3 part d. Various scenarios can be used in the industry, including allocating design QC responsibilities to the Design-Builder and retaining QA responsibility for the agency; or allocating both QA and QC to the Design-Builder; or securing the services of an independent QA firm as shown in Figure 6.2 that depicts the project organization for the Texas State Highway 130 DB project completed in 2007. More rarely, the agency retains full QA/QC responsibility, which increases its ability to check closely for the design quality but also increases inefficiencies, risk of disputes, and may slow down the design review schedule. Under any approach, the line of communication between designers and the agency goes through the Design-Builder who is often a design-build firm or a joint venture made up of builders and design firms.

Historically, DB contracts have been awarded competitively, based on best value formulas that heavily weighted price. As a result, DB contracts are strongly built around this price component that later impedes owner-builder collaboration on design. This has caused some transportation agencies to seek alternatives to DBB and DB for project delivery. A promising alternative that has generated interest in the highway sector, CMGC may offer some of the same advantages as DB related to expediting projects, and involving contractors in project design, while allowing the agency to retain control of design (through a separate contract with the designer) as in DBB. Previous studies have found that adding CMGC to a

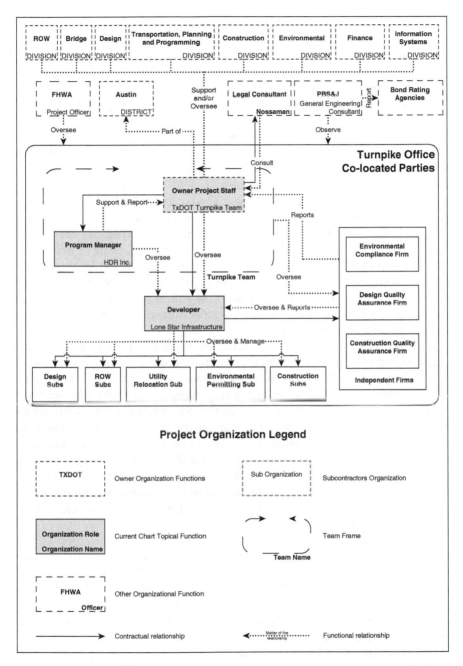

Figure 6.2 State Highway 130 project organization

DOT's delivery toolbox provides several benefits (Gransberg and Shane, 2010; McMinimee *et al.*, 2009). First, CMGC provides DOTs with an alternative option when DB and DBB are not able to satisfy contrasting project objectives. CMGC is an integrated team approach applying professional management during the planning, design, and construction of a project. The team consists of the owner, the designer, sub-designers, the builder, and subcontractors. As in the case of DBB, the owner contracts separately for design and construction services. However, CMGC enhances early contractor involvement (ECI) by retaining a builder with the role of a construction manager (CM) about the same time as the designer, typically through a qualifications-based or best-value selection process (see Figure 6.3 part b). During preconstruction, the CM acts as an advisor, providing professional services to the owner. A CM performs constructability reviews, cost estimates, construction phasing and schedules, and budget recommendations to determine the best options for the owner based on the project budget. The CM also may perform duties not typically performed by builders, such as assisting in securing financing or selecting or helping in the selection of designers. Once design activities mature, the level of scope definition improves to a level that would provide an efficient pricing of the construction activities. At this time, the CM role evolves into a role similar to that of a traditional GC, and the construction entity becomes "at risk" during the construction phase.

This is the two-step nature of CMGC contracting with the second phase usually starting within the 60–90 percent design complete range as shown in Figure 6.3 part c. In this second phase, the construction entity awards subcontracts in either a lump sum, or cost-reimbursable format with a guaranteed maximum price (GMP) contract. When the construction entity is bound to a GMP, the most fundamental character of the relationship is changed. In addition to acting in the owner's interest, the CM must manage and control construction costs to not exceed the GMP to protect its own interest (AIA-MBA Joint Committee, 2014). The two most important advantages of CMGC are: (1) It strongly and tangibly encourages innovation by parties to the contract; and (2) it offers a flexibility that allows the parties to easily allocate and re-allocate risk throughout the life of the contract.

More recently, some industry sectors are introducing a variation to DB called progressive DB (PDB) that attempts to overcome some of the issues with traditional price-competitive DB. PDB adopts the two-step nature of CMGC contracting. Under PDB, the agency first selects a Design-Builder based on qualifications, then, it works jointly with all the parties associated with this entity, including the designer, to evolve the design to the point that the Design-Builder can price it. At this point, a new contract, either lump sum or cost-plus-fee with guaranteed maximum price can be signed.

PDB has only been recently introduced in some industry sectors, such as the water, mass transit and building sectors; but its implementation may still be considered at the experimental stage, as any assessment of its

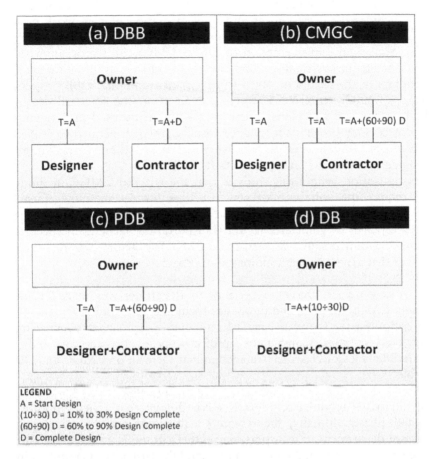

Figure 6.3 Timed contractual frameworks

performance is purely anecdotal. Therefore, PDB was excluded from the scope of this chapter. The chapter, instead, continues with a description of common issues and knowledge gaps that led to the identification of a set of research objectives and a methodology for a case study analysis of transportation projects completed through DB and CMGC with the intent of identifying common RC features of design management under these project delivery methods in the United States.

Research motivation and methodology

Changes in the approach for delivering public works are often a response by public agencies to changes in their organizational environment. Once an agency has decided to pursue DB or CMGC to deliver its projects, it is forced to seek ways to adapt to the new approach as there are certain broad concepts that must be understood by all parties involved. This adaptation process includes the development and implementation of new work processes within new project organizational structures (Migliaccio *et al.*, 2008, p. 486). Successful implementation of DB or CMGC in a programmatic fashion often requires a significant and aggressive change in the culture and philosophies of the parties involved from that of traditional DBB projects (Migliaccio, 2007). In terms of design management, the standard design methods, schedules, and plans review stages frequently used in designing DBB projects may prove inadequate or insufficiently accelerated to realize the advantages of these alternative delivery methods, making the task more challenging for designers and agency staff.

The motivation behind this study is that the DB and CMGC methods would require the utilization of new practices for design management. Therefore, the authors focused their study on the identification of practices and development of guidelines to aid transportation agencies in successfully implementing design management on CMGC or DB projects. Research objectives that drove the methodological approach were:

- to review owners' recent experience in design services management under CMGC and DB, and synthesize them into case studies;
- to critically assess the relative merits of alternative approaches to managing key aspects of the design that affect project scope, quality, and cost; and
- to highlight lessons learned from design management on CMGC and DB projects.

These objectives required a research methodology that relied on three sequential phases. Initially, the research team contacted, by telephone, every state department of transportation (DOT) in the United States. This included 52 agencies including Puerto Rico and the District of Columbia. In addition, 13 non-DOT public transportation agencies were included in the study due to their expertise with DB and CMGC. An initial round of phone interviews was carried out with the personnel identified by the agency as the individuals most knowledgeable on the agency's design process under DB and/or CMGC. This first round of interviews was performed using a structured questionnaire that included strategic, exploratory questions regarding the agency's recent experience with design services under DB and CMGC. While it was found that not all DOTs have enough experience with either system, 18 agencies were deemed to have the most experience and information to offer and were asked to participate in a second round of data collection, which included the performance of in-depth phone interviews and the establishment of an additional e-mail correspondence aimed at obtaining data and information from their projects. From the in-depth questions, critical assessments were made regarding the relative merits of alternative approaches to managing key aspects of the design that affect implementation, project scope, quality, and cost.

The results of these two-phased phone interviews guided the selection of programs and projects that could provide an in-depth diverse portfolio of sample case study implementations of design management procedures. Agencies chosen for case studies were visited by one or two team members. During these visits, the team conducted detailed interviews and gathered specific information from various parties, including agency staff and consultants, designers and builders. Between 6 and 20 individuals were interviewed at each of the programs visited. Interview findings and analyses of documentation were then used to create a series of program-level or project-level case studies.

Design-Build case studies

Program-level

- Maryland State Highway Administration (MSHA)—MSHA started using DB in 1998. Since then, the MSHA has delivered almost 40 highway and bridge projects using this delivery method. This case describes the programmatic effort of SHA in implementing DB for highway projects.
- North Carolina Department of Transportation (NCDOT)—NCDOT started using DB in 1999. To date, NCDOT has delivered almost 90 highway and bridge projects using the DB system. This case describes the programmatic effort of NCDOT in implementing DB for highway projects.
- Utah Department of Transportation (UDOT)—UDOT has been one of the pioneer states in terms of implementing innovative delivery methods. Starting with the successful implementation of DB for the I-15 Corridor Reconstruction project (1996–2001; $1.6B), this delivery method has been institutionalized and extensively used by UDOT as shown in Figures 6.4, 6.5, and 6.6. This case describes the programmatic effort of UDOT in implementing DB for highway projects. In addition to analyzing UDOT documentation about the program, four DB projects were analyzed: (1) Pioneer Crossing, Lehi—15 American Fork Interchange (2008–10; $175m); (2) SR-154 Bangerter at 7800 S, 7000 S, and 6200 S (2010–12; $40m); (3) I-15 at 11400 South Interchange (2008–11; $245m); and (4) I-15 South Layton Interchange (2009–11, $95m).

Project-level

- Bundle 401 Project, Oregon Department of Transportation (ODOT)— Bundle 401 was one of the bundled contracts awarded during the delivery of the OTIA III State Bridge Delivery Program. This bundled contract consisted of replacing five concrete bridges on Oregon Route 38 between Elkton and Drain. Oregon Route 38, also known as Umpqua Highway No. 45, is a state highway connecting the city of Reedsport on the Pacific coast with Interstate 5. The total bundle cost was $46,390,721. The

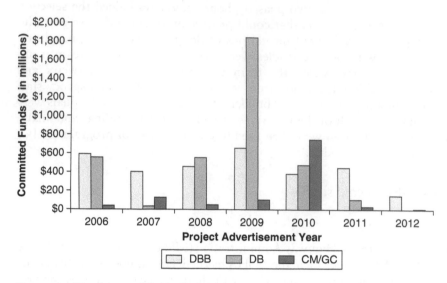

Figure 6.4 Total annual funds committed to projects using each delivery system

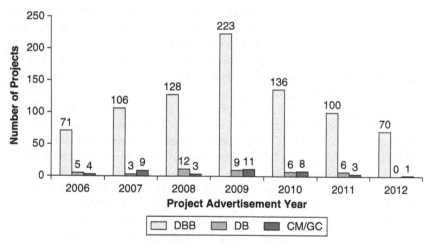

Figure 6.5 Comparison of number of advertised projects per year

Notice to Proceed (NTP) for DB procurement was issued in November 2005 and the project was completed in June 2009.

* I-15 Core Project, UDOT—In 2004, UDOT initiated the process to expand I-15 in Utah County. In March 2008, the Legislature authorized funding for the project and directed UDOT to complete the project scope and assemble a management team. Due to the 2008 financial crisis, the Legislature lowered the budget from $2.63 billion

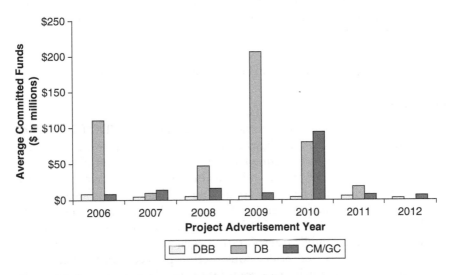

Figure 6.6 Average size of projects using each delivery system

to $1.73 billion still mandating the reconstruction of at least 15 miles of highway. To meet these requirements, UDOT selected DB as the project delivery method and fixed price/best design as its procurement approach. The agency established the contract value, challenging proposers to submit a design providing the highest value, while meeting the schedule deadline and minimizing inconveniences for the public. The fixed price/best design procurement approach proved extremely successful. The selected Design-Builder proposed to reconstruct 24 miles of the corridor, whereas the agency only expected reconstruction of 15 miles. Construction operations began in April 2010 and concluded December 2012, two years ahead of schedule.

- SR 99 Bored Tunnel Project, Washington State Department of Transportation (WSDOT)—The Alaskan Way Viaduct is a two-mile-long, double-deck, elevated section of State Road 99. Second in traffic volume to only I-5 in the state of Washington, the viaduct is a major north-south corridor through downtown Seattle carrying about 110,000 vehicles per day. In January 2009, WSDOT, King County, and the City of Seattle agreed to replace the Viaduct with a single bored tunnel under downtown Seattle, the SR 99 Tunnel. The procurement process started in October 2009. In December 2010 WSDOT awarded the contract to build the SR 99 Bored Tunnel Alternative for over $1 billion. WSDOT issued a first NTP for the preliminary design while waiting for the completion of the environmental permitting process. Afterward, the agency issued a second NTP in August 2011 for the final design and construction. Construction activities started at the end of 2011 and were expected to finish by the end of 2015. Recently, the project has encountered a drawback due to the failure of the Tunnel Boring Machine, which is expected to significantly delay its completion.

Construction Manager/General Contractor case studies

Program-level

*   Utah Department of Transportation (UDOT)—The UDOT has a long history of innovation in highway and bridge construction contracting. Among its many such accomplishments is the execution of the largest (up to that time) DB project in US history. The I-15 reconstruction project, built for Salt Lake City's 2002 Winter Olympics, was also the largest project ever undertaken by the state of Utah. The success of this high-visibility project gave UDOT the reputation as one of the nation's most innovative public transportation agencies and showed other agencies that highway and bridge construction projects can be successfully completed using a delivery system other than DBB. Having proven the viability of DB, UDOT turned its sights on developing a new system providing contracted parties the benefits of DB along with the benefits of DBB. The system they turned to was the CMGC delivery system. UDOT now has built more than 25 projects with CMGC since 2005, and is, therefore, the state agency most experienced in using this method on a large variety of projects.
*   City of Phoenix, Arizona—The City of Phoenix has built more than 200 projects using what they call a CMR construction project delivery system since initiating the system in 2000. Only recently has the city commenced using CMR for horizontal construction, totaling 12 horizontal CMR projects since their first project, let in 2008.
*   Osceola County, Florida—The CMR program in Osceola County was initiated under great controversy due to the long-term instability of the county road building program, and political pressures to complete and execute a major infrastructure plan. As a result, the program was under an ultimatum from the County Commission to have nine projects under contract within one year, when only one was under contract at the time. The new administration boldly decided to implement CMR, an untried delivery system to meet the target, adding to the controversy. To begin this program, a tremendous training effort was initiated, focusing first on the design community.
*   Utah Transit Authority (UTA)—UTA has used the CMGC construction project delivery system on five major projects since 2002. At the time of the case study, the $2.5 billion cost of these projects may be more than any other agency has spent on CMGC projects.
*   Oregon Department of Transportation (ODOT)—ODOT has used the CMGC construction project delivery system on three projects since 2011. When using this system, ODOT employs several methods of managing

March 2009
Funding

May 2009
Program Manager

June 2009
Designer

September 2009
Contractor

February 2010
Construction Starts

Figure 6.7 Project timeline from funding through commencement of construction

post-award design activities. Their process allows design professionals to adjust their plans with "real-time" information provided by the CMGC firm. There are written standard operating procedures for the design of CMGC projects, and the agency now utilizes these contracts, because while the agency had only worked on one CMGC project at the time of the study, this delivery method had worked well.

Project-level

- Mountain View Corridor (MVC) Project, UDOT—The MVC is a 15-mile "planned" freeway in western Salt Lake County and northwestern Utah County servicing 13 municipalities. There actually were three contracts on this project. A small one upfront included early order items such as girders and some canal crossings that had to be done at certain times of the year, a flexibility made possible because of the CMGC process. This process also allowed all contractual parties to get onboard quickly after funding was secured, as shown in Figure 6.7. Eventually, the information available was enough for the development of a complete set of final plans, but that was deemed unnecessary since the project operated on a system of continuing pricing.

Findings

This section summarizes the most important facts and recommendations that were generated by the NCHRP study. Many of the lessons learned and best practices for DB and CMGC are common to both systems; however, many are appropriate for only one. This is not unexpected, since both are fast-tracked construction delivery systems that are only as similar to the traditional contraction project delivery system as they are to each other.

Overall, the fast-track nature of both DB and CMGC methods leads to a short-term need for increased plan production rates. This places additional requirements on the designers, such as extended work hours, to keep pace with the acceleration and changes proposed by the builder. Successful implementation also often requires that a project be broken into additional multiple "mini" phases, enabling the constructor to start work early in areas where right-of-way (ROW) permits have been obtained and utilities relocations have been completed. Early work packages can be broken down into such items as retention ponds, partial clearing and grubbing, constructing on friendly parcel takes, which requires more design effort than traditional "station-to-station" designs. Standard items under the designer's oversight, such as utility coordination during design, partially transfer to the constructor to accelerate utility relocations, advance order long-lead items, have one "point" of responsibility with the utility companies. These shifts in responsibilities are often required for the builder to take responsibility for the overall project schedule and budget.

Design-Build

After almost two decades of use in transportation, DB is a tested delivery system that is most often preferred when delivery time is critical. A majority of state transportation agencies have already used DB, even if several agencies have only used it for a few projects. It seems that a few common barriers existed for all agencies in their first-time use. First, public agencies often needed to obtain legislative authority to employ the system and to use procurement approaches specific to DB. In addition, the approach to manage design activities is also different from traditional DBB delivery. As a result, different approaches to design management under DB were developed that were specific to the context and constraints of a specific agency. Since, a large majority of local transportation agencies and a small group of state transportation agencies have not used DB, the authors only highlighted these different approaches without recommending one over the other to provide a diverse range of practices as an outcome of the study. The rest of this section outlines a few of the design management features typical to DB and describes how an agency has approached them, which seems crucial to effective design management.

As explained in the introductory section, lump sum contracts are prevalent in DB delivery. Under these contracts, designers lack a contractual relationship with the agency. With few exceptions, they are usually consultants to the Design-Builder even if the agency often contracts with a separate designer for preparing the conceptual design to be included in the tender. Similarly to CMGC, designers are required under DB to take a much more active role in working with the builder (almost always the leading DB entity), but their ability to communicate directly with the agency is limited when compared to DBB and CMGC. This is due to the nature of a lump sum contract where the price is tied to certain design assumptions, so the Design-Builder may tend to be defensive regarding these assumptions during the contract administration. Under these circumstances, agencies may still be able to implement effective design management by setting certain boundaries in the specifications and through an appropriate and clear approach to design QA and QC. However, these approaches need to be identified early and conveyed to proposers during the proposal preparation phase. Often, proposers' specific approaches to quality management for both design and construction are used as a component of the best value evaluation.

At the program level, the authors found that a series of factors affect the agency's approach to design management and its effectiveness. First, to effect meaningful organizational change as required by the specificity of DB, agencies must address their formal and informal cultures alike. Without attention to aligning these two organizational realities, agencies are likely to see opposition to new processes. Agencies should never underestimate the positive or negative impact of the informal culture on the implementation of new delivery methods, such as DB. For instance, the successful implementation of DB for the I-15 Corridor Reconstruction project created a positive environment for evaluating and adopting innovative contracting methods by the Utah Department of Transportation. Other agencies have instead encountered significant challenges in their initial DB projects, with mixed results. Within these agencies, an initial negative perception of DB slowed down or stopped its use.

Second, the approach to assigning personnel as well as the availability and capability of the agency's staff in return influenced the need for training and/or utilization of external consultants. For example, one of the early adopters of DB, UDOT now provides guidelines on assembling the project team. Similarly, individuals with DB experience are often used on an as-needed basis for DB projects. For example, several of the WSDOT employees assigned to the SR 99 project had been involved in other critical DB projects and carried this tacit knowledge with them. When using external consultants, a series of sub-factors arise. First, if the agency decides to develop ad hoc engineering standards for the program, the role of consultants would need to be well-defined. Then, it would need to define how these consultants will fit into operating processes as well as organizational structures. To manage the OTIA III State Bridge Delivery Program, ODOT selected Oregon Bridge Delivery Partners (OBDP), a joint venture of two large private firms. ODOT and OBDP collaborated closely to administer a programmatic delivery toolbox that included DBB, DB, and CMGC for bridge projects in the program. The Bundle 401 project case study well exemplified how ODOT had addressed issues related to consultant utilization.

The next crucial factors include the agency's approach to seek participation by stakeholders as well as how it sets communication between proposers and stakeholders. For the SR 99 bored tunnel project, the Design-Builder learned that other stakeholders may need to review design submittals. In this case, WSDOT entered into several Memoranda of Agreement (MOAs) with the City of Seattle, detailing oversight requirements and expectations. These MOAs were needed, since much of the project work was on and under City of Seattle property and, thus, might have affected city-owned infrastructure. But, the MOA conditions complicated design review time-lines and expectations because the Design-Builder had to coordinate with many stakeholders within the city and its utility subsidiaries. It was also necessary to meet City of Seattle standards for certain work efforts and WSDOT standards for others.

How the agency allocates design management responsibilities among its units and how the project delivery process is managed by units dealing with phases adjacent to post-award design (i.e., DB contract procurement and construction) are both critical to design management. At an extreme, three different units will manage design during (1) pre-award design (during procurement), (2) post-award design, and (3) post-award construction. However, two approaches are common depending on the type of project and the level of maturity of an agency's DB programme. When projects are particularly unique (e.g., SR99 Bored Tunnel Project) or when an agency is not expert with DB delivery, the same group manages the design process from procurement and throughout post-award delivery. This approach creates a continuity of design information. In case the project lacks uniqueness and the agency is highly versed with DB, it seems common that pre-award design administration is assigned to agency staff (and units) that specialize in DB procurement. Often, an individual from the same unit who is involved in the procurement is later assigned to manage design activities in post-award jointly with a construction project manager. The role of this design manager decreases as the project moves from a release for construction to completion. This industry practice attempts to "bridge" project responsibility from pre-award to post-award while recognizing specific knowledge requirements for each phase.

At the project level, a similar set of factors affect design management, including the approach to collect project data and information that would feed the design process, the existence of particularly complex situations or restrictions to design or construction alternatives. Furthermore, more traditional parameters, such as the procurement selection approach and criteria, bidding contingencies, and completion date determination were found to shape design management practices. In terms of practices, the approach to deal with pre-award design activities substantially affects post-award design management. In fact, a significant part of design is concurrently carried out during the DB contract procurement by all the competing teams. Design alternatives are generated at this point and incorporated into the final design. When the procurement process allows for submittal of alternative technical concepts (ATC) by competing teams, a significant feature of pre-award design management is how the agency will handle the selection, approval, and incorporation of these alternative design ideas. For readers unfamiliar with ATCs, they are a way to allow the value engineering process to occur during procurement. During procurement, each competing team can submit for approval by the agency these value engineering proposals. Some agencies require that these proposals be approved before being included in the final proposal package. Once a team obtains approval for an ATC, it can include it into its proposal and price the project accordingly. In some jurisdictions, some of these design alternatives can be incorporated into the final design even if the team proposing the idea is not selected. This approach is often the object of claims and disputes by teams that are not selected. Therefore, it is clearly regulated in the state procurement laws. Commonly, intellectual property rights on these ATCs are acquired from the agency through the payment of stipends to teams that are not selected. If the agency likes some of the ideas from teams that are not selected, the pricing of these design alternatives often occur after the winning team is selected and before a contract is awarded, so that they can be incorporated into the contract. Otherwise, they can be handled as change orders post-award. Again, the state procurement law would dictate the approach. The case studies provided a comprehensive variety of approaches to deal with pre-award design management, either when ATCs are allowed or not. When ATCs are not allowed, agencies rely only on post-award value engineering proposals by the Design-Builder.

During post-award, an agency's approach to design management is mostly shaped by how it establishes a collaborative partnering environment, how it handles communications and coordination on matters that contribute to design development, how it handles value engineering proposals, how it handles interdependencies between design and other activities, and, especially, how it handles formal design management processes. The set of case studies highlighted different approaches to these features of a design management process. In addition, constraints within agencies and projects that may motivate the selection of one feature over another were provided in the full report, together with a set of guidelines (Minchin et al, 2014).

Construction-Manager-as-General-Contractor

CMGC is a delivery system with some history in commercial and industrial construction, but is new to most of the transportation construction industry. In the early days of the new century, portions of the industry tried to establish an early form of CMGC, called Construction-Management-at-Risk (CMR) as a fast-tracked alternative to the more established fast-tracked DB delivery system. CMR offered all the speed inherent to a fast-tracked system, but also offered the owner more control over the design process than DB. However, builders fought the system almost everywhere it was implemented because CMR laws generally either forbade the CM from self-performing any work at all, or required that the CM bid for any work against qualified subcontractors. This, the builders feared, would eliminate smaller, local builders from ever being awarded any project large enough to attract the larger national or international builders or CM firms. The logic of this was faulty on the surface, as any builder or CM firm, no matter how large, that was awarded such a contract would have to find someone, probably local, to perform the actual construction. Regardless, CMR never gained any momentum as a national delivery system option for highway construction, and CMGC, which had been used in Utah and a few other places for a time, was embraced as the fast-tracked alternative to DB because CMGC either allows, or in most cases requires, the CM to self-perform a set percentage of the work. The Federal Highway Administration (FHWA) has supported the implementation of CMGC from the time that it was introduced to the transportation construction industry, and has made that support tangible through the Every Day Count (EDC) initiatives (EDC-1 in 2010 and EDC-2 in 2012).

Under CMGC, the designer has a contractual relationship with the agency. However, it is required to take a much more active role in working with the owner and the CM (builder) throughout the entire design process, including early and continuous use of value engineering concepts (without the constraints of the accepted process), ROW acquisition, real-time pricing, increased coordination meetings, and accelerated design. Designers must budget additional funding and management personnel for frequent team meetings and binding decisions while working with both the owner and CM. They also need to be educated in the process of receiving real-time input from the builder as well as being flexible in modifying standard items such as traffic control plans to best fit the chosen approach to construction.

Whereas CMGC projects rely on a set of two separate two-party agreements between the owner and the designer and the owner and the builder, it promotes project partnering. The CMGC partnership, or team, is comprised of the owner, the CM, the designer, the designer's specialty consultants, the subcontractors, and any other party that would be beneficial to include. The CM is best retained at the same time as the designer, very early in the process. Assuring transparency throughout the process is the most important priority for the owner. To that end the committee that chooses the designer and the CM often includes a design consultant and a builder—either active or retired. Fortunately, CMGC facilitates openness and trust, providing real-time costs, schedule, and constructability input.

During preconstruction, the CM acts as an advisor to the designer, providing pre-construction professional services to the owner. Plans reviews and constructability reviews are the two most discussed of these services, but the two most important benefits of CMGC are made possible by the early involvement of the CM: innovation and the flexibility to allocate and re-allocate risk until relatively late in the construction phase. These two benefits are not unrelated. When UDOT took on an inordinate and unbalanced share of the risk on the Mountain View Corridor project, it not only brought the construction's prices down by millions of dollars as a natural reaction to suddenly not having to add contingency to the contract price, it also freed the builder to implement several innovative construction methods which eliminated some work and lowered the cost to perform other work, saving millions of additional dollars.

It is very important for those considering implementing CMGC to consider the cultural shift that has to take place if one (agency or individual) has never worked on a CMGC project before. The importance of this is most manifested in the importance of choosing the right people to lead the CMGC team, as well as who makes up the team. The agency can pretty well assemble its own partnering team, and it needs to take maximum advantage of that opportunity. A fundamental question to ask when evaluating potential team members is: Will this individual advance the CMGC process or impede its application? If there is any doubt that the individual will impede the process, that person should be eliminated from consideration as a team member. However, nothing should be done in secret. The most important characteristic of putting the CMGC team together is transparency. Transparency in the placing of in-house personnel on the team, transparency in hiring new personnel, transparency in procuring the CM, the designer, the designer's specialty consultants, the owner's independent cost engineering consultant, and all other team members.

To help assure overall success, the owner should spend great volumes of time, resources, and effort early in the process, planning in detail the entire CMGC operation/project for its entire service life; and one of the initial actions for the agency, even before selecting outside team members, is to reach out to the community and all stakeholders and begin the process of educating them. Very few areas of the country know enough about the CMGC culture or process for the agency to skip the vital step of aggressive public relations and education effort, and the effort must contain a message that is consistent, no matter who is sharing or receiving the message.

Chapter summary

Based on the findings reported, some RC norms seem more present than others in design management. First, Role Integrity prescribes design management behavior in both DB and CMGC projects. For instance, designers are subject to short-term needs for increased plan production rates under these delivery methods. To achieve project goals under these methods of delivery, designers adapt to flexibility requirements, such as extended work hours, to keep pace with the acceleration and changes proposed by the builder. This same practice is also representative of Reliance and Expectations behavior on the part of builders in the DB projects as they select the designer consultant under the assumption that these fast-forward plan production requirements will be met. Given that DB has been shown to outperform other delivery methods in terms of schedule, this reliance is mostly well placed (Hale *et al.*, 2009; Minchin *et al.*, 2013).

Flexibility also permeates behavior on both DB and CMGC projects. A striking example is the practice of breaking down the project into multiple "mini" phases forcing all contractual parties to blend and sometimes loop their standard work processes. Other examples include the practice of allocating and re-allocating risk until relatively late in the construction phase under CMGC. Also, agencies seem flexible in adapting their behavior to maximize chances to succeed when they select different units to manage DB or CMGC projects in all or some of the lifecycle phases. Under CMGC, whereas the designer has a contractual relationship with the agency, it is required to take a much more active role in working with the owner and the CM throughout the entire design process. This norm has shaped CMGC practice with designers budgeting additional funding and management personnel to be able to fulfill these requirements. This same example is also representative of Contractual Solidarity due to the seemingly seamless approach to work together. Contractual Solidarity and Harmonization of Conflict are both infused into the DB and CMGC projects by the standard practice of setting up a collaborative partnering environment to better handle communications and coordination in design activities.

Reciprocity seems to affect broadly the parties' behavior to include external parties. For instance, owners often attempt to share equally and fairly risks associated with third-party input on design by entering into MOAs with permitting agencies and other stakeholders to mitigate some of the known impacts of these external players on the project delivery. Similarly designers and builders seem to find a reciprocally neutral ground though the handover of the utility coordination task from design to construction in the utility relocation stage.

Findings from the NCHRP study also suggest the action of two other RC norms on design management: Restraint of Power and Propriety of Means. Clear examples of the former include the self-imposed approaches by agencies to purchase intellectual property rights on innovative ideas generated during procurement in the form of ATCs and to guide the design management process by an early identification of certain boundaries in the specifications and through an appropriate and clear approach to design QA and QC. Both of these norms seem to guide agencies in their continuous push to transparency in the procurement process and competency in using expert consultants during the evaluation of proposals and the oversight of design activities. To a lesser extent, Propriety of Means can be seen in the agency's behavior in slowly handing over the project management role from the design/procurement manager to the construction manager. In fact, procurement, design and construction functions tend to be very specialized in the transportation sector requiring different professionals to be involved at different stages from the agency's standpoint.

Revisiting the findings of the NCHRP study seems to confirm that RC norms are present in DB and CMGC contracts, which was one of the conclusions reached by Harper and Molenaar (2014). In addition, some findings suggest that these norms often infuse standard design management practices on projects that rely on these contracts. Where Harper and Molenaar based their conclusions on a content analysis of standard contracts that are usually used for the building sectors, the NCHRP study was based on data and information from the transportation infrastructure sector and relied on a more comprehensive data exposure that included surveys, phone interviews, and extensive case studies based on agency's homebrewed contracts and other project documents as well as on extensive interviews of participants.

Acknowledgments

This chapter summarizes and revisits results from NCHRP Report 787, Guide for Design Management on Design-Build and Construction Manager/ General Contractor Projects, which was conducted as part of National Cooperative Highway Research Program (NCHRP) Project 15–46. The NCHRP is supported by annual voluntary contributions from the state Departments of Transportation. The report was prepared by the authors of this chapter with the support of several researchers and consultants. In particular, Lourdes Ptschelinzew and Umberto Gatti were involved in the data collection and case study write-ups; Tom Warne and Ken Atkins participated in some of the case study visits, and respectively in writing some of the sections of the report; Gregg Hostetler and Sylvester Asiamah contributed to writing some of the sections of the report.

The work was guided by a task group chaired by Rodger D. Rochelle, North Carolina DOT, which included Reuel S. Alder, Utah DOT, Baabak Ashuri, Georgia Institute of Technology, Henry I. Chango, D'Ambra Construction Co., Jon M. Chiglo, Minnesota DOT, Robert Dyer, Washington State DOT, Arunprakash M. Shirole, S & A Shirole, Inc., Richard Duval, FHWA Liaison, Frederick Hejl, TRB Liaison. The project was managed by Andrew Lemer, NCHRP Senior Program Officer.

Reflections

1 The National Cooperative Highway Research Program (NCHRP), referred to throughout this chapter, has numerous construction-related resources available at: http://www.trb.org/NCHRP/NCHRP.aspx.
2 Design-Build (or variations such as Design-Bid-Build) has been the predominant form of construction procurement in the United States for a number of years. The American Institute of Architects (AIA) has what it calls 9 "families" of construction contracts, one of which includes the variations of the Design-Build method. More information is available at: http://www.aia.org/contractdocs/aias076693.
3 Public infrastructure projects in the United States have historically been funded by various levels of Government. However, like elsewhere, recent fiscal pressures have led to the use of alternate procurement methods

such as PPPs. The National Council for Public-Private Partnerships (NCPPP) is the main advocate and facilitator for the formation of PPPs at the federal, state and local levels: http://www.ncppp.org.

4   With the findings of this chapter in mind, and indeed the Internet resources highlighted above, compare the aspects of construction procurement in the United States with those in other parts of world.

5   The Construction Industry Institute (CII), based at the University of Texas at Austin, is a consortium of more than 130 leading owner, engineering-contractor, and supplier firms from both the public and private arenas. Examples of best practice examples, including procurement, can be found at: https://www.construction-institute.org.

References

AIA-MBA Joint Committee (2014). *Construction Management: CM as Constructor.*

CII (1995). *Pre-project Planning Handbook* (Special Publication, pp. 1–137). Austin, TX: The Construction Industry Institute.

Gransberg, D., and Shane, J. (2010). *NCHRP Synthesis 402: Construction Manager-at-Risk Project Delivery for Highway Programs.* Transportation Research Board.

Hale, D., Shrestha, P., Gibson, G. E., and Migliaccio, G. C. (2009). Empirical comparison of design/build and design/bid/build project delivery methods. *Journal of Construction Engineering and Management,* 135(7), 579–587.

Harper, C., and Molenaar, K. (2014). Association between construction contracts and relational contract theory. *Construction Research Congress 2014* (vol.1, pp. 1329–1338). Atlanta, GA.

Knight, A., Griffith, A., and King, A. (2002). Supply side short-circuiting in design and build projects. *Management Decision,* 40(7), 665–662.

McMinimee, J., Schaftlein, S., Warne, T., Detmer, S. S., Lester, M. C., Mroczka, G. F., and Yew, C. (2009). *Best Practices in Project Delivery Management* (pp. 1–137). Transportation Research Board.

Migliaccio (2007). Planning for strategic change in the project delivery strategy, Doctoral Dissertation, The University of Texas at Austin. https://repositories.lib.utexas.edu/handle/2152/3370.

Migliaccio, G. C., Gibson, G. E., and O'Connor, J. T. (2008). Changing project delivery strategy: an implementation framework. *Public Works Management and Policy,* 12(3), 483–502.

Minchin, E., Migliaccio, G., Ptschelinzew, L., Gatti, U., Atkins, K., Warne, T. R., and Asiamah, S. (2014). *Guide for Design Management on Design-Build and Construction Manager/General Contractor Projects* (pp. 1–226). Transportation Research Board.

Minchin, R. E., Li, X., Issa, R. R., and Vargas, G. G. (2013). Comparison of cost and time performance of design-build and design-bid-build delivery systems in Florida. *Journal of Construction Engineering and Management,* 139(10), 04013007.

Molenaar, K. R., Songer, A. D., and Barash, M. (1999). Public sector design-build evolution and performance. *ASCE Journal of Management in Engineering,* 9(2), 54–62.

Rein, C., Gold, M., and Calpin, J. (2004). The evolving role of the private sector in the U.S. toll road market. *Journal of Structured and Project Finance,* 9(4), 27–33.

7 How do construction firms learn on collaborative infrastructure projects?

Le Chen and Karen Manley

Chapter introduction

In recent years, collaborative procurement approaches such as Partnering, Early Contractor Involvement (ECI) and Project Alliances have been increasingly used for infrastructure construction (Lahdenperä 2012; Chan *et al.* 2010), which reflects the advantages of these approaches in managing high levels of complexity in project delivery (Morwood *et al.* 2008). At the same time, the turbulent global economic environment has created pressure for these collaborative procurement approaches to evolve to suit changing market conditions (Mignot 2012; Hartmann and Bresnen 2011; Bresnen 2010, 2009). From a learning capability perspective (Lewin *et al.* 2011; Zollo and Winter 2002), the joint learning of participant construction organisations during collaborative project delivery plays an essential role in driving the evolution of procurement methods.

Learning capability is a special type of dynamic capability, and is defined as the capacity of an organisation to align its knowledge-base purposefully with the needs of an evolving market in order to secure superior performance in the long term (Lewin *et al.* 2011; Lichtenthaler and Lichtenthaler 2009). Organisations achieve this alignment through reconfiguring learning routines that explore, retain, and exploit knowledge inside and outside organisational boundaries (Lewin *et al.* 2011; Lichtenthaler and Lichtenthaler 2009). Learning capability literature proposes that organisations use three major types of learning routines to reconfigure their knowledge-base: these are exploratory, transformative and exploitative learning routines (Lewin *et al.* 2011; Lane *et al.* 2006; Zahra and George 2002).

- *Exploratory learning* routines identify, acquire, analyse, and understand critical external knowledge (Lichtenthaler 2009; Lane *et al.* 2006; Zahra and George 2002), and help the members of an organisation to externalise knowledge for new knowledge creation (Kale and Singh 2007; Zollo and Winter 2002; Nonaka 1994).
- *Transformative learning* routines select, retain, disseminate, and codify both internally generated and externally acquired new knowledge (Lichtenthaler and Lichtenthaler 2009; Lane *et al.* 2006).

- *Exploitative learning* routines integrate newly acquired and generated knowledge into the existing operating routines, so as to refine and extend those existing routines and technologies (Lane *et al.* 2006).

Learning routines, especially their interdependence and complementarities, are idiosyncratic, by virtue of the specific ways they are applied by organisations, and are dependent upon organisations' evolutionary paths (Lewin *et al.* 2011). Hence, the combination of learning routines applied by organisations varies between organisations, thus causing heterogeneity of dynamic learning capability among organisations. The imperfect imitability of the learning routines for an organisation makes the resultant dynamic learning capability a source of competitive advantage and performance heterogeneity between organisations.

Applying this theoretical perspective to the infrastructure construction context, collaborative learning capability (CLC) can be perceived as a special dynamic learning capability that construction organisations specifically develop to improve the performance of collaborative project delivery. CLC builds on the micro-foundations of exploratory, transformative and exploitative learning routines, which are used by construction organisations to carry out joint learning with other participant organisations during collaborative project delivery. The shared learning of participant organisations modifies and improves the project operating routines applied by each organisation, and helps to develop new collaborative procurement models that ultimately improve project performance (Hartmann *et al.* 2010; Bresnen 2009; Leiringer *et al.* 2009). Figure 7.1 illustrates the routine-based conceptualisation of CLC and the three learning routine dimensions that underpin it.

The study of organisational CLC in the collaborative infrastructure construction is in its infancy in the literature. The CLC concept has not yet been clearly operationalised and measured. The absence of CLC measurement scales leads to difficulties in clearly estimating the impact of CLC on the performance

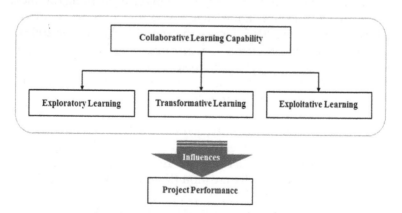

Figure 7.1 Conceptualisation of collaborative learning capability

of collaborative projects, a topic which is facilitated by the current research and which will be examined in future research by the authors. In order to facilitate the investigation of CLC on project performance in the future, this study applies a routine-based conceptualisation of CLC in the empirical context of collaborative infrastructure projects in Australia. The study has two research objectives:

1   The study sought to develop valid and reliable measurement scales to evaluate the CLC of construction organisations.
2   The study used the CLC measurement scales to evaluate the learning routines of construction firms.

The scales are developed based on a survey sample that comprised three main types of construction organisations: clients, contractors, and consultants. This study seeks to use the scales to measure the CLC of the contractors and consultants only, excluding the client portion of the sample. Hereafter in the chapter, contractors and consultants will be referred to as "construction firms". The research scope was designed to focus on the CLC of construction firms since their learning needs and approaches are different from those of client organisations in collaborative project delivery (Hartmann *et al.* 2010; Bresnen 2009; Leiringer *et al.* 2009). As these firms play the essential roles of design and construction it is important to understand their learning routines. A study on the CLC of client organisations will be reported in a subsequent publication. The study first reviews the learning activities of construction firms from the perspective of dynamic learning capability theory, and subsequently reports a survey study that measured the CLC of construction organisations in the Australian infrastructure sector.

Background: the learning of construction firms

The delivery of a collaborative infrastructure project requires construction-related services provided by a mix of different private sector firms, such as engineering consultants, designers, contractors, cost planners, and facilities managers (Chan *et al.* 2010; Miller *et al.* 2009). When construction firms such as these participate in collaborative styles of project delivery, this enables them to reconfigure their resources and capability, thus improving their competitive advantage in a dynamic market (Leiringer *et al.* 2009; Green *et al.* 2008). The objectives of capability reconfiguration vary from firm to firm, ranging from the desire to achieve new market penetration (Green *et al.* 2008), to technology acquisition and innovation (Rose and Manley 2012; Gluch *et al.* 2009). The learning that drives the reconfiguration is also firm-specific and path-dependent (Green *et al.* 2008). In spite of the diversity of firms' learning paths, one of the common intentions of participant organisations is to take advantage of the knowledge-supporting network that is specifically created during collaborative styles of project delivery (Ruana *et al.* 2012), where individual interactions and dialogue across the organisational boundaries of project participants are facilitated (Carrillo *et al.* 2006). This knowledge network not only enables construction firms to acquire knowledge from other participant organisations, but also provides new stimulus for firms to overcome learning inertia and to establish sustainable organisational learning (Love *et al.* 2010). The longer the history a construction firm has with collaborative project management, the more knowledge and competence the firm is likely to have built into their project routines (Jin and Doloi 2008).

Learning routines are higher-order routines that are underpinned by lower-order operating routines which directly influence project performance. Operating routines need to be developed by learning routines that are purposely deployed to capture, share, and apply collaborative project know-how; this is particularly true for routines associated with risk management, contractual management, and relationship building (Abdul-Rahman *et al.* 2008; Green *et al.* 2008; Jin and Doloi 2008; Senaratne and Sexton 2008). Exploratory, transformative and exploitative learning routines play a unique and essential role in the reconfiguration of the operating routines employed by construction firms, and the subsequent development of new, more effective procurement models.

Exploratory learning routines

During collaborative infrastructure construction, exploratory learning occurs when construction managers and professionals identify, acquire, analyse, and process new knowledge which is critical for project delivery. The exploratory learning routines within construction firms are carried out at both a meso-level (e.g. within project teams and communities of practice), and at firm level. Face-to-face interactions and project team meetings are common routines of collective learning (Love *et al.* 2005). The discussion of problems, searching for solutions, sharing and exchange of learned mistakes, knowledge and experience often lead to effective communication by storytelling and brainstorming (Abdul-Rahman *et al.* 2008; Australian Constructors Association 1999). Wenger (Wenger *et al.* 2002) advocates communities of practice as an important learning mechanism for construction firms (Sage *et al.* 2010; Yu *et al.* 2009). Bound by close working relationships, common interest and ownership of knowledge, communities of practice facilitate articulation and assimilation of tacit knowledge (Love *et al.* 2002).

A knowledge broker can serve as a transferral mechanism between departments, projects, and communities of practice, as well as between individuals and professionals (Sage *et al.* 2010; Styhre *et al.* 2004). Project managers who are involved in the design phase and are familiar with all the formal documentation and decision-making processes are best suited for the knowledge broker role (Morwood *et al.* 2008). The outcomes of meso-level learning are ideally fed back into learning mechanisms at firm level (Sexton and Lu 2009; Carrillo *et al.* 2006). The communication mechanisms for this exchange can be formal, such as regular firm-level meetings, workshops, and video conferencing; or informal, such as an "away day" during which managers from different business units share knowledge through non-work related activities (e.g. golf) away from formal working environments (Sexton and Lu 2009; Carrillo *et al.* 2006).

In terms of knowledge identification and acquisition, most construction firms now have structured relationship management mechanisms to identify the clients that are aligned with the firms' strategic business goals (Morwood *et al.* 2008). During project delivery, these mechanisms help to build several healthy point-of-contact relationships across the client's organisation and with the client's advisors (Morwood *et al.* 2008). This relationship building ensures good communication that provides the foundation for a thorough understanding and appreciation between the client and the construction firm with regard to project delivery drivers, deliverables, and expectations (Love *et al.* 2010; Hauck *et al.* 2004). The knowledge acquired not only enables a construction firm to deliver the project on hand, but also gives the firm a competitive edge when bidding for future collaborative projects (Miller *et al.* 2009).

Transformative learning routines

Transformative learning routines enable the exchange and dissemination of knowledge through interpersonal interactions, and the selection of relevant new knowledge that needs to be codified for ongoing project operations. The knowledge acquired and created through exploratory learning is shared within firms through transformative learning routines. Meetings and workshops at both meso- and firm level help to disseminate this knowledge (Sexton and Lu 2009). Project managers and construction professionals normally share learning from previous collaborative projects with respect to mistakes, and good and bad practices, through informal face-to-face communications during the preparation for new collaborative projects (Abdul-Rahman *et al.* 2008; Australian Constructors Association 1999). Regular formal project reviews are conducted during project delivery to assess project status during the different delivery phases, so as to: identify mistakes and problems, highlight optimal solutions, and make decisions about courses of necessary subsequent actions (Carrillo *et al.* 2006; Love *et al.* 2002). The learning is much more effectively disseminated if the review process and project performance evaluation embrace benchmarking and other continuous improvement approaches (Bresnen 2007; Robinson *et al.* 2005; Love *et al.* 2002).

Converting tacit knowledge into explicit concepts involves repeated time-consuming dialogue (Sexton and Lu 2009; Love *et al.* 2002). The knowledge gained through this evaluation and sharing is filtered, compiled, and stored in a standard template for future reference (Abdul-Rahman *et al.* 2008). Best practice and problem-solving solutions are ideally codified into explicit forms such as best practice guidelines, procedures, manuals, and checklists to guide the improvement of project management routines (Senaratne and Sexton 2008).

The knowledge codification routines are best supported by IT applications such as intranet, groupware, skills yellow pages, expert systems, intelligent systems, data mining tools, document management systems, virtual reality tools, and search engines (Carrillo *et al.* 2006; Peansupap and Walker 2006).

Construction firms also manage to retain knowledge and learning in their inter-firm relationships. For example, the infrastructure sector often sees repetitive partnerships formed by some contractor and consulting firms to enable collaborative bidding for infrastructure projects (Morwood *et al.* 2008). The collaborative relationships and trust built through prior joint projects provide these firms privileged access to their partners' knowledge-base (Roehrich and Lewis 2010), especially when special and unique technologies are involved (Morwood *et al.* 2008).

Exploitative learning routines

The application of both externally acquired and internally generated knowledge is achieved through use of key exploitative learning routines (Lewin *et al.* 2011). Exploitative learning routines ensure that knowledge gained from previous collaborative projects through exploratory and transformative learning can be used to match governance mechanisms and operating routines with the needs of the present market (Leiringer *et al.* 2009; Abdul-Rahman *et al.* 2008; Carrillo *et al.* 2006). In the exploitative learning stage, internal training is the common approach adopted by construction firms to internalise knowledge (Morwood *et al.* 2008; Australian Constructors Association 1999). Some large construction firms integrate collaborative contracting content into the formal training programmes of corporate universities (Carrillo *et al.* 2006). Human-centred approaches are essential for internalising tacit knowledge. In collaborative project teams, experienced staff members provide on-the-job training as mentors to the members who have less experience with collaborative delivery methods (Morwood *et al.* 2008). Lessons learned from previous projects are used as guidance in decision making, for instance during the planning of project activities (Abdul-Rahman *et al.* 2008). Through applying transformed knowledge during the operations of collaborative projects, new ideas and innovations are created (Love *et al.* 2010; Sexton and Lu 2009).

Methods

Concept operationalisation

CLC is based on the three learning phases described above. In order to operationalise the CLC concept, an in-depth literature review was conducted to identify the routines underpinning each of the three learning phases. A directed content analysis approach (Krippendorff 2004) was used to explore the literature. Well-established theories and findings of prior research provided guidance to define the coding categories of learning routines

(Krippendorff 2004). The literature review involved two steps. The first step of the review drew on recent research advances in (1) knowledge management (e.g. Nonaka 1994), (2) absorptive capacity (e.g. Lewin *et al.* 2011; Lane *et al.* 2006; Zahra and George 2002; Cohen and Levinthal 1990), (3) dynamic capabilities (e.g. Teece 2007; Zollo and Winter 2002), and (4) strategic alliance management (e.g. Kale and Singh 2007; Grant and Baden-Fuller 2004). These research works provided the solid theoretical foundation for CLC. The measurement scales developed by Lichtenthaler (2009) for managing external knowledge, and those developed by Kale and Singh (2007) for managing internal knowledge, provided coding categories to enable the operationalisation of the CLC concept.

The second step of the review focused on literature related to the empirical context of the infrastructure industry, particularly in Australia, to contextualise the CLC concept. The learning practices of both client organisations and construction firms were reviewed. The review covered construction management journals, books, government documents, industry reports, and guidelines. The two-step review derived a total of 19 learning routine items to operationlise the CLC concept: six items for exploratory learning dimension, seven items for tramformative learning dimension, and six items for exploitative learning dimension, as shown in Table 7.1. These categories are the result of triangulation across four theoretical disciplines, two types of literature (academic and industry), and two author perspectives, where each author undertook independent coding that was later compared and refined.

Validation and use of measurement scales to evaluate construction firm CLC

Data collection

The next step was both to test the value of the learning scales developed in the operationalisation stage, and to use these scales to assess the learning routines of construction firms (consultants and contractors). Data for this study was gathered through the conduct of a quantitative survey that sought to characterise the learning routines of organisations involved in collaborative infrastructure construction in Australia. Respondents were required to indicate the degree to which each of the 19 identified learning routines were implemeted by their respective organisations during a recent collaborative project they had worked on. A seven-point Likert scale was used to measure the learning routine implementation (1 = strongly disagree, 7 = strongly agree). The survey was distributed to the contact database of the Alliancing Association of Australasia (AAA), a total sampling frame of 1,688 prospective respondents, comprising construction sector practitioners representing public and private sector clients, contractors, consultants, and suppliers with experience in collaborative contracting. Following the advice of Neuman (2003), a pilot study was carried out to ensure the face validity of measurement items of learning routines.

Table 7.1 Operational items of CLC

Exploratory learning routines	Transformative learning routines	Exploitative learning routines
We liaise with external partners to collect information about market developments.	We regularly update guidelines for staff behaviour during collaborative projects.	In our organisation it is well known who can best exploit new knowledge to collaborative projects.
We liaise with external partners to collect information about technological advancements.	We maintain a database of learnings from our collaborative projects.	We regularly apply new knowledge to collaborative projects.
We liaise with external partners to collect information about staff skill enhancement.	Staff regularly use a benchmarking approach in collaborative project review for continuous improvement.	We constantly consider how to better exploit the organisation's knowledge-base during collaborative projects.
We maintain a database of individuals who can help us with collaborative projects.	Staff regularly engage in informal information sharing about collaborative projects.	We incentivise managers' use of organisational databases on collaborative project experience.
We document the development of different types of collaborative governance arrangements.	Staff regularly participate in formal forums, such as meetings, seminars, or retreats, to exchange information about collaborative project implementation.	We use external behavioural coaches to improve staff skills in relation to collaborative project delivery.
We regularly debrief staff on collaborative projects in formal meetings.	Staff with substantial experience in managing collaborative projects are rotated across our key collaborative projects.	Staff regularly attend training programmes on collaborative project management.[b]
	Staff incentives are used to encourage information sharing about collaborative projects.[a]	

a: This item is assigned to the exploitative learning dimension by the factor analysis.
b: This item is deleted by the factor analysis due to low factor loading (< 0.50).

The survey was distributed by email as a link to an online form, and was open for a response period of 12 weeks which ended in early 2013. At closure of the survey, 320 valid responses were received, providing an overall response rate of 19 per cent. Applying the sample size estimation formula recommended by Bartlett *et al.* (2001), this response rate ensured statistical rigour of the data at an alpha level of 0.05 with a 3 per cent margin of error. Responses were approximately equally distributed between representatives of client, contractor, and consultant organisations (34 per cent, 34 per cent, and 31 per cent, respectively) while sub-contractor and supplier organisations were infrequently represented. There are 1 per cent missing values on the learning routine items in the data set, which is less than 5 per cent threshold as stipulated by Tabachinick and Fidell (2001). Further details of the survey process are reported by the authors (Chen and Manley 2014).

Data analysis

Data analysis comprised two steps, and was performed using IBM SPSS 20.0. The first step was to undertake Exploratory Factor Analysis (EFA) on the whole data set of 320 cases. This confirmed the reliability and validity of the scales used to measure the CLC of construction organisations, and also identified the factorial structure of the CLC concept. Various parameters were reported to demonstrate the overall validity of the EFA results. The EFA resulted in the grouping of learning routine items into "like" groups (factors). The authors applied their understanding from the literature to interpret which of the three learning phases each factor was most closely aligned with, to provide a more strategic indication of the structure of the concept. The EFA analysis also identified which of the 19 learning routine items identified from the literature was *not* considered by survey respondents to be a significant component of their CLC.

The second step was conducted on the 207 responses received from contractor and consultant firms, as this chapter is concerned with highlighting the learning behaviour of the service providers in the supply chain. This step used the CLC measurement scales to evaluate the degree to which the construction firm respondents perceived that each learning routine was applied in the collaborative projects they reported on in the survey. The mean degree of implementation across the 207 construction firm respondents was reported for each learning routine. Similarly, the mean degree of implementation was determined for each of the factors and for each of the overarching learning phases. Finally, the means for each routine were averaged to determine a single mean degree of overall learning routine implementation, which was to be representative of the overall degree to which construction firms implement CLC.

Results

The results are reported in Table 7.2.

Table 7.2 Factorial structure of the CLC concept and construction firms' learning routine implementation level

Factorial structure of construction organisations' CLC	EFA factor loadings	Construction firms' learning routine implementation	
		Mean	SD*
Exploratory learning factors		4.63	1.47
Factor 1–1 External knowledge exploration		**4.91**	**1.71**
We liaise with external partners to collect information about market developments.	0.87	4.90	1.93
We liaise with external partners to collect information about technological advancements.	0.85	4.94	1.87
We liaise with external partners to collect information about staff skill enhancement.	0.81	4.89	1.78
Factor 1–1 Internal knowledge exploration		**4.35**	**1.66**
We maintain a database of individuals who can help us with collaborative projects.	0.75	4.20	2.00
We document the development of different types of collaborative governance arrangements.	0.72	4.29	2.04
We regularly debrief staff on collaborative projects in formal meetings.	0.56	4.59	1.93
Transformative learning factors		4.64	1.38
Factor 2–1 Explicit knowledge transformation		**4.61**	**1.57**
We regularly update guidelines for staff behaviour during collaborative projects.	0.77	4.60	1.75
We maintain a database of learnings from our collaborative projects.	0.71	4.70	1.83
Staff regularly use a benchmarking approach in collaborative project review for continuous improvement.	0.70	4.60	1.86
Factor 2–2 Tacit knowledge transformation		**4.68**	**1.51**
Staff regularly engage in informal information sharing about collaborative projects.	0.78	4.97	1.72
Staff regularly participate in formal forums, such as meetings, seminars, or retreats, to exchange information about collaborative project implementation.	0.76	4.59	1.80
Staff with substantial experience in managing collaborative projects are rotated across our key collaborative projects.	0.55	4.49	1.90

Exploitative learning factors		4.34	1.12
Factor 3–1 Knowledge application		**5.20**	**1.31**
In our organisation it is well known who can best exploit new knowledge to collaborative projects.	0.84	5.00	1.66
We regularly apply new knowledge to collaborative projects.	0.78	5.64	1.24
We constantly consider how to better exploit the organisation's knowledge-base during collaborative projects.	0.73	4.99	1.72
Factor 3–2 Knowledge internalization		**3.47**	**1.37**
Staff incentives are used to encourage information sharing about collaborative projects.	0.75	2.89	1.70
We incentivise managers' use of organisational databases on collaborative project experience.	0.72	2.94	1.71
We use external behavioural coaches to improve staff skills in relation to collaborative project delivery.	0.56	4.61	2.00
Construction firms' CLC		4.50	1.16

Reliability			
Total variance explained (rotation sums of squared loadings)	74.3%		
Cronbach's Alpha (α)	0.92		

* SD: Standard deviation

Validation of the measurement scales

The EFA results were found to be significant, thus supporting the validity of the measurement scales derived from the literature review. The EFA reported a significant Bartlett test of sphericity. The assessment of Kaiser-Meyer-Olkin measure of sampling adequacy (0.91 > 0.60) and the inspection of the anti-image correlation matrix established the factorability of the correlation matrices. Following the advice of Hair *et al.* (1998: 110), principal component analysis and Varimax rotation were adopted in the EFA to derive a clear separation of the factors. As reported in Table 7.2, the cumulative percentage of total variance extracted by the factors in the EFA was 74.3 per cent, which is much higher than the 60 per cent minimum threshold of significance proposed by Hair *et al.* (1998). Cronbach's Alpha (α) value of 0.92 indicates a very good reliability of the scale configuration.

Factorial structure and composition of CLC

EFA was used to validate and clarify the components of the CLC concept derived from the literature. All of the 19 learning routine items presented in Table 7.1 were inputted into the EFA. Each learning routine item was assigned a factor loading, which was indicative of whether the item was considered to be a significantly discrete and influential contributor to CLC. In the analysis, a factor loading of 0.50 and above was considered significant at the 0.05 level (α) to obtain a power level of 80 per cent with the sample of 320 cases (Hair *et al.* 1998: 112), and the evidence of satisfactory convergent validity (Bagozzi and Yi 1988). Table 7.2 shows the resultant factor loadings for each learning routine item. The analysis resulted in the removal of an exploitative learning item due to a low factor loading (< 0.50); thus only 18 of the 19 learning routines considered were confirmed as significant components of CLC.

Table 7.2 also shows the factorial groupings of the learning items. Six factors were identified by the EFA, each of which are respectively underpinned by three learning routine items. The authors applied their understanding of the literature to nominate which of the three overarching learning phases each factor is associated with, resulting in two learning routine groups being associated with each learning phase. Further, an item that was originally associated with transformative learning was reassigned to an exploratory learning factor.

Table 7.2 confirms the propositions in Figure 7.1, that CLC is underpinned by three types of routines to carry out exploratory, transformative and exploitative learning. The exploratory learning factors are "external knowledge exploration" and "internal knowledge exploration". The transformative learning factors are "explict knowledge transformation" and "tacit knowledge transformation". The exploitative learning factors are "knowledge application" and "knowledge internalisation". The EFA results indicate that according to the perspectives of the respondents, construction organisations: explore knowledge from both external and internal sources; transform both explicit and tacit knowledge; exploit new knowledge in collaborative project delivery; and encourage internalisation of newly assimilated knowledge.

CLC and the learning of construction firms

As shown in Table 7.2, the values of the mean and standard deviation (SD) indicate the degree to which each learning routine was implemented by the consultant and contractor firms represented in the survey. The overall degree to which construction firms implemented CLC was a value of 4.50 on the seven-point Likert scale, with a SD of 1.16. This equates to the degree to which respondents agree that CLC is implemented sits between "neutral" and "slightly agree", indicating that CLC is not implemented intensively by the construction firms represented in the survey. Whilst the standard deviation of this overall result suggests a relatively small variation in the different degrees of implementation between the three learning phases, the factors into which they are categorised, and the underlying individual learning routine items, some differences of note will be briefly discussed.

The results in Table 7.2 show that construction firms conduct exploratory and transformative learning to a slightly higher degree than exploitative learning. The factor results in Table 7.2 also suggest that construction firms conducted more intensive external knowledge exploration than internal knowledge exploration. Liaising with external partners to collect information about market development, technological advancements, and staff skill enhancement are common practices. Respondents were less likely to use internal knowledge exploration routines, such as database maintenance, documentation of collaborative governance arrangements, and regular formal meetings on collaborative projects. The results similarly indicate that the firms were less likely to implement explicit than tacit knowledge transformation routines.

There was similarly little variance in the implementation of individual routines, with the most intensely used scoring 5.64 out of 7; "We regularly apply new knowledge to collaborative projects". The least used routine was "Staff incentives are used to encourage information sharing about collaborative projects" at a score of 2.89, indicating that whilst such incentives are important enough to be considered a significant component of CLC, they are not intensively employed in the construction industry.

These results provide a descriptive indication of the degree to which the learning routines identified from the literature are implemented by construction firms during collaborative project delivery in Australia. This provides an indication of the composition of CLC in this context. Whilst the project observations and case study findings in the literature suggest that these learning routines are important for construction firms to improve project operating routines, the results of this study indicate that most of the routines were not implemented to a high degree by the construction firms in the survey sample. Given that the learning routines identified from the literature were those that were reported to be applied during successful projects that achieved excellent performance outcomes (e.g. Department of Infrastructure and Transport 2010; Morwood *et al.* 2008; Hartmann *et al.* 2010; Love *et al.* 2010; Hauck *et al.* 2004), it is likely that a high intensity of learning routine implementation is associated with high project performance. It then follows that the construction firms in the Australian infrastructure sector that were captured in this survey have the potential to improve their project performance by improving the degree to which they implement CLC and its underlying routines. Therefore it would be beneficial to undertake further study to improve our understanding of the impact of learning routines and CLC on project performance outcomes.

Chapter summary

The study fulfilled the two research objectives. The study conceptualised and operationalised the concept of CLC of construction organisations based on learning capability theory. The study developed reliable and valid measurement scales for the concept, and used the scales to evaluate the learning of construction firms when applied during collaborative infrastructure projects in Australia. The analysis demonstrated that during collaborative infrastructure delivery construction organisations explore knowledge from both internal and external sources, transform both explicit and tacit knowledge, and apply and internalise assimilated new knowledge. The study also confirmed that the construction firms in the Australian infrastructure sector do implement the three types of learning, as found in earlier qualitative studies (Love *et al.* 2010; Hauck *et al.* 2004) and industry publications (Kelly 2011; Morwood *et al.* 2008). Despite these positive findings, the study also revealed that collaborative learning was not practised to a very high degree.

The findings suggest that there is potential for further improvement of collaborative learning by construction firms. In particular, more systematic learning approaches deserve attention from the business leaders of construction firms. The findings suggest corporate databases could be further developed to maintain information about key personnel and governance methods related to collaborative projects. Further, greater use of incentive schemes would encourage continual database development and knowledge sharing. The findings imply that construction firms in the Australian infrastructure sector would benefit from improving their learning behaviour during collaborative project delivery. The learning practices of high performance firms may be used as a benchmark.

Based on the findings of this exploratory study, the authors have planned further studies to investigate the performance implications of CLC and the underlying learning routines of construction organisations, including identification of which learning routines are most influential in optimising collaborative project performance. Further validation of the measurement scales for different populations would also be useful, based on new locations, industry sectors, and stakeholders.

Acknowledgements

This study is supported by the Alliancing Association of Australasia (AAA) and the Australian Research Council (ARC) under the Linkage Project scheme (LP110200110). The authors gratefully acknowledge the assistance provided by Joanne Lewis and Deborah Messer in managing the data collection process during the survey, and by Joanne Lewis in the editing of the research findings.

Reflections

1   Innovation and knowledge sharing is explored in both theoretical detail and by using supporting case studies in Rowlinson, D. and Walker, D.H.T. (2008). "Innovation management in project alliances". In (eds) Walker, D.H.T. and Rowlinson, S. *Procurement Systems: A Cross-Industry Project Management Perspective*. Taylor & Francis, London and New York, pp. 400–422. This publication provides an excellent additional resource to this chapter.

2   The Australian Government's Department of Infrastructure and Regional Development published, amongst other useful resources, its "Infrastructure Planning and Delivery: Best Practice Case Studies" in 2010 and 2012. These publications provide leading practice guidance to clients and contractors on several forms of procurement methods, including Alliance Contracting, and can be accessed at: https://www.infrastructure.gov.au/utilities/publications.aspx.

3   Are there any successful examples of collaborative learning capability (CLC) in other sectors from which the construction industry can learn?

4   This chapter explores the findings of a research project that was supported by the Alliancing Association of Australasia (AAA) who provide further information on collaborative management issues at: http://projectmanager.com.au/tag/alliancing-association-of-australasia.

5   Another publication that includes aspects of knowledge management in Alliancing and also draws on the findings of an ARC research project is that of Walker, D.H.T. and Harley, J. (2014). *Program Alliancing in Large Australian Public Sector Infrastructure Projects*, RMIT University, Centre for Integrated Project Solutions, Report for ARC LP110200110.

References

Abdul-Rahman, H., Yahya, I. A., Berawi, M. A., and Wah, L. W. (2008). "Conceptual delay mitigation model using a project learning approach in practice". *Construction Management and Economics*, **26**(1): 15–27.

Australian Constructors Association. (1999). *Relationship Contracting: Optimising Project Outcomes*, Australian Constructors Association, Sydney.

Bagozzi, R., and Yi, Y. (1988). "On the evaluative of structural equation models". *Journal of the Academy of Marketing Science*, **14**(3): 74–94.

Bartlett, J. E., Kotrlik, J. W., and Higgins, C. C. (2001). "Preview organizational research: determining appropriate sample size in survey research". *Information Technology, Learning, and Performance Journal*, **19**(Spring)(1): 43–50.

Bresnen, M. (2007). "Deconstructing partnering in project-based organisation: seven pillars, seven paradoxes and seven deadly sins". *International Journal of Project Management*, **25**(4): 365–374.

Bresnen, M. (2009). "Living the dream? Understanding partnering as emergent practice". *Construction Management and Economics*, **27**(10): 923–933.

Bresnen, M. (2010). "Keeping it real? Constituting partnering through boundary objects". *Construction Management and Economics*, **28**(6): 615–628.

Carrillo, P. M., Robinson, H. S., Anumba, C. J., and Bouchlaghem, N. M. (2006). "A knowledge transfer framework: the PFI context". *Construction Management and Economics*, **24**(10): 1045–1056.

Chan, A. P. C., Chan, D. W., and Yeung, J. F. (2010). *Relational Contracting for Construction Excellence: Principles, Practices and Case Studies*, Spon Press, Abingdon, UK.

Chen, L. and Manley, K. (2014). "Validation of an instrument to measure governance and performance on collaborative infrastructure projects". *Journal of Construction Engineering and Management*, **140**(5): 04014006.

Cohen, W., and Levinthal, D. (1990). "Aborptive capacity: a new perspective on learning and innovation". *Administrative Science Quarterly*, **35**: 128–152.

Department of Infrastructure and Transport. (2010). *Infrastructure Planning and Delivery: Best Practice Case Studies*, Department of Infrastructure and Transport, Australian Government, Canberra, Australia.

Gluch, P., Gustafsson, M., and Thuvander, L. (2009). "An absorptive capacity model for green innovation and performance in the construction industry". *Construction Management and Economics*, **27**(5): 451–464.

Grant, R. M., and Baden-Fuller, C. (2004). "A knowledge accessing theory of strategic alliances". *Journal of Management Studies*, **41**(1): 61–84.

Green, S. D., Larsen, G. D., and Kao, C. C. (2008). "Competitive strategy revisited: contested concepts and dynamic capabilities". *Construction Management and Economics*, **26**(1): 63–78.

Hair, J. F., Anderson, R. E., Tatham, R. L., and Black, W. C. (1998). *Multivariate Data Analysis*, Prentice-Hall International, Upper Saddle River, NJ.

Hartmann, A., and Bresnen, M. (2011). "The emergence of partnering in construction practice: an activity theory perspective". *Engineering Project Organization Journal*, **1**(1): 41–52.

Hartmann, A., Davies, A., and Frederiksen, L. (2010). "Learning to deliver service-enhanced public infrastructure: balancing contractual and relational capabilities". *Construction Management and Economics*, **28**(11): 1165–1175.

Hauck, A. J., Walker, D. H. T., Hampson, K. D., and Peters, R. J. (2004). "Project alliancing at national museum of Australia: collaborative process". *Journal of Construction Engineering and Management*, 130(1): 143–152.

Jin, X.-H., and Doloi, H. (2008). "Interpreting risk allocation mechanism in public-private partnership projects: an empirical study in a transaction cost economics perspective". *Construction Management and Economics*, 26(7): 707–721.

Kale, P., and Singh, H. (2007). "Building firm capabilities through learning: the role of the alliance learning process in alliance capability and firm-level alliance success". *Strategic Management Journal*, 28(10): 981–1000.

Kelly, J. (2011). *Cracking the VFM Code: How to Identify and Deliver Genuine Value for Money in Collaborative Contracting*, Big Fig Publishing, under the imprint of Intelligentsia Press, Australia, New Zealand.

Krippendorff, K. (2004). *Content Analysis: An Introduction to Its Methodology*, Sage, London.

Lahdenperä, P. (2012). "Making sense of the multi-party contractual arrangements of project partnering, project alliancing and integrated project delivery". *Construction Management and Economics*, 30(1): 57–79.

Lane, P. J., Koka, B. R., and Pathak, S. (2006). "The reification of absorptive capacity: a critical review and rejuvenation of the construct". *The Academy of Management Review*, 31(4): 833–863.

Leiringer, R., Green, S. D., and Raja, J. Z. (2009). "Living up to the value agenda: the empirical realities of through-life value creation in construction". *Construction Management and Economics*, 27(3): 271–285.

Lewin, A. Y., Massini, S., and Peeters, C. (2011). "Microfoundations of internal and external absorptive capacity routines". *Organization Science*, 22(1): 81–98.

Lichtenthaler, U. (2009). "Absorptive capacity, environmental turbulence, and the complementary of organizational learning processes". *Academy of Management Journal*, 52(4): 822–846.

Lichtenthaler, U., and Lichtenthaler, E. (2009). "A capability-based framework for open innovation: complementing absorptive capacity". *Journal of Management Studies*, 46(8): 1315–1338.

Love, P. E. D., Tse, R. Y. C., Holt, G. D., and Proverbs, D. G. (2002). "Transaction costs, learning, and alliances". *Journal of Construction Research*, 3(2): 193–207.

Love, P. E. D., Fong, P. S. W., and Irani, Z. (2005). *Management of Knowledge in Project Environments*, Elsevier Butterworth-Heinemann, Boston, MA.

Love, P. E. D., Mistry, D., and Davis, P. R. (2010). "Price competitive alliance projects: identification of success factors for public clients". *Journal of Construction Engineering and Management*, 136(9): 947–956.

Mignot, A. (2012). "Who moved my cheese? Adapting to the changing nature of collaboration in infrastructure". Alliancing Association of Australasia, http://www.a3c3.org/ (June 2012).

Miller, G., Furneaux, C., Davis, P., Love, P., and O'Donnell, A. (2009). *Built Environment Procurement Practice: Impediments to Innovation and Opportunities for Changes*, Curtin University of Technology, Perth, WA.

Morwood, R., Scott, D., and Pitcher, I. (2008). *Alliancing A Participant's Guide: Real Life Experiences for Constructors, Designers, Facilitators and Clients*, AECOM, Brisbane.

Neuman, W. L. (2003). *Social Research Methods: Qualitative and Quantitative Approaches*, Allyn and Bacon, Boston, MA.

Nonaka, I. (1994). "A dynamic theory of organizational knowlege creation". *Organization Science*, 5(1): 14–37.

Peansupap, V., and Walker, D. H. T. (2006). "Innovation diffusion at the implementation stage of a construction project: a case study of information communication technology". *Construction Management and Economics*, 24(3): 321–332.

Robinson, H. S., Anumba, C. J., Carillo, P. M., and Al-Ghassani, A. M. (2005). "Business performance measurement practices in construction engineering organisations". *Measuring Business Excellence*, 9(1): 13–22.

Roehrich, J. K., and Lewis, M. A. (2010). "Towards a model of governance in complex (product–service) inter-organizational systems". *Construction Management and Economics*, 28(11): 1155–1164.

Rose, T. M., and Manley, K. (2012). "Adoption of innovative products on Australian road infrastructure projects". *Construction Management and Economics*, 30(4): 277–298.

Ruana, X., Ochiengb, E. G., Pricec, A. D. F., and Egbud, C. O. (2012). "Knowledge integration process in construction projects: a social network analysis approach to compare competitive and collaborative working". *Construction Management and Economics*, 30(1): 5–19.

Sage, D. J., Dainty, A. R. J., and Brookes, N. J. (2010). "Who reads the project file? Exploring the power effects of knowledge tools in construction project management". *Construction Management and Economics*, 28(6): 629–639.

Senaratne, S., and Sexton, M. (2008). "Managing construction project change: a knowledge management perspective". *Construction Management and Economics*, 26(12): 1303–1311.

Sexton, M., and Lu, S. L. (2009). "The challenges of creating actionable knowledge: an action research perspective". *Construction Management and Economics*, 27(7): 683–694.

Styhre, A., Josephson, P.-E., and Knauseder, I. (2004). "Learning capabilities in organizational networks: case studies of six construction projects". *Construction Management and Economics*, 22(9): 957–966.

Tabachinick, B. G., and Fidell, L. S. (2001). *Using Multivariate Statistics*, Allyn and Bacon, Boston, MA.

Teece, D. J. (2007). "Explicating dynamic capabilities: the nature and microfoundations of (sustainable) enterprise performance". *Strategic Management Journal*, 28(13): 1319–1350.

Wenger, E., McDermott, R., and Snyder, W. M. (2002). *Cultivating Communities of Practice*, Harvard Business School Press, Boston, MA.

Yu, W.-D., Chang, P.-L., Yao, S.-H., and Liu, S.-J. (2009). "KVAM: model for measuring knowledge management performance of engineering community of practice". *Construction Management and Economics*, 27(8): 733–747.

Zahra, S. A., and George, G. (2002). "Absorptive capacity: a review, reconceptualization, and extension". *Academy of Management Review*, 27(2): 185–203.

Zollo, M., and Winter, S. G. (2002). "Deliberate learning and the evolution of dynamic capabilities". *Organization Science*, 13(3): 339–351.

8 PPP procurement
Adding value through relationship development

Peter Davis

Chapter introduction

Initially this chapter will establish the influence that relationship concepts have on construction supply chains, endeavouring to show where thoughtful use of relationship marketing techniques can benefit the construction supply chain (Davis 2008). Recognising the need to build and sustain relationships in procurement generally and most particularly in PPP type projects, this chapter will present a model that has been developed and tested in research undertaken with industry practitioners. Contemporary validation of the model will be drawn from other researchers that study PPP and integrated-type project delivery. The models and their constituent parts can be used as a template by researchers and practitioners alike to encourage a culture of relationship building/maintenance, reflective learning and mutual trust that may be maintained beyond project specific delivery dates well into the operational phase of the PPP project (Davis and Love 2011).

Relationship marketing (RM) has been investigated widely in the normative literature, from which a plethora of definitions that endeavour to capture its scope have arisen. Suffice to say there is a lack of consensus on the definition of relationship marketing; however, several concepts come to the fore. These themes marry with those central to the Alliancing and Partnering literature in construction, which has been equally and extensively researched – for example see Jefferies, *et al.* (2014) and Davis and Love (2011). Primarily RM has the potential to provide organisations with significant supply chain benefits, particularly long-term value to clients and key stakeholders in construction procurement.

Built on traditional marketing theory RM provides a change in focus toward service industries similar to construction as distinct from transactional processes associated with products. RM benefits will come from enhanced commitment, trust and performance satisfaction. These benefits form the basis of the model that is developed and discussed in this chapter.

Upon reading this chapter it will be found that relationship marketing techniques are relevant to the construction project environment, relationship development techniques are a major contributor to successful contracting, and both are able to add considerable value to a PPP's supply chain. There is a recognised structure to relationship development that is underpinned by specific themes; these warrant careful consideration when managing relationships in construction procurement. Commitment and trust are shown as explicit elements that should be continually maintained in PPP procurement, they significantly contribute to joint learning, joint problem solving and reduced ambiguity.

PPP background

The meaning of PPP varies between countries, with different governments focusing on specific attributes (see also Chapters 1, 3 and 9 for further discussion on this topic). In the United Kingdom a Public-Private Partnership is defined as an arrangement typified by joint work between the public and private sectors. PPPs can be very broad and cover many types of collaboration and risk sharing to deliver infrastructure. Infrastructure, for example roads and hospitals or school buildings, are often purchased over a 30-year concession period during which time a private sector partner together with the government takes responsibility for providing the original public service and maintaining it (for further details see the work of Akintoye, *et al.* 2003). This whole-life perspective is important for PPP procurement, relationship development and most particularly relationship maintenance. Management of risk is an essential component of all procurement whether it is traditional design bid build, or as identified in Figure 8.1, at the other end of the procurement continuum PPP. The essential difference being that well-thought-out risk transfer is largely limited in traditional procurement and rarely extends beyond the construction phase of the project.

In a review of PPP Liu, *et al.* (2014) identifies a comprehensive list of critical success factors that includes accelerated infrastructure provisioning with reduced whole-life costs and enhanced financial management, improved service quality and innovation, timely implementation and managed risk exposure that

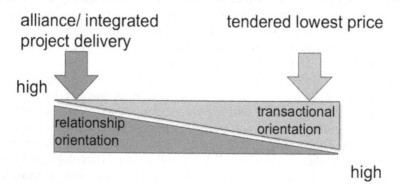

Figure 8.1 Strength of relationship in the procurement continuum

increases accountability and transparency. They suggest that long concession periods ensure PPPs have a stronger reliance on relationships than traditional infrastructure development routes. They also suggest that this factor alone adds to the complexity of the resultant relationship, its development and maintenance. In his PhD, Davis (2005) describes permeable boundaries that support coordination responsibilities shared between contracting parties as one aspect of relationship development; this is supported by Liu, *et al.* (2014) in their research that investigates PPP performance and critical success factors.

The shape of relationships in a PPP project

To introduce relationship development in PPP it is worthwhile considering some background, for many years marketing has been seen as an avenue to develop strong business outcomes. Some years ago the notion of relationship marketing, as an independent strand of marketing thinking, was embarked upon (Gronroos 1990; and Payne and Ballantyne 1991). It was found to be materially different from the marketing concept upon which it is based. The idea behind relationship marketing is to create value for customers or clients and share that value with key stakeholders in a project venture (Davis 1996). Strikingly the value created in relationship marketing is not tangible and appears quite subjective, for example the various parties working together to drive innovation in project PPP delivery will design and realign processes that support value. Similarly continuous and collaborative effort together with mutual understanding founded on long-term commitment will create increasingly tight bonds in the team and continually escalate value. This value may manifest as something quite tangible, for example an innovative process of working, or safer practices, or in the alternative it may be an aspect of 'double loop learning' derived from 'war stories' (Davis and Walker 2009) conveyed through communities of practice – something relatively intangible. Some writers identify this as a chain of relationships, which is a complex network that incorporates both upstream and downstream interactions within the delivery team.

There are several key principles of relationship marketing that are pertinent to procurement in construction. These are adapted from Davis' work (2008) and set out below with examples provided that focus on contemporary PPP procurement.

- A focus on *client retention* – rather than chase new clients it is suggested that a strong focus is maintained on keeping clients. More recent research in the context of safety suggests that strong bonds or a 'personal deal' created over time potentially limits damage caused by adversarial episodes that may eventuate within a procurement process (Walker 2013).
- Orientation on *product benefits* – as opposed to features of the project, product benefits are highlighted in RM research. This feature enhances

the contemporary approach to the whole-life cost/risk evaluation of PPP procurement in as much as the product of the procurement – as opposed to the process of time, cost and quality – becomes the focus.

- Long *timescale* – relationships and relationship development is contingent on a long time scale. This aspect supports the focus on product benefits and is crucial to relationship development and its ongoing maintenance in PPP-type projects. Due to the long concession periods associated with PPP projects broad stakeholder management is essential.
- High *customer emphasis* – promises and commitment help create an enduring relationship. Several researchers suggest a focus on relationship marketing will enhance understanding of an organisation's customers' needs whether they are upstream or downstream in the supply chain.
- High *customer commitment and contact* – this is manifest both up and down the structure of the organisation in both development and operational phases of a PPP. Several researchers refer to core management driving this. However, to be successful the commitment and contact must come from all levels of participating organisational entities within the delivery and operational phases.
- Concern for *quality* – as in the foregoing commitment and contact a concern for quality in both process and product must be embedded in the entire special purpose vehicle commissioned for the project to ensure success.

These six key principles form the basis of discussion for this section of the chapter, they have been reworked in a project environment as critical success factors in recent times and the reader is recommended to review Love, *et al.* (2010), Love, *et al.* (2011) and Jefferies, *et al.* (2014) for an insight to their particular approaches. Underpinning these six principles are three defining attributes of relationship development in PPP procurement. They are commitment, trust and performance satisfaction.

There are many attributes to consider in the subject of developing relationships and applying it to innovative and emerging styles of procurement. Primarily commitment, trust and performance satisfaction have been found to represent the headline constituent parts that attract consistent focus (Davis and Walker 2008).

Committed parties believe the relationship that they have is worth working on to ensure that it endures, and this desire to continue a relationship displays three characteristics: inputs to the relationship (there is a bond), the relationship's durability (conviction) and its ongoing consistency (better off remaining in the relationship than not).

A construction service is intangible by design and this intangibility provides little scope for early objective measurement by clients or their representatives in the construction supply chain; essentially the final result is unknown as there are little or no templates from which to make direct comparison. Considerable research has been carried out over a long time encompassing many industry sectors in an effort to determine an appropriate procurement method for project delivery, which is followed by equal consternation regarding contractor selection (for an elaboration on these points papers written in the recent past by authors such as Gary Holt, Peter Love and Martin Skitmore are commended to the reader).

Trust appears consistently in contemporary literature in association with relationships and accordingly should be given a priority in PPP selection initiatives. Essentially trust is a belief in a promise of another and understanding that an obligation will be fulfilled, providing confidence that the recipient will benefit and be helped to attain a desired goal. It also overcomes the intangibility inherent with construction procurement and at the same time builds personal and group behaviours in a project team – for an interesting perspective on team trust as a predictor of performance see Zhang and Zhang (2015). Trusting behaviour may manifest in psychological contracts or the 'personal deal' that was mentioned earlier (Dabos and Rousseau 2013), and exist between an individual and groups or organisations. In a trusting relationship such as a PPP, stakeholders are able to focus on essential long-term benefits. These long-term benefits are suggested to provide, amongst other things, enhanced competitiveness/innovation, and to lift productivity.

Expanding on the foregoing, a buying firm's trust in a supplier would be enhanced across three important areas: people, organisation and process. Following arguments made in several published works, the intertwined threads of people, organisation and process (POP) can be found in the relationship development phases described above. In research by Davis and Love (2011), these were catalogued as individual relationships, engendering trust and organisational development. Later models of equal rigour can be seen in a recent book from Walker and Lloyd-Walker (2015) and Ibrahim's (2014) PhD study. All three models are mutually supportive of one another. The Davis and Love (2011) model is reproduced for ease of reference (Figure 8.2), while the alternative models are recommended for independent review.

In each case a PPP procurement example is provided where trust may be enhanced.

1   People – the extent to which the parties are prepared to share private information and the extent and nature of the information disclosed is a signal of good faith. By providing confidential information contracting parties in a PPP procurement relationship provide tangible evidence that they are willing to make themselves vulnerable in their attempt to help their client obtain a desired goal.

2   Organisation – the willingness of a supplier/contractor to provide specialist equipment or adapt existing processes to meet a buyer's needs is fundamental to PPP procurement. This action of enhancing close engagement cements the relationship required between the stakeholders. The cost of changing to meet a buyer's needs may be considerable and customisation provides evidence that the supplier of the service can be relied upon.

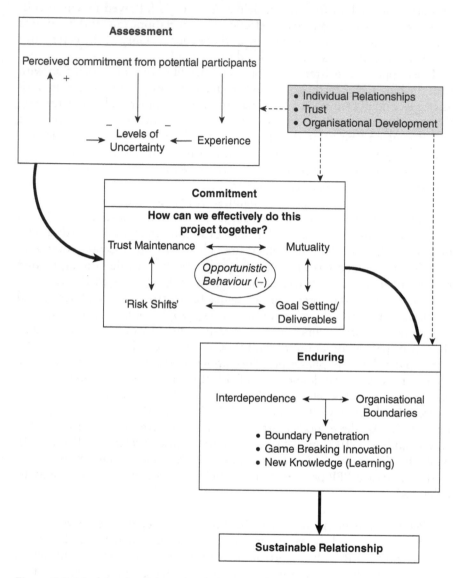

Figure 8.2 Models of collaboration in integrated project teams contracting
(Source: Davis and Love, 2011)

3   Process – the outcome of past ventures, the length of a relationship, together with document business records provide predictability to the process engaged in by the partners in the relationship and provides a framework for ongoing mutually beneficial interaction.

It may seem simple but many firms have trouble identifying clients' expectations. Mistakenly they think that they already know what their client wants. Research has identified that clients prefer companies with a reputation for fairness and an established track record. Importantly, despite a vast array of literature over a considerable period of time, clients do not appear to rely wholly on official criteria in supplier selection. So in order to be successful in the relationship it is necessary for organisations to analyse their client's need and determine what satisfaction would look like. Satisfaction may fall into several categories including: a demonstrated understanding of the problem, needs or interests, an interactive and communicative relationship, consistency in time budget, meeting expectations and matching previous favourable experience together with predictability in the process. There is considerable evidence to suggest that customer satisfaction is positively associated with the receipt of added value over time.

Having established that commitment, trust and performance satisfaction are recognised key attributes in a procurement relationship, the following explores their development, which is known to proceed in a fairly structured and defined way.

The development of relationships in PPP procurement

Many writers suggest relationship development can be conceptualised as a series of iterative phases. Typically writers identify partner selection, purpose definition, boundary setting, value creation and relationship maintenance as stages where commitment, trust, corporate and mutual goal development – individually or collectively – become an operational focus, while major components within other phases remain somewhat dormant. The phases are typically given various names, however Davis and Love (2011) distilled these to three: assessment, commitment and enduring phases, which lead to an ongoing relationship. The following analyses each of these within the context of PPP procurement.

Assessment phase

Generally it is a client who initiates the use of a particular procurement method for infrastructure development. The project team then becomes embedded within the preordained organisational structure with little or no thought as to the implication of the relationship with which it will be involved. There are some initial questions that should be asked: Should a collaborative PPP approach be pursued? Which aspects of the relationship warrant development? And how should the new organisational structure be developed to manage the ensuing collaborative relationship? Macbeth and Wagner (2000) suggest that organisations need to be aware of several factors when embarking on a collaborative strategy.

1   Historical actions of staff hindering collaborative relationships should be identified, as the intra-organisational context will have an effect on the initiating behaviours that are displayed at the outset of the relationship.
2   To militate this, frameworks that encourage cooperative behaviour with the firms involved in the collaborative relationship should be developed.
3   Norming behaviour, where early encounters within relationship development create positive project-specific objectives, should be nurtured and captured.

In essence this assessment phase is one of strategic formulation, where potential partners will look for organisational alignment and strategic fit, as early determination of goals and objectives at an institutional or project level will influence the way forward. To enable this an organisation should be able to analyse and describe its goals in terms that its prospective partners can comprehend, and most importantly relate to.

At this point in time a psychological contract or personal deal may come into play. The term 'psychological contract' became popular within HR studies in the 1990s. The earliest studies identified in construction occurred in the mid-2000s when Dainty *et al.* (2004) wrote about psychological contracts with respect to project managers in construction. The basis of the psychological contract relationship is reciprocity between organisation and employees on their perceived obligations to and expectations of one another. Current research on psychological contracts seems to be able to explain some of the important actions that happen in the early assessment phases of the team. A psychological contract is described as an internal motivator that drives individuals to determine the perceived or institutional contract – in this assessment phase motivation is high, although commitment remains in the development stage. Despite the scope of the relationship being fluid and lacking definition certain consideration would be given to key benefits and requirements, for example finance, technology platforms and managerial expertise. In this phase of the relationship partners may choose to proceed with caution, gauging reciprocal commitment and reflecting on exchanges made. Interim exchanges may be made with multiple partners initially as this represents a typical risk reduction strategy. As familiarity increases through the development of established norms and behaviours cultural and power-distance factors would also decrease. The development of trust mitigates high levels of uncertainty more quickly with some potential partners than with others. As a consequence multiple partners will be reduced as certain parties will not be considered appropriate for forming a relationship with. As Macbeth and Wagner (2000) indicate, selection disqualification can occur if displayed behaviours or competencies appear to be less than expected. While organisational alignment may be becoming established, individual commitment to one another within the newly forming team will be contingent on psychological contract and trust assessment, as guarded exchanges of information take place. Perceived investments from individuals, however, will likely increase trust; these investments may be of an economic or social nature for example. Typically individuals would be looking to determine if these investments add value to the project from their perspective. In this phase comparative analysis with other potential relationships may take place; however, eventually a decision based on the limited information will allow a team to continue to the commitment phase.

Commitment phase

In reality there are no distinct boundaries between any of the phases that are described in this section of the chapter. For example the transition from the assessment to commitment phase may occur earlier in the development or later contingent on the team, the project and the procurement solution being considered. Indeed many academic writers discuss decision-making models in a similar context using the terms people, organisation and process as the variables most intertwined. It is fair to say that the boundaries are blurred and fuzzy and in actuality they may be crossed a number of times. Considering the psychological contract or personal deal once again, the unwritten agreement referred to in the assessment phase continues to be formalised in the team members and their constituent organisational minds. Mutually accepted promises and obligations between the organisation and the employees with loose parameters are formed. It is noteworthy that PC writers, for example Rayton and Yalabik (2014), refer to PC contract breaches that can have adverse effects on the organisations in this phase in a similar way that occurs in standard contract violations where sanctions may be levied. The parties to the potential relationship display serious interest and consider relationship obligations in order to overcome a propensity to depart. Trust and commitment go hand in hand and through their synergistic formation become fundamental to the relationship interaction. Trust affects a buyer's behaviour and attitude, it impacts on negotiation and bargaining. It is fair to say that social bonding and trust development underpin relationship development. Incompatibility or a lack of personal chemistry are often blamed for failure in this phase; however, a focus on the psychological contract or personal deal can provide a means through which the team leadership can establish and maintain healthy relationships with increased levels of commitment and satisfaction (Guest 1998). In this formative phase of relationship developmental risk is prevalent because partners are assessing their strategic, operational and tactical positioning in the project. Dealings between the organisations become more direct as the relationships develop, and as trust and the desire to work together increases, the potential partners' shifts in risk thinking become more marked. For example a large concession that requires reciprocation, a proposal for compromise, a unilateral action of tension reduction, a candid statement concerning motives and priorities all represent risk shifts (Ford 1998). In this phase these may be either formal or informal. Several writers have demonstrated that informal cooperation or working together translates into new roles as the relationship development progresses. Furthermore new models of working together and co-operative behaviours embed new values in the wider context of associated organisations and the people that operate within them. Despite the foregoing in this phase the relationships will remain fragile with limited commitment and they can end relatively easily. Simultaneously there will be growing trust, though there may be some conflict over what should be done for mutual achievement. Comparisons and measurements against predetermined benchmarks will be made although performance satisfaction between the parties will reduce reliance on this. Many writers suggest a committed environment would include:

- longer-term focus on the strategic goals of the stakeholders;
- relationship agreements without guarantees in terms of workload and resource transfer;
- reduced duplication and process improvements;
- shared authority with open and honest risk sharing.

Similarly to the earliest stages there is no set pattern; however, as the actors within the relationship develop, an emergent culture begins to form, which leads to the enduring phase.

Enduring phase

Several writers describe a hybrid team that evolves as the relationship development process begins to acquire communal assets. The team begins to become more interdependent and any organisational demarcations or organisational boundaries disappear throughout this phase. Increased organisational interactions take place that surpass expectations, attractiveness increases and enhances goal congruence, and deeper cooperation emerges. All partners within the hybrid team tend to alter their procedures and are prepared to make informal adaptations. Davis and Love (2011) refer to people, organisation and process to describe a learning organisation where sharing collaborative experiences or 'war stories' as described by Davis and Walker (2009) become prevalent in the people. They explain common performance measurement systems adopted by the hybrid organisation and finally provide an example of a process where transfers of adopted asset-specific resources will incur significant cost. All of these factors coalesce to form an environment of implicit trust, mutual recognition and sharing of risk.

Relationship development: an industry perspective

Davis (2008) undertook an analysis of construction supply chains reviewing upstream and downstream relationships in construction organisations. Responses were sought from nearly 900 industry practitioners in Western Australia. A good response provided interesting results that were analysed using descriptive statistics and exploratory and nonparametric confirmatory statistical techniques. Of the responses it was found over 50 per cent of the respondents were consultants, over 20 per cent were general contractors and around 10 per cent subcontractors. The sample was stratified into three groups: contractors, clients and design professionals. Questions in the survey covered attitudes to relationships, commitment, trust and satisfaction, and were used to test the principles and attributes discussed in the earlier sections of this chapter.

Attitudes to relationships

When asked about attitudes to relationships the respondents indicated that they would endeavour to foster both upstream and downstream relationships. Upstream relationships are defined as those between your company and a controlling party (for example your client), downstream relationships are defined as those in which your company has control over another company in their business dealings (for example your specialist contractors). Respondents agreed there were no real differences between them.

Several benefits were suggested to come from upstream relationships – they were categorised in the research as related either to relationship marketing (RM) i.e. long-term and product-focused, or transactional marketing (TM) i.e. short-term and project-focused. They included:

- alignment of organisational objectives
- constructability benefits
- value-engineered solutions benefits.

Alignment of organisational objectives provided the largest upstream benefit to the respondents followed by constructability and value-engineered solutions. Further work to distil the respondents' attitudes to relationships was undertaken. Some interesting results were discovered and the sample strata of contractor, client and design professional were found to be divergent in their views. Both the contractor and design consultant respondents identified product issues (RM), which are long-term considerations that generally exist far in excess of the actual project duration (alignment of organisational objectives, value-engineered solutions, constructability) and process issues (TM), which are short-term and confined to the project duration (cost reductions, schedule reductions). Interestingly from an RM perspective this shows that construction procurement personnel intuitively differentiate between relationship and traditional marketing. They have a tendency to focus on RM in their construction supply chains. This is a positive behaviour in the context of PPP procurement due to its inherent long-term nature.

Similarly, several benefits were suggested to accrue from downstream relationships. These included:

- knowledge-based benefits
- cost benefits
- time benefits.

The research outcome indicated that knowledge-based benefits provided the largest advantage, followed by cost benefits and time benefits downstream. Additional analysis provided some interesting results. Again there was divergence in the sample strata with contractors identifying RM benefits as knowledge-, process-, resource- and skill-based benefits that enable contractors to add client value via relationship development with others. These benefits would regularly manifest in strategic relationships with organisations downstream and benefit PPP procurement. Design respondents grouped the benefits in a similar way; however, quality benefits produced unclear results that were explained by an unresolved distinction between quality activities incorporated in the project and functional qualities in the final product. Clients varied from the contractors, grouping quality benefits with relationship attributes. This was ascribed to a lack of distinction between functional quality in the final product and quality activities incorporated into the project delivery. Two other variables, value adding benefits and risk allocation benefits, were found to be complex and difficult to interpret. Moreover it was difficult to differentiate them in terms of TM and RM.

Summarising this, the majority of respondents concurred that they endeavour to foster both upstream and downstream relationships; perhaps understandably upstream relationships are paid marginally more attention than downstream relationships – we should always be keen to meet the needs of our clients in whatever form they present themselves. In upstream relationships, the primary determinants were found to be relationship issues for both contractors and clients. Design professionals recognised transactional issues as the primary factor – perhaps this is a consequence of their focus being primarily on the project (the processes involved in building the structure). All respondents – clients, contractors and design professionals – identified cost benefits as a process-based or transactional marketing attribute.

Importantly perhaps, cost or price becomes a secondary criterion; its focus is moderated by the intangible outcomes that build capacity for the future, being commitment, trust and satisfaction. There is ample research that supports these arguments: for example Day and Barksdale (1992), in a report concerning architects and engineers, found that firms rarely spoke about price as part of the selection process, and Patterson (1995) identified cost and the value of fees as secondary to reputation.

Managerial implications for a PPP team

The industry respondents in the sections above identified underlying factors of product and process. In doing so they differentiated between relationship and transactional marketing, identifying important factors in the relationships and relationship development.

Key reports arising from the UK concerning construction, for example Latham (1994) and Egan (1998), are supportive of these findings in as much as both reports identify a desire to move away from adversarial transactional relationships towards mutual goal achievement and the alignment of organisational objectives that engender long-term relationships.

Despite these and other reports of significance, adversarial relationships continue to reside in the thoughts of academic writers. For example a review of several sources of academic work identifies many papers that highlight problems derived from the adversarial nature of construction (for example see Eriksson 2010; Boukendour and Hughes 2014; Rose and Manley 2014; Smiley, *et al.* 2014).

Other research, from Baccarini (1999) and Cooke-Davies (2002), helps us understand the importance of the differences between relational and transactional dealings. They use the term 'product success', which matches long-term RM criteria, involving satisfying user and stakeholder needs relating to products and meeting project owners' strategic organisational objectives. This can be contrasted against the concept of 'process success', which is short-term and transactional in marketing terms, involving meeting time, cost and quality objectives and satisfying project stakeholders' needs as they relate to project management processes. In effect process success is transactional marketing; however, it is argued that transactional marketing is not conducive to organisational achievement or success in a PPP context, as relationships in this environment are short-term and essential elements of relationship development, such as commitment, trust and satisfaction, cannot be nurtured (Davis 2005). These three aspects are explored in the closing sections of this chapter.

Commitment – trust – satisfaction

As we saw in the earlier sections of this chapter, commitment has been identified by several writers as a factor that is founded on an enduring relationship. Other writers suggest that supply chain management is built around intra- and inter-organisational relationships that require a strategic approach. In the research example the sample, while considering whether the level of conviction of remaining in a relationship would yield higher benefit than from terminating it, found upstream commitment was marginally greater than that displayed for downstream commitment. In this regard there were no significant differences between any of the sample strata.

Relationship-marketing literature has great depth in the area of trust. Prior research found several useful indicators of trust that are pertinent to procurement, supply chain management and PPP initiatives. There were no differences between groups in their reporting of a hierarchy of trust factors' (confidential information sharing, a willingness to customise and predictability in the process) impact on relationships. The research also highlighted a propensity toward an expectation of trust from both upstream and downstream organisational relationships; the strongest association was with the sharing of confidential information. However, it was identified that respondents were marginally more cautious with regard to downstream relationships; clearly sharing confidential information with a partner downstream in a relationship without careful consideration would be a significant risk. In summary, it was found that the entire sample had high expectations that their dealings with other stakeholders would be underpinned by trust.

Attributes that satisfy clients were identified as: understanding problems, needs and interests; interactivity in the relationship and being communicative, on time and on budget; additional cost providing value; meeting clients' expectations and matching previous favourable experiences. In upstream relationships the concept of being interactive and communicative in the relationship ranked most highly, but this was ranked fourth when respondents considered downstream relationships. The highest ranked variable in downstream relationships was understanding problems, needs and interests. In both upstream and downstream relationships, matching previous favourable experiences was ranked in second position. In upstream relationships understanding problems, needs and interests was ranked third, while in downstream relationships meeting clients expectations was ranked third. Testing found that there were no differences between the groups surveyed.

Chapter summary

Building relationships in a PPP project environment is risky, as in the assessment period all parties jockey for strategic position and endeavour to define their place in the context of the project's scope. Generally, because of the nature of projects and their finite timeframe, a relationship development process is novel, requiring a considerable amount of planning and investigation. At the same time this process could be used to identify and manage risks in a collaborative way. When thought is given to the operational and maintenance phase of a PPP project and its extended timeframe, individual project relationships should be replaced with organisational relationships that are supported or structured with appropriate organisational governance protocols. Individual relationships are framed from past experience and are prioritised through favourable experience, indeed previous experience reduces risk and provides a more tangible assessment from which to build. Despite the fact that the relationship development model described earlier appears to be somewhat contrived, in reality using management games and selection themes has been shown to facilitate and nurture the fast-track development of relationships in alliance projects. This fast tracking enables the individuals to assess associates within a much-reduced timeframe. Previous work by Davis and Love (2011) shows that in relationship development in alliance projects partners attempted to attain relationships akin to respectful personal relationships, for example contractors have an objective of maintaining a position of high regard in the eyes of their clients. In this regard relationship tests sought to identify contractors' positions and whether they determined a stance of remaining adversarial, conducting businesses as usual or adopting best practice.

Relationship-based supply chain activities that are critical to PPP projects have been identified. Improved supply chain management is manifest through more satisfying longer-term business partnerships and the development of social capital that provides goodwill and commitment to the project. This goodwill is a more effective governance mechanism than that which is presented in more traditional contract-based supply chains. Strong relationship-based supply chains that appear in PPP projects that adopt this approach tend to generate additional tangible assets, such as knowledge-base benefits and improved joint problem solving. These assets focus upon teams finding more holistic solutions that better satisfy a broader range of upstream and downstream constituencies. Commitment is required by stakeholders in order to meet project objectives; furthermore showing concern and portraying a positive attitude, fostering upstream relationships to obtain work and indicating that remaining in a relationship would yield significant benefits upstream are attributes required in successful PPP-type projects. Similarly value adding, innovation and organisational achievement are derived from collaboration and commitment. These benefits become leading indicators; in fact they are the seeds for future success.

Reflections

1   PPP is becoming a commonly accepted term for partnerships between the public and private sectors; however, various models still exist, so can you provide an overview of the different approaches to this form of procurement?

2   In their seminal publication on PPPs, Grimsey and Lewis include an investigation into the governance of contractual relationships in PPPs.

This section of the book is a useful resource that supplements this chapter and focuses on the changing public sector administrative requirements for PPPs. See Grimsey, D. and Lewis, M.K. (2007). *Public Private Partnerships: The Worldwide Revolution in Infrastructure Provision and Project Finance*, Cheltenham, UK, Edward Elgar.

3   What are the key principles of relationship marketing that are pertinent to the PPP procurement method?

4   The work of Walker, *et al.* in Chapter 2 further explores the issues of relationships and in particular the shared characteristics between PPPs and Alliances.

5   Is relationship building any different in procurement approaches such as PPPs or Alliances when compared to more traditional methods such as Construct Only or Design and Construct?

References

Akintoye, A., Beck, M. and Hardcastle, C. (2003). *Public-Private Partnerships: Managing Risks and Opportunities*. Oxford: Blackwell Science.

Baccarini, D. (1999). 'The logical framework method for defining project success'. *Project Management Journal* 30(4): 25–32.

Boukendour, S. and Hughes, W. (2014). 'Collaborative incentive contracts: stimulating competitive behaviour without competition'. *Construction Management and Economics* 32(3): 279–289.

Cooke-Davies, T. (2002). 'The "real" success factors on projects'. *International Journal of Project Management* 20(3): 185–190.

Dabos, G.E. and Rousseau, D.M. (2013). 'Psychological contracts and informal networks in organizations: the effects of social status and local ties'. *Human Resource Management* 52(4): 485–510.

Dainty, A.R.J., Raiden, A.B. and Neale, R.H. (2004). 'Psychological contract expectations of construction project managers'. *Engineering, Construction and Architectural Management* 11(1): 33–44.

Davis, P.R. (1996). *Marketing Project Management Services: A Review and Assessment of Some Possible Approaches Used by Project Management Organisations in Perth, Western Australia*. Report presented as part of the requirements of the award for the degree of Master of Project Management, Curtin University of Technology, Perth, WA.

Davis, P.R. (2005). *The Application of Relationship Marketing to Construction*. Unpublished PhD Thesis, Royal Melbourne Institute of Technology.

Davis, P.R. (2008). 'A relationship approach to construction supply chains'. *Industrial Management and Data Systems* 108(3): 310–327.

Davis, P.R. and Love, P.E.D. (2011). 'Alliance contracting: adding value through relationship development'. *Engineering Construction and Architectural Management* 18(5): 444–461.

Davis, P.R. and Walker, D.H.T. (2008). 'Case study: trust, commitment and mutual goals in alliances'. In (eds) Walker D.H.T. and Rowlinson, S. *Procurement Systems: A Cross Industry Project Management Perspective*. London, Taylor & Francis, pp.378–399.

Davis, P.R. and Walker, D.H.T. (2009). 'Building capability in construction projects: a relationship-based approach'. *Engineering, Construction and Architectural Management* 16(5): 475–489.

Day, E. and Barksdale, H.C. (1992). 'How firms select professional services'. *Industrial Marketing Management* 21: 85–91.

Egan, J. (1998). *Rethinking Construction*. The report of the Construction Task Force to the Deputy Prime Minister, John Prescott, on the scope for improving the quality and efficiency of UK construction. London, Department of Trade and Industry.

Eriksson, P.E. (2010). 'Partnering: what is it, when should it be used, and how should it be implemented?' *Construction Management and Economics* 28(9): 905–917.

Ford, D. (1998). *Managing Business Relationships*. Chichester, UK; New York, J. Wiley.

Gronroos, C. (1990). 'Relationship approach to marketing in service contexts: the marketing and organisational behavior interface'. *Journal of Business Research* 20: 3–11.

Guest, D.E. (1998). 'Is the psychological contract worth taking seriously?' *Journal of Organizational Behavior* 19(S1): 649–664.

Ibrahim, C.K.I.B.C. (2014). *Development of an Assessment Tool for Team Integration in Alliance Projects*. Unpublished PhD thesis, University of Auckland, NZ.

Jefferies, M., Brewer, G., and Gajendran, T. (2014). 'Using a case study approach to identify critical success factors for alliance contracting'. *Engineering, Construction and Architectural Management* 21(5): 465–480.

Latham, M. (1994). *Constructing the Team*. Joint Review of Procurement and Contractual Arrangements in the United Kingdom Construction Industry, Final Report. London, HMSO.

Liu, J., Love, P., Davis, P., Smith, J. and Regan, M. (2014). 'Life-cycle critical success factors for public-private partnership infrastructure projects'. *ASCE Journal of Management in Engineering* 31(5): 4014073.

Love, P., Mistry, D. and Davis, P. (2010) 'Price competitive alliance projects: identification of success factors for public clients'. *Journal of Construction Engineering and Management* 136(9): 947–956.

Love, P.E.D., Davis, P.R., Chevis, R. and Edwards, D.J. (2011). 'Risk/reward compensation model for civil engineering infrastructure alliance projects'. *ASCE Journal of Construction Engineering and Management* 137(2): 127–136.

Macbeth, B.D.D. and Wagner, B. (2000). 'Implementing collaboration between organisations: an empirical study of supply chain partnering'. *Journal of Management Studies* 37(7): 1003–1017.

Patterson, P.G. (1995). 'Choice criteria in final selection of a management consultancy service'. *Journal of Professional Services Marketing* 11(2): 177–187.

Payne, C.M.A. and Ballantyne, D. (1991). *Relationship Marketing: Bringing Quality Customer Service and Marketing Together*. Oxford, Butterworth-Heinemann.

Rayton, B.A. and Yalabik, Z.Y. (2014). 'Work engagement, psychological contract breach and job satisfaction'. *International Journal of Human Resource Management* 25(17): 2382–2400.

Rose, T.M. and Manley, K. (2014). 'Revisiting the adoption of innovative products on Australian road infrastructure projects'. *Construction Management and Economics* 32(9): 904–917.

Smiley, J.P., Fernie, S. and Dainty, A. (2014). 'Understanding construction reform discourses'. *Construction Management and Economics* **32**(7–8): 804–815.

Walker, A. (2013). 'Outcomes associated with breach and fulfillment of the psychological contract of safety'. *Journal of Safety Research* **47**: 31–37.

Walker, D.H.T. and Lloyd-Walker, B.M. (2015). *Collaborative Project Procurement Arrangements*. Newtown Square, PA, Project Management Institute.

Zhang, L. and Zhang, X. (2015). 'SVM-based techniques for predicting cross-functional team performance: using team trust as a predictor'. *IEEE Transactions on Engineering Management* **62**(1): 114–121.

9 Relationship Contracting in a Local Government Public-Private Partnership (PPP)

Marcus Jefferies and Denny McGeorge

Chapter introduction

There is general acceptance that PPPs are part of the procurement landscape in Australia. However, a strong body of opinion exists to support the concerns of the private sector that current social infrastructure projects in Australia are not true partnerships (Curnow *et al.*, 2005; Jefferies and McGeorge, 2009; Jefferies *et al.*, 2013). The public sector needs to make PPPs more attractive to the private sector and clarify the risk identification in order to transfer more responsibility to the private sector. This issue is supported by recent industry criticism of PPPs concerning the 'narrowness' of the scope of work that is offered to the private sector. In reality, PPP project costs relating to finance, building design, construction, maintenance and waste management amount to less than 15 per cent of the total life-cycle cost of the enterprise. Figure 9.1 illustrates the limited nature of current private sector participation in PPPs.

As a result, the private sector is frustrated with the high transaction costs of PPPs, which offer only a marginal increase in scope of business opportunity. This is in stark contrast to opportunities that are available in the much lower

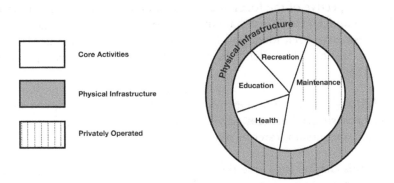

Figure 9.1 Current private sector participation in PPPs (Source: adapted from Curnow *et al.*, 2005 and Jefferies *et al.*, 2013)

cost-to-bid ratio of more traditional procurement models. Governments are looking for significant increases in efficiency through the PPP process, but no matter how well the 15 per cent of the enterprise available to the private sector is organised, it is not going to make up for inefficient management in the remaining 85 per cent. This has led to the decision by a number of major construction contractors to withdraw from the PPP process.

Additionally Shepherd (1999) and Jefferies and McGeorge (2008) argue that there are fundamental reasons for the need to review the PPP process, these reasons include:

- Lack of flexibility in the evolution of the project where the host authority must juggle competing bidders and keep them on the same baseline, often while simultaneously processing an environmental impact statement.
- Current arrangements lack flexibility in operation whether it be extending or widening a tollway, converting a power plant or extending a hospital.
- High transaction/tender costs in taking at least two fully developed and underwritten bids to the finishing line (e.g. Melbourne City Link incurred external tender costs alone of $24 million at financial close).
- PPPs need to allow the private sector to utilise its expertise and gain a broader scope of work with an increased transfer of risk and responsibility.

There is some evidence to suggest that the growth of PPPs in Australia would be accelerated if the concerns highlighted above could be addressed. The inability of the private sector to win enough projects to offset the significant tender costs for more complex PPPs appears to have had an adverse influence on the construction industry (e.g. reduction of company share values, company mergers and take-overs). PPPs are normally linked to large-scale projects that, in many cases, have a high public profile. Because of this, the risks associated with PPPs are often perceived to be correspondingly higher than for more conventional forms of contractual relationships. However, irrespective of the form of contractual relationship there are two critical risk management issues that emerge during the bidding stage of any project. These have been identified from previous work by Tiong (1990), Walker and Smith (1995), Jefferies and McGeorge (2008) and Jefferies *et al.* (2013) as:

1   *Legal Risk* – legislative framework, project agreement, tax, laws; and
2   *Financial Risk* – bid process, form of financing, evaluation, commercial investors, ownership, rates of return.

The aim of this chapter is to investigate current approaches to risk management of infrastructure projects using an Australian mixed-use development, procured using the PPP approach, as a case study. The results of the case study were developed from an analysis of project documentation and a semi-structured interview process and they explore issues that are essential for the successful delivery of a PPP such as the legal structure, stakeholder relationships and the management of risks – which are part of five themes that emerged which are discussed in more detail later in the chapter. This chapter complements others in this book that take a similar PPP focus. But whereas other authors contextualise their research in other parts of the world – such as the UK, Europe, Asia, US, Africa and so on, this one very much focuses on Australia.

Defining Public-Private Partnerships (PPPs)

One of the problems with a PPP is with its very definition. Definitions tend to depend on a commentator's own particular perspective and range from the very general to the quite particular. A general definition is provided by Akintoye *et al.* (2003), where Public-Private Partnerships (PPPs) are defined as a long-term contractual arrangement between a public sector agency and a private sector concern whereby resources and risk are shared for the purpose of developing a public facility. PPPs are considered to be a form of Relationship Contracting, which, according to Chueng *et al.* (2005), is based on a recognition of and striving for mutual benefits and win-win scenarios through more cooperative relationships between the parties. Relationship Contracting embraces and underpins various approaches, such as partnering, alliancing, joint venturing, PPPs and other collaborative working arrangements and better risk-sharing mechanisms. Relationship contracts are usually long-term, develop and change over time and involve substantial relations between the parties.

The focus of the case study in this chapter is a public sector local government client (City of Ryde Council). Thus, the definition of PPPs adopted for this research project was: 'An arrangement between a council and a private person for the purposes of providing public infrastructure or facilities and/or delivering services in accordance with the arrangement' (NSW Department of Local Government, 2005).

The *Local Government Amendment (Public Private Partnerships) Act 2004* and the Local Government (General) Regulation 2005 commenced on 1 September 2005. The Act brings into effect the recommendations of the Emeritus Professor Maurice Daly, Commissioner of the Liverpool City Council Public Inquiry, in relation to Public-Private Partnerships (PPPs). The Act defines PPPs, requires councils to follow the procedures set out in guidelines and establishes the Local Government Project Review Committee (NSW Public Accounts Committee, 2006).

The origins of PPPs in Australia

Most Australian commentators such as Jones (2003), Malone (2005), Walker (2003), Jordan and Stilwell (2004), English (2005) and Jefferies

(2006) trace PPPs back to the late 1980s as a natural progression from Build Own Operate (BOO) contracts such as the Gateway Motorway and Bridge, Brisbane (completed 1986) and Build Own Operate Transfer (BOOT) contracts such as the Sydney Harbour Tunnel (completed 1992). Quiggin (2005) subscribes to the view that the 'Partnerships Victoria' policy document was a watershed in the development of PPPs in Australia, and that, by and large, this document is representative of the approach adopted by other states. Duffield (2005) classified Australian PPPs into 'first' and 'second' generation with the release of 'Partnerships Victoria' being the watershed between the two generations. The difference between first- and second-generation PPPs in Duffield's (2005) view is:

- *First-generation PPPs:* primarily motivated by the public sector gaining access to private capital and the transfer of new full project risks.
- *Second-generation PPPs:* State Governments sought to retain direct control of 'core' services and to involve the private sector in, amongst other things, value-for-money outcomes.

Table 9.1 presents the work of Jefferies (2014) and is a summation of key events in the development of Australian PPPs from the 1980s, which updates the work of the above authors and further illustrates the first- and second-generation divide.

Characteristics of PPP projects

In discussion about the nature of social PPP projects, Jefferies and McGeorge (2009) state that comparisons were typically made against economic infrastructure projects. Key features and differentials are identified in Table 9.2.

The public sector must make PPPs more attractive to the private sector and clarify the identification of risk in order to transfer more responsibility to the private sector. In relation to commercially viable value-adding, Government typically restrict the outsourcing of services to 'non-core'

Table 9.1 Key events and initiatives in the development of PPPs in Australia

1980s–1990s: 1st-generation PPPs	2001 to date: 2nd-generation PPPs
1980s: Pressure due to poor balance of trades, excessive high debt, Government borrowing limit capped by loans council, poor fiscal management – Australian Government seeks alternative methods for development without further reducing credit ratings.	**2001:** PPP manual 'Partnerships Victoria' released. Queensland Government follows with 'Public Private Partnership (PPP) Policy' released in September.

1983: Australian dollar floated on international money markets – first step to deregulating the national economy.

1987: NY stock market crash – ripple effect in Australia ends the speculation boom that had followed the deregulation of the economy.

1988: NSW first documented formal procedures and controls governing private sector participation.

1990s: Corporate liberalism emerges in Government. An ideological shift towards Government playing more of a managerial role leading to a number of privatisations and outsourcing taking place across Australia.

1990s: Victoria takes lead with privatisation, outsourcing and BOO and PPP projects.

1996: National Competition Policy, supported by Competition Principles Agreement endorsed by all Australian Governments.

2000: Airport Link Company collapses six months after Sydney's airport rail link is opened, becoming one of Australia's first PPP projects to fail.

2000: TAS. Government releases a policy statement, and guidelines, on private sector participation in the provision of public infrastructure.

2002: NSW Government publishes a 'State Infrastructure Strategic Plan'. SA Government releases PPP policy and establishes PPP Unit in Treasury. WA releases 'Partnerships for Growth' – was released as the Policies and Guidelines for Public Private Partnerships.

2002: Intergenerational report released with the Budget papers (Treasury 2002) warned that net Government spending will need to rise by 5% of GDP by 2041–2 to fund the same standard of services and level of benefits.

2003: National PPP Forum held; Victoria 'Fitzgerald' review; NSW 'Parry Inquiry' recommends public debt used only when all other funding options have been fully explored.

2005: Local Governments propose to use PPP model for a number of urban revitalisation projects, such as Parramatta, Liverpool in NSW.

2006: NSW Parliamentary Inquiry into Cross City Tunnel and PPPs. NSW and Victoria announce continued use of PPPs as well as increase use of public debt to meet infrastructure shortfall.

2008: National PPP Guidelines prepared and endorsed by Infrastructure Australia and various State, Territory and Commonwealth Governments. The Guidelines set a national framework for the procurement of PPPs.

2012: NSW Government produces PPP guidelines to complement National Guidelines but also to provide stakeholders with NSW-specific requirements for PPP procurement.

2013: Proposed PPP reforms in response to changing market conditions, new fiscal challenges, private sector concerns and public's increasing demand for new infrastructure.

Source: Jefferies (2014)

services such as administration, catering and cleaning. There appears to be some conflict in the division of 'core' and 'non-core', the example being given – the employment of nurses, many of whom are employed in hospitals

Table 9.2 Key features and differentials between social infrastructure projects and economic infrastructure projects

PPP type	Social infrastructure (e.g. schools and hospitals)	Economic infrastructure (e.g. motorways, tunnels and bridges)
Scale of project	Smaller	Larger
Examples	Schools and hospitals	Motorways, tunnels and bridges
Complexity	More complex – especially ongoing involvement with community	Less complex
Risks	Associated with performance of the facility, e.g. major failure in air-conditioning systems	High construction risk engineering projects (e.g. collapse of Lane Cove Tunnel during tunnelling phase). Financial risks can also be high.
Revenue generation	Rental streams via the Government. Value-adds are sought, e.g. rental space, service contracts and other additional means	Direct payments, e.g. from tolls

Source: adapted from Jefferies and McGeorge (2009)

via private enterprise agencies. A branch of a consortium could undertake such a function (Jefferies and McGeorge, 2009).

Figure 9.2 illustrates the complexity of the relationships within a PPP.

Social PPPs: emergence and barriers for successful implementation

Jefferies and McGeorge (2009) established a compilation of all PPP projects from 1986 to date in Australia. Their data demonstrates that the application of the PPP approach to social infrastructure PPPs is a relatively recent trend. Jefferies *et al.* (2013) indicate that many of the PPP projects currently under consideration are for social projects and whether or not these projects progress to fruition will largely depend on the perceived risks and returns to both the public and private sectors. There are many barriers when considering the use of a PPP strategy. The following list has been developed using issues identified by the Australian Council for Infrastructure Development Limited (2003), Curnow *et al.* (2005) and Jefferies and McGeorge (2009):

- *Tax reform:* section 51AD of the Australian Tax Act is a serious barrier to many PPP projects. Rectification of this could pave the way for further use and implementation of PPP arrangements, e.g. shadow tolling as a form of payment for infrastructure services is restricted, if not prohibited, under 51AD as it currently stands.

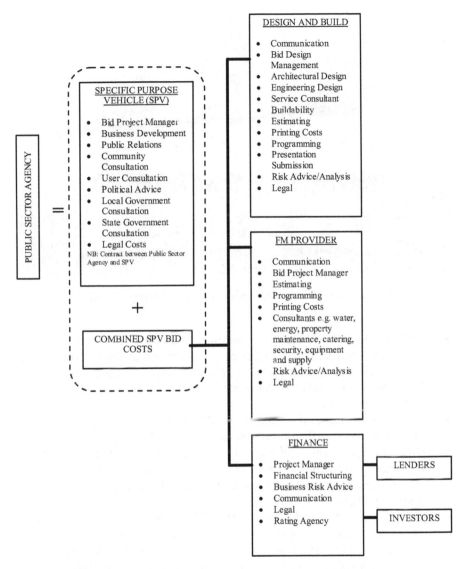

Figure 9.2 The complexity of costs and commercial relationships within a typical PPP consortium structure (Source: adapted from Jefferies and McGeorge, 2008; 2009)

- *Whole-of-Government approach:* strong central Government control has led to a lack of consistency in PPPs. To ensure PPPs succeed the consistency must be driven at the highest levels of Government.
- *Lack of suitable skills in Government:* public sector agencies may not have the skills or experience to ensure a successful project. Greater training and experience will overcome these.

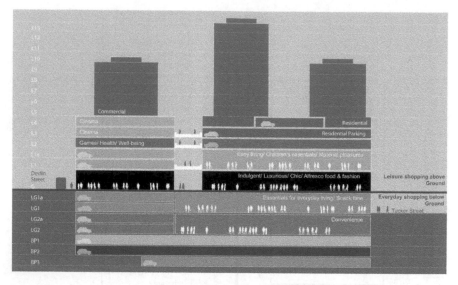

Figure 9.3 Cross section of Top Ryde redevelopment (Source: City of Ryde, 2015)

- *High participation costs:* high bidding costs are viewed as the most prominent barrier to entry. It has been indicated that tender costs are up to six times higher than that of traditional procurement arrangements. Therefore it is unlikely smaller contractors will be able to participate due to high tender costs.
- *High project values:* the majority of PPP projects are larger than those for which many small contractors can realistically aspire to bid.
- *High risk:* one of the fundamental requirements of a PPP is that the private sector must genuinely assume risk. Many smaller companies would not be equipped to handle such large risk, which would prevent them from entering the PPP market.
- *Lack of credibility and contacts:* the PPP process invariably involves a contractor working in a consortium with partners from outside of the construction industry. This is an effective barrier to smaller contractors as they are less likely to have external contacts to form a PPP consortium.

Case study: the Top Ryde PPP

Project background

The Top Ryde PPP project is an $800 million mixed-use development of retail, commercial, residential and public services over 78,000m². The original Top Ryde Shopping Centre was built in 1957, being New South Wales' first regional shopping centre and one of Australia's earliest. The role of

Top Ryde as an important retail and social centre declined over the 1980s and 1990s as its infrastructure became out-dated with insufficient parking and a limited choice of food and retail outlets. The City of Ryde recognised that the redevelopment was required in order to reduce escape expenditure outside the suburb, increase employment and restore the Town Centre as a social and civic hub.

The Beville Group, i.e. the private sector contractual partner with the City of Ryde (CoR), purchased the Centre in 2000 and had approached the City of Ryde to obtain approval to expand the old Top Ryde Shopping Centre on several occasions. Prior to 2004, the Beville Group had a poor relationship with the City of Ryde but again initiated an approach to negotiate planning approval with the new GM of CoR. The proposed expansion of the Town Centre was of such a significant size that an Integrated Traffic Solution (ITS) was required. From that ITS, some of CoR land was required for the underpass system to access the new Shopping Centre and CoR was approached to sell the land. Eventually planning documents moved through Local Government to State Government, where the new access routes were a key issue for planning approval under a Local Environment Plan (LEP). Defined Developments (DD), established by Bevillesta Pty Ltd (as a subsidiary of the Beville Group) as the Developer to deliver Top Ryde Shopping Centre, signed a 49 by 50 year lease for a 'peppercorn' rent of $1. Under the terms of the lease, DD wouldn't pay any significant rent for the land providing they successfully linked CoR Civic Precinct land to the ITS in the form of bridges and underpasses.

Under the *Local Government Amendment (Public Private Partnerships) Act 2004*, the NSW State Government subsequently deemed the project a 'PPP' as the Developer was providing free infrastructure to the City of Ryde in the form of rights of way through the underpasses and over the bridges linking CoR land from west to east and enhancing the access solution for CoR ultimately allowing for future redevelopment to be undertaken by CoR. In March 2005, Bevillesta entered into a management services agreement with Lend Lease, requiring them to provide design master planning services that combined their in-house architectural expertise together with their cost planning programming and resources. In 2006, the LEP Number 143 was gazetted, which allowed for future growth of Top Ryde City and the surrounding suburbs. Lend Lease was appointed on a guaranteed maximum price contract to undertake the design and construction of the project. Construction on the Centre commenced in September 2007.

Planning approval was based on the best traffic solution to minimise impact to the community but at the same time maximising customer efficiency. By working with the State and Local authorities, DD were able to develop the best outcomes for all parties and execute long-term agreements with the Government departments that benefited the Centre owner supplying the community with facilities and infrastructure for the future growth of CoR's land. It is also important to note that redevelopment of Top Ryde must be viewed in a broader context of establishing a momentum for revitalising the whole of the Town Centre over time.

The results of the case study research were developed from an analysis of project documentation and semi-structured interviews with senior management stakeholders representing the various public and private sector partners within the Top Ryde PPP. The following five themes emerged:

1   contractual arrangements: the Tripartite Agreement
2   legal and administration costs
3   standardisation of documentation
4   relationship management
5   risk management.

1. The contractual arrangements: Tripartite Agreement

Fundamentally, the main participant roles bound by the Tripartite Agreement involved:

- City of Ryde (Local Government) – administration and review
- Bevillesta (private sector developer) – design, financial and delivery risk
- Roads and Traffic Authority (State Government) – approval process for relevant works

Under the Tripartite Agreement, the City of Ryde is the roads authority and the owner of the surrounding main roads (Devlin Street and Blaxland Road). As owner, the City of Ryde agreed to lease a portion of these roads, comprising the site for Top Ryde, to the Developer under the Agreement. The Developer undertook the works on the site and agreed to own and operate the works (finance, design, construct and operate). Under section 138 of the Act, the consent of Council, as roads authority, with concurrence of the Roads and Traffic Authority (RTA), was required. A condition of the Development Application (DA) consent is that the three parties enter in a Tripartite Deed of Agreement. The Tripartite Deed served to clarify the roles and responsibilities of the three key players, i.e. City of Ryde, RTA and Bevillesta, at the beginning of the project. The tone of the Tripartite Deed was not adversarial. The roles and responsibilities of each party were discussed, agreed and formalised at the start of the project.

A three-layer communication model was established and embraced enthusiastically by all the parties to manage the PPP work. The three levels consisted of:

i   *PPP communication meeting*
    Hosted weekly by the Main Contractor (Lend Lease) and attended by CoR, and often RTA, with stakeholders such as State Transit Authority

and the Project Verifier as necessary. This meeting served to advise and discuss detailed works progress, certification issues, focus points for co-operation, feedback from the local community and communications required with local residents about future work. Weekly communication meetings were structured to review the works programme and the community's interest together to manage expectations and minimise inconvenience.

ii *PCG (Project Control Group) meeting*

Initiated and hosted by CoR weekly as appropriate, this brings together the project managers operating on behalf of the CoR, the Developer, Main Contractor and RTA to overview general progress, identify and mitigate risks and issues, ensure information flow is timely, agree points of collaboration, resolve contentious issues, defuse potential problems and agree action points for all participants to ensure the project proceeds as smoothly as possible. Weekly PCG meetings enhanced communication and collaboration between the three PPP stakeholders' representatives, the effectiveness of which has been reinforced by the inclusion of the Main Contractor in this forum.

iii *High-level PCG meeting*

This was held every two or three months according to need and, as specified in the Tripartite Deed is chaired by an independent person. Given RTA's concurrent role in the PPP, the principal attendees had been the CoR and the Developer. These meetings provided a platform of supervision, negotiation and control of the direction of the PPP works and design intent of the overall project. This forum was principally concerned with policy, direction and the progress of the project at a strategic level. Collaboration at quarterly high-level meetings between the Heads of the CoR and Bevillesta organisations provided a strategic control upon the relationship between the PPP Agreement and the DA process. The nominated project managers for the CoR, RTA, Bevillesta and Lend Lease (LL) operated as consistent principal contacts for their organisations facilitating resources as required and ensuring that communication was not weakened or diluted. The Tripartite parties and LL each have large organisations with many 'interested' members.

The Tripartite Agreement brought together the owner (CoR) and regulatory body (RTA) together with the private sector developer (Bevillesta in the form of Defined Developments). The PPP model supplied a very-low-risk solution for the Local Government (CoR). In removing the likes of design and maintenance risk, CoR were able to supply infrastructure to the community at merely an 'administration cost'. The PPP, via the Tripartite Agreement, also provided a framework for the RTA to work within, streamlining approvals and providing an efficient way of achieving the infrastructure that was such a key component of delivery of Top Ryde.

2. Legal and administration costs

The legal aspects of a PPP include the contracts, the legal entity and the law and regulations that the PPP will be working under (Bult-Spierinh and Dewulf, 2006). Whilst all project participants acknowledged that legal costs were an inevitable consequence of the construction industry's highly litigious environment, the dominant view was that PPP legal and administration costs were 'excessively high'. These costs can act as a deterrent to the private sector at tender stage and the public sector at development stage.

There were a number of legal and administration costs identified as part of the Top Ryde project, and indeed unique to PPPs. These include:

- legal advice regarding the establishment of the PPP;
- the legalities of setting up the PPP and also arrangements with the contractor(s);
- liaising with project stakeholders to work through the contract and assessing the risk in the contract;
- costs incurred as part of the Independent Verifier process stipulated by the Department of Local Government (NSW State Government requirement of PPP projects);
- the lack of standardised contract documentation for projects of this nature;
- efficiency and effectiveness on the focus on the 'finer points', given the economics of long-term legal obligations contained in many PPPs, particularly in contracts of over 20 years, and in this case the 49 by 50 year lease arrangement.

Upfront costs incurred under a PPP

The 'upfront' or bid costs (legal fees, consultant fees etc.) incurred by the Developer on Top Ryde were initially minimal and were thought of as being similar as if the project was delivered as a standard Design and Construct (D&C) contract as opposed to a PPP. However, additional costs were incurred once the project was classified as a PPP by the NSW State Government. This additional cost wasn't envisaged upfront as Top Ryde was classified as a PPP late in the procurement process as the project itself was already established and many of the costs had already been incurred. One of the additional costs, was that of independent verification, where the Verifier was to be paid $500,000 for their services. This cost was wholly incurred by CoR but after negotiations between the Council's General Manager and the Developer, however, the Developer did contribute a substantial amount of $300,000 towards the fees. There were some other costs incurred once the project was confirmed as a PPP, such as the running costs that were involved in the monitoring of the management framework system where an independent PCG committee had to be established and the representatives subsequently had to be paid for their time.

Independent verification

A key issue in a PPP is independent verification/certification. For example, as far as the road services on Top Ryde are concerned, the RTA stated that CoR should be responsible for this area and should lead the certification process. CoR took responsibility for this and this speeded up certification and reduced costs. Initially, CoR thought this additional certification process would cost $300,000 (hence the contribution from the Developer), however, these costs 'blew out' to $500,000 due to a lack of communication and an understanding of what the certifier's role would actually be. Subsequently, the relationship with the certifier became very 'contractual' and 'adversarial'. This was not part of the collaborative style that was embraced by all other project stakeholders and this had a negative impact on project morale. Often, consultants who provide engineering services and certification are excellent engineers, but they have an organisational culture where they place project administration with their highest qualified engineers who receive little to no admin support. Subsequently, the certifying engineer on Top Ryde was distracted and overworked and became bogged down in micro issues. On Top Ryde, the engineer responsible for certification was a global specialist in tunnel design but CoR had to wait for his specific approval which was frequently delayed due to the extent of his global portfolio and the nature of his role. Certification fees were also very expensive in terms of consultancy rates with engineers costed at very high hourly rates. If independent certification has to occur as part of the PPP process then it needs to include greater value for money with significantly more administrative support.

The nature of the verification process is also somewhat questionable in its current format. For instance, the Contractors (LL) were uncomfortable with this process as they paid their own consulting engineers to design the tunnels. The appointed engineer had significant expertise in tunnel design yet another consultant (project certifier) is then engaged to assess their work yet bear no responsibility over that work. Therefore, if anything goes wrong with the design and construction of the tunnel then it is the Main Contractor/engineer who bears responsibility regardless of certification. This issue had a negative impact on project culture. The cost of independent certification is enormous to the public sector, and in its current format is another unnecessary cost of PPPs not incurred in traditionally procured projects.

In future, if a certifier has to be used, then a tender process could achieve a successful outcome by calling for an emphasis on the certifier's ability to become part of a collaborative team and demonstrating an ability to carry out certification with the necessary administrative support without burdening the specialist engineer. If independent certification is not stipulated then certification could be done internally via in-house expertise (self-certification process) or via a contracted private certifier that could be engaged by the Main Contractor or the client.

The current process is a duplication where the certifier excludes responsibility with statements such as 'it appears to comply'. Therefore, under the current process, what is the real value of independent/third-party certification? The trepidations conveyed reflect concerns expressed by Evans and Bowman (2005) who point out that the legal framework in which the PPP project operates will be a crucial factor to the success of a PPP model: 'the legal framework within which a PPP project operates will also be a determinant of the optimal PPP model' (Evans and Bowman, 2005, p.63).

3. Standardisation of documentation

A lack of standardised PPP documentation was cited as a core reason in driving high legal and administration costs. All project participants agreed there was a lack of consistent principles and practices across the various State jurisdictions in terms of guidelines, and that generally drove costs up. Disappointment was expressed that the potential of the National PPP Forum established in 2003, which aimed to standardise many PPP processes, had not been realised. It is hoped that the recent initiative of the current State Government in establishing 'Infrastructure NSW' will continue to explore this further.

After the Development Application (DA) for Top Ryde was lodged an independent assessment panel was established which recommended to CoR that the project should proceed after ratification. A Project Manager was then employed in late 2006 to oversee all infrastructure and ensure that CoR expectations were met. This became a form of 'due diligence' so that CoR could avoid problems that had occurred on similar projects led by Local Government. As CoR has an obligation to avoid unnecessary expense and risk, this methodology could become standard procedure for future similar projects. The Department of Local Government were supportive of this, particularly as this was the first PPP of this type involving CoR and there was no existing standardised templates or management models to follow.

The Fitzgerald Report (Fitzgerald, 2004) recommended that PPP processes should undergo streamlining in order to reduce the costs of tenders and encourage wider bidder participation to increase competition. Divergent views existed on how best to standardise PPP processes and which model to follow. There was a general concern that international precedents may not reflect the smaller scale of the Australian PPP market. Differences of opinion existed in regards to whether national standardisation could ever be achieved considering the nature of the Australian federal system (in contrast to the UK where standardisation was implemented under a more centralised Government system).

However the UK PPP model, in which legal/contractual documents have been standardised, was viewed as the reference point for those who recommend standardising documentation for Australian PPP models. There was some consensus that a reduction in transaction costs could result from standardised templates such as those in the UK. It is worth noting, however, that despite the introduction of templates in the UK, the transaction costs are 'high and appear likely to remain relatively so despite the development of templates' (Public Accounts and Estimates Committee Report on Private Investment in Public Infrastructure 2006, p.84). This was also reflected in experiences of the UK PPP market where, despite the standardised documentation being in place, legal costs did not reduce. These experiences raised the question, would the associated cost of implementing standardised documentation be too prohibitive considering the relatively limited Australian PPP market?

A form of standardised documentation was adhered to on the Top Ryde PPP. According to the Department of Local Government the project was determined as meeting the definition of a PPP under s400B of the Local Government Act 1993. Subsequently, after further assessment of project documentation by the Department of Local Government it was deemed that the project did not represent a high risk to CoR and there was no requirement to submit project processes for review to the Local Government Review Committee. However, the Department reminded CoR that the processes outlined in the standardised document, *Guidelines on the Procedures and Processes to Be Followed by Local Government in PPPs*, must still be followed even though the project was assessed as non-reviewable. The Department of Local Government could always change their assessment if the risk profile of the project changed over time.

On a positive, during the period of this research, recognition needs to be given to the significant work done by Government agencies to standardise some of their PPP processes based upon previous experiences and feedback from the private sector. All were in accord that all levels of Government appear to be committed to refining contractual negotiations, in order to reduce extended legal debates which are not only costly for the contractors involved but also for Government. Examples of initiatives included work undertaken by the NSW Department of Health to standardise legal documentation based on previous PPP project experiences that included the Newcastle Mater Hospital, Long Bay Hospital and Correctional Facility projects. Despite general consensus from interview participants that the lack of standardisation was often costly, particularly at the initial bidding stage, there were some concerns that a tightly standardised approach could restrict flexibility and innovation, features often cited as major strengths of PPPs.

4. Relationship management

Relationships

All senior project stakeholders within CoR were involved in the negotiation and planning process and each of them developed successful relationships with senior stakeholders from the other private and public sector partners. Initial commercial negotiations were carried out on CoR's behalf by both the GM and Manager of Buildings and Property. The GM, in tandem with the Group Manager of Environment and Planning also developed the planning instruments. The GM and Group Manager of Public Works negotiated with the transport authorities (State Transit and the RTA). A 'softly, softly' approach to negotiations had to be adopted by CoR stakeholders as CoR had a vested interest in the project and a great deal of patience had to be shown to the other parties/stakeholders. This type of approach was replicated by the main private sector stakeholder, the Developer, as their Director of Development embodied similar traits which helped foster a successful working relationship with CoR where negotiations were performed in a very positive and 'can-do' manner.

Project relationships were driven by a 'win-win' culture and the successful management of what were very diverse expectations of each stakeholder. The cultural issue of relationship management (teamwork, trust, mutual goals etc.) helped to drive the project away from typical adversarial contracting. On the odd occasion when the Contractor, LL, had to be reminded of their contractual obligations, e.g. if the programme was falling behind schedule, then reminders were all carried out at the appropriate level, the highest level if possible, and then subsequently filtered down to onsite operations. Excellent holistic management of Top Ryde has no doubt helped to deliver a successful project. Obligations under each party have been met in a timely manner and the process of agreements made monthly, weekly and daily under different management structures from the top down have enabled outstanding results. This was achieved in practice at the likes of the 'PCG high-level meeting' where obligations were agreed and passed on down through the chains of management to those responsible for the daily delivery on site.

Commitment

Commitment is another key aspect of building successful stakeholder relationships. All stakeholders must want to 'travel the journey'. They must realise and assess the risks involved, accept that there will be problems, but must be committed enough to be able to work around the problems and subsequently accept that for every risk successfully managed there will be mutual rewards. Projects such as Top Ryde are ever evolving and fluid and this must be recognised by sharing challenges and risks. There is an onus on the broader project stakeholders, such as the regulatory bodies, to understand the dynamic nature of PPPs and the complexities of this type of project. Often, however, regulatory bodies tend to avoid relationship-based contracting approaches and still want to refer specifically to a contract, which in some cases could have been signed several years earlier, leading to micro levels of contract compliance which in turn leads to adversarial and poor relationships and ultimately variations and increased costs.

If stakeholders have, to use an analogy, 'skin in the game' (in this case those parties bound by the Tripartite Agreement – CoR, RTA and the Beville

Group) then they were 'in' and extremely supportive and keen to solve problems. But, if stakeholders didn't have this direct 'involvement' (as is the case with the likes of the Department of Local Government, the Independent Commission Against Corruption *et al.*) then they would only be part of the project if things were successful, and if things went wrong, or for political reasons it was not deemed appropriate to be 'part of the project', then they didn't want to be seen to have any involvement. This led to some negativity and meant that some of the regulatory bodies sat somewhat removed from the project. As discussed previously, this also led to some over expenditure in order to cover for personal or hardship risk (i.e. Independent Verifier).

Communication

At a project operation level, the Project Control Group (PCG) process was introduced and implemented by CoR's Project Manager. This led to significant improvements in the process and culture of face-to-face contact with key project players. This process was implemented after the initial meeting between CoR, Defined Developments, RTA and LL and enabled a very positive project culture with successful communication occurring at diverse levels. There is no doubt that the project's culture of open and collaborative discussion led to the avoidance of contract disputes. Key issues were identified at PCG level and senior project players could then act by solving problems almost immediately after the meetings. Establishing the correct collaborative atmosphere ensures the spirit of intent of contract is fulfilled without going into dispute.

Communication channels were open and collaborative at all times. All project participants, from senior management through to site supervisors, were free to contact CoR's Project Manager at any time. Open lines of communication and access to data were also reciprocal as CoR's Project Manager had continuous availability to the Contractor's (LL) 'Project-Web' data-base. Project-Web is LL's web-based project management tool that is used to link the site-based team, remote designers, clients, consultants and local contractors for document control, drawing management and project updates. Project-Web worked very well for fostering open communication and collaboration allowing freedom of access on a daily basis.

Open and collaborative communication also enabled project stakeholders to talk to senior people at very short notice. Top Ryde project stakeholders actively embraced collaborative communication approaches and this was enhanced by the fact that these senior and middle management players were all diversely experienced and practical thinkers. This level of collaborative communication was particularly espoused by the main players within CoR, RTA, Beville Group and LL. A major achievement of this project is that the key communication team was kept small and consistent and this has worked successfully. Project managers from each of the main stakeholders have remained unaltered during the project and as a small team remained focused and consistent facilitators for their organisations.

Community support

The issue of developing a successful relationship with the broader community of the City of Ryde was initially sought during the public notification process during the planning stages. There was an unprecedented community support of 95 per cent for the planning instruments to develop the site. The community was aware from day one both of the need to create a certain number of storeys for the Top Ryde precinct to make it commercially viable and that the real issue regarded more holistic planning to create a 'real' township for Ryde. This meant the right civic and governance precincts, the right retail precincts, the right commercial and the right residential premises, the right open spaces, the right alfresco dining spaces etc. It wasn't just about the one precinct but the coming together of several.

Previous applications for planning approval to extend Top Ryde had been rejected by Council. What made a significant difference with the approved DA was the overwhelming support of the community. Ryde residents helped to drive project support as they had got to the stage where they wanted to 'just make it happen' as they effectively had nowhere to go for acceptable commercial and retail activity. Residents were faced with going to other suburbs for the likes of cinema and extensive shopping activities. So, by 2004 the old 1950s Centre was very out-dated with a huge amount of public support for redevelopment. Therefore, during the early stages of construction, when many disruptions occur – noise, demolition, increased traffic, dust, trucks etc. – the community was very forgiving as they desperately wanted Top Ryde to progress.

Dealing with the community on a daily basis was successfully managed by CoR's Project Manager, with communication channels established between the public and key project stakeholders. A policy was put in place that the Project Manager would communicate with a member of the community within 24 hours of their initial enquiry. This not only helped to hasten the addressing of any community complaints but also created a happy ongoing 'working' relationship between CoR and the community of Ryde. The levels of community engagement need to be maintained by all relevant stakeholders from the inception and planning stages through to project completion and building operation and evolve during that period to reflect the changing expectations and concerns of the community.

5. Risk management

Compliance, due diligence and the Tripartite Agreement

The basis of the model used for the delivery of Top Ryde involved the establishment of the risk profile, internal and consultant reports, deeds, marketing plans, statutory compliance and Project Control Groups as the foundations for a successful project structure. This process was authorised, with subsequent project approval to proceed, by the State Department of Local Government. In order to mange this process, and particularly the process of engaging consultants, commissioning external reports and studies, especially where in-house expertise was lacking, the GM put together a cash surplus of A$9m to support and fund these issues for the duration of the project. At the pre-delivery stage A$800,000 was spent with LL under a consultancy agreement to mitigate risk in the ground (sub-soil). A detailed geotechnical and structural analysis was carried out by LL under contract with the Developer. A quantity surveying and cost consultant, was engaged to provide a benchmark review of all of the Contractor's (LL) costs. These tests enabled a 'no latent conditions' clause under the contract and LL did all the necessary research to price and deliver the contract. This level of geotechnical research enabled LL to price work more accurately.

The Tripartite Agreement succeeded in bringing the owner (CoR) and regulatory body, RTA, together with the Developer (DD). CoR and RTA mitigated some of their risk by having input into the design by ensuring compliance with design standards. CoR were allowed to enforce their rights under the design standards and DD, as Developer, had to deliver this under the contract. CoR also mitigated some of their risk by putting bonds into place under the Tripartite Agreement that were relevant to significant stages of project delivery. Further risk was mitigated by not issuing occupancy until the Integrated Traffic Solution (ITS) was complete and approved in accordance with the DA, i.e. in accordance with Local Government planning law and State Law (LEP), which meant that the Centre could not expand until the ITS was developed. Maintenance risk was managed by using an ongoing clause in the Centre's lease agreement that DD had to maintain and certify annually.

Financial risk

The Developer mitigated financial risk by establishing funding arrangements with a syndicate of six lending institutions (banks). Finance could be effected if one or more of the banks withdrew from the deal, but this in turn was mitigated by a 'no reason' clause, e.g. there had to be a significant event such as a dispute or extension of time submission for them to do so, and therefore allow financiers to enforce step-in rights.

The risk of ensuring Centre occupancy (tenants) was managed by DD as the financiers would only fund the project if budgets were correct. This involved a significant projection and feasibility study of project finance costs, leasing plans and agreed revenue figures for the Centre. The financiers also undertook their own finance checks, assessed by an independent cost consultant, who checked costs, values variations etc. The banks also engaged independent retail experts every month who sat on an agreed leasing panel to assess if DD were meeting budgets and adhering to the project's programme. Checks were also carried out on demographic studies to ensure demand was there for a centre of this size and nature and DD had to sign up major tenants before finance was approved.

Community risk

As discussed elsewhere in this report, community support for the development of Top Ryde was significantly high. This risk was initially managed at approval stage by engaging a full public review and debate. One frequent topic was the issue of pedestrian access to the Centre. Access to the Centre was proposed to change via the use of pedestrian bridges over a main arterial road as opposed to the historical pedestrian crossings that were integrated directly onto the road which led to the slowing down of through traffic. The community raised the question: 'Why do we have to cross the bridge, why can't we cross the road like we used to?' These issues change over time, as do community members, so referring them back to a public consultancy process years earlier doesn't always lead to a positive outcome for the public. These risks must be managed by a process that continuously responds to public questions so a community interface, e.g. newsletters, progress meetings, consultation etc., was established in order to provide a method for problem solving and to keep the community in a positive frame of mind regarding the project.

Greater security measures were also introduced to prevent ongoing vandalism to lifts that served the pedestrian bridges. This also helped to enforce that the biggest risk was managing the change for people (community). Change can be perceived as simple to some, yet significant to other members of the community. The simple fact that they can no longer walk across the road and must now use a bridge, or that a bus stop is moving 50m down the road can lead to a ferocious community backlash. Constant assessment on the impact on community is important to ongoing project success. The issue of the pedestrian bridges was a significant contractual issue as the RTA made it a condition that a pedestrian crossing on one of the surrounding streets – a main arterial route with over 90,000 daily traffic movements – be removed to improve traffic flow.

Chapter summary

A PPP consortium is a temporary organisation with a complex network of players with competing goals and objectives, many of whom never get to see the complete picture. Inevitably the group operates under pressure, particularly the members of the SPV (Special Project Vehicle) who are the drivers of the process. Social, as opposed to economic, infrastructure PPPs are more complex. Much of the negativity and adversarial environment which surrounds PPPs is due to a lack of transparency not just in terms of the bidding process, but also with regards to the identification of risk, opportunity and success factors. PPPs act as an essential but relatively minor part of Governments' asset acquisition programmes. However, as they tend to be large, complex projects that can affect people's lives for a very long time, PPPs arouse a great deal of interest and passion.

The PPP market in Australia, particularly for social PPPs, is gathering momentum but is less mature than markets such as the UK. The rate of maturation may be inhibited to some extent by the scale of the available market (in comparison to the UK) and the compartmentalisation of the market on a State-by-State basis. However, the existence of both a National PPP Forum and the recent establishment by the State Government of 'Infrastructure NSW' is very encouraging.

Reflecting on the case study analysis of the Top Ryde project, five features appear crucial issues in the success of PPPs: the delivery model (contractual arrangement), legal/admin costs, standardised documentation, relationship management and risk management. The main recommendations of this research are outlined below:

- Public sector clients must be precise as to their outcomes, with a clear vision of what they want to deliver to ensure project clarity.
- The project must be feasible, with appropriate market research performed to enable the project to be market tested.
- The public sector, be it at State or Local Government level, must be able to bring in expertise where needed. If this can be delivered in-house by Government then 'external' costs could be reduced.
- Project partners must be of the same culture in order to embrace the concept of risk-reward and fully encompass open lines of collaborative communication in a trusting relationship. The public sector must perform 'due diligence' on potential partners as the relationship must 'fit' in order to ensure the right partners are selected with both the project and broader stakeholders in mind.
- A 'can do' attitude among public and private sector stakeholders must be fully embraced, especially during the negotiation stages and open lines of collaborative communication must be adopted at all project levels from senior management to operations on site.
- A new department or unit, within the Local, State or Federal Government, must be created for delivering this type of project. If at Federal or State level then this unit could be 'ballooned' into Local Government as and when needed and if necessary be supported by a pool of experts or external consultants.

Ongoing research into PPPs is vital to ensure the development of sustainable procurements methods, the continued funding of a nation's infrastructure, successful operational viability, fair risk distribution, financial success and that greater rewards are provided for all stakeholders, particularly the community at large. The growing acceptance of alternative project delivery and finance methods, such as PPPs, implies that Governments will be increasingly faced with strategic choices whether to use 'public' or 'private' mechanisms, or a combination of the two, in the provision of infrastructure facilities. PPPs should provide a means for developing infrastructure without directly impacting on the Government's budgetary constraints. The principles embodied in PPPs are now established worldwide as a significant means of developing public services.

Reflections

1   Standardised procurement models for the delivery of certain PPP projects have been suggested by several authors and practitioners. What are the challenges faced by the public sector client in making standard contracts for relationship-based procurement models successful?

2   Much criticism is still aimed at various public sector benchmarking tools for assessing PPP tenders. The Public Sector Comparator (PSC) model in Australia attracts negative feedback from the private sector as it is seen as a means of adjusting project scope as opposed to contract value.

3   The National PPP Policy and Guidelines are applicable to all Australian, State and Territory Government agencies. It effectively replaces previously existing policy and guidelines in those jurisdictions. The likes of policy framework, procurement analysis, public sector comparator guidelines and so forth can all be found on the Australian Government's 'Infrastructure Australia' website: http://infrastructureaustralia.gov.au/.

4   Routledge has recently published *Public Private Partnerships: A Global Review*, edited by Akintoye, Beck and Kumaraswamy (2016) which supplements the contents of several chapters in this book. It focuses purely on the PPP approach, as opposed to other forms of Relationship Contracting, and it provides an international perspective through the use of 21 case studies. The book is available from the publisher at: https://www.routledge.com/.

5   The NSW State Government Department of Treasury provides an excellent online resource in PPPs as part of its commitment to provide infrastructure and services to the people of NSW. Their website contains information on PPP policy, projects, funding and supporting publications. A direct link to this PPP resource is: http://www.treasury.nsw.gov.au/ppp.

References

Akintoye, A., Beck, M. and Hardcastle, C. (2003). *Public-Private Partnerships: Managing Risks and Opportunities*. Oxford: Blackwell Science.

Australian Council for Infrastructure Development (2003). *Public Private Partnerships: A Brief Summary*. http://www.infrastructure.org.au [Viewed on 19th June 2015].

Bult-Spiering, M. and Dewulf, G. (2006). *Strategic Issues in Public-Private Partnerships: An International Perspective*. Oxford: Blackwell Publishing.

Cheung, F. Y. K., Rowlinson, S., Jefferies, M. C. and Lau. E. (2005). 'Relationship Contracting in Australia'. *Journal of Construction Procurement*, 11: 123–135.

City of Ryde (2015). Development Application 672/2006: Mixed Use Development – Ryde Town Centre. http://www.ryde.nsw.gov.au/Business-and-Development/ Major-Development/Top-Ryde-City/Redevelopment-2006 [Viewed on 1st September 2015].

Curnow, W., Jefferies, M. C. and Chen, S. E. (2005). 'Unsustainable Bidding Costs: A Critical Issue for Public Private Partnerships'. *Proceedings of Public Private Partnerships: Opportunities and Challenges*. Hong Kong.

Duffield, C. F. (2005). 'PPPs in Australia: Public Private Partnerships'. *Proceedings of Public Private Partnerships: Opportunities and Challenges*. Hong Kong.

English, L. (2005). 'Using Public-Private Partnerships to Deliver Social Infrastructure: The Australian Experience'. In: Hodge, G. and Greve, C. (eds) *The Challenge of Public-Private Partnerships: Learning from International Experience*. Cheltenham, UK: Edward Elgar, pp. 290–304.

Evans, J. and D. Bowman (2005). 'Getting the Contract Right'. In Hodge, G. A. and Greve, C. (eds) *The Challenge of Public-Private Partnerships: Learning from International Experience*. Cheltenham, UK: Edward Elgar, pp. 62–80.

Fitzgerald, P. (2004). *Review of Partnerships Victoria: Final Report to the Treasurer*, Department of Treasury and Finance, State Government of Victoria. Melbourne: Growth Solutions Group.

Jefferies, M. C. (2006). 'Critical Success Factors of Public Private Sector Partnerships: A Case Study of the Sydney SuperDome'. *Engineering, Construction and Architectural Management*, 13(5): 451–462.

Jefferies, M. C. (2014). *An Analysis of Risk Management in Social Infrastructure Public-Private Partnerships (PPP's)*. Unpublished PhD Thesis, University of Newcastle, Australia.

Jefferies, M. and McGeorge, D. (2008). 'Public-Private Partnerships: A Critical Review of Risk Management in Australian Social Infrastructure Projects'. *Journal of Construction Procurement, Special Edition: Building across Borders*, 14(1): 66–80.

Jefferies, M. and McGeorge, W. D. (2009). 'Using Public-Private Partnerships (PPPs) to Procure Social Infrastructure in Australia'. *Engineering, Construction and Architectural Management*, 16(5): 415–437.

Jefferies M. C., McGeorge, D., London, K. and Rowlinson, S. (2013). 'Relationship Contracting: A Case Study of the Top Ryde Public Private Partnership (PPP)'. *Proceedings of the 19th CIB World Building Congress*, Brisbane 2013: Construction and Society, QUT, Brisbane.

Jones, D. (2003). 'Evaluating What Is New in the PPP Pipeline'. *Building and Construction Law*, 19: 250–270.

Jordan, K. and Stilwell, F. (2004). 'In the Public Interest? A Critical Look at Public-Private Partnerships in Australia'. *Arena Magazine*, 1 August, 9–13.

Malone, N. (2005). 'The Evolution of Private Financing of Government Infrastructure in Australia: 2005 and Beyond'. *The Australian Economic Review*, 38: 420–30.

New South Wales Department of Local Government (2005). *Guidelines on the Procedures and Processes to Be Followed by Local Government in Public-Private Partnerships*. Sydney.

New South Wales Public Accounts Committee (2006). *Inquiry into Public Private Partnerships*, Legislative Assembly, NSW, Sydney. http://www.parliament.nsw. gov.au/prod/parlment/committee.nsf/0/6FB3D448CE8BF349CA25707000159 52F [Viewed on19th June 2015].

Public Accounts Committee (2006). *Report on Private Investment in Public Infrastructure*. Parliament of Victoria, State Government of Victoria, No. 240, Session 2003–06.

Quiggin, J. (2005). 'Public-Private Partnerships: Options for Improved Risk Allocation'. *Australian Economic Review*, 38: 445–450.

Shepherd, A. F. (1999). 'Project Alliancing on a Public/Private Partnership Project'. *Building Australia*, June.

Tiong, R. L. K. (1990). 'BOT Projects: Risks and Securities'. *Construction Management and Economics*, 8: 315–328.

Walker, R. G. (2003). 'Public-Private Partnerships: Form over Substance?' *Australian Accounting Review*, 13(30): 54–59.

Walker, C. and Smith, A. J. (1995). *Privatized Infrastructure: The BOT Approach*. London: Thomas Telford.

10 The growth and emergence of PPPs in Asia

Steve Rowlinson

Chapter introduction

Modern developments in procurement systems and procurement strategy have been based on a series of principles that guide clients in their choice of approach (see Figure 10.1). Key issues, from the client viewpoint and a governance perspective, are as follows. The procurement system adopted must be efficient and the process of procurement must be auditable. Furthermore, a clear line of accountability is important and the systems adopted must be effective. The old adage about performance still holds true: efficiency is about doing things right and effectiveness is about doing the right things. This leads to a contingency view of procurement systems in that we need to apply an appropriate system to the circumstances in which we find ourselves. In the case of Public-Private Partnerships (PPP) this means that we should choose first of all to go down the PPP route only when circumstances indicate that it is best or fit for purpose.

Once the PPP route has been chosen the key, overriding principles focus on finance and integration systems. The special purpose vehicle and the public demand must be properly integrated throughout the whole life-cycle of the concession and the sources of finance must be appropriate. This also

Efficiency	**Innovation**
Audit	**Change**
Accountability	**Relationship**
Effectiveness	**Partnering**
	Alliance

Finance

Figure 10.1 Prerequisites for successful procurement strategies

applies to the returns and rewards of all parties, the public, stakeholders and the private special purpose vehicle.

In addition, we are looking for something extra from the process compared with an ordinary 'infrastructure development'. What is required is innovation, in terms of product and delivery and in the process of use. There also needs to be recognition that change may be necessary throughout the life of the concession and there must be flexibility within both the design and use processes to accommodate this. In terms of project management there needs to be a focus on relationship, between public and private organisations, but also with the stakeholders involved in the process throughout the life-cycle. Furthermore there is a need for consideration of relational methods of design and construction which include the concepts of partnering and alliancing. PPPs are generally a 30-year concession and so relationship is vitally important from the outset. Consequently, this chapter investigates how these processes and principles have coalesced in different ways within Asia and also explores the different procurement routes, therefore providing an introductory Asian perspective ahead of the more focused Chapters from 11 to 14.

Governance

Figure 10.2 illustrates a conceptual view of governance within any construction industry and we will use this model to assess the different levels of compliance within the Asian countries we are examining. The way the construction industry is organised varies from location to location and culture to culture. The Hong Kong construction industry is characterised by a 'can do' attitude and the achievement of stretch targets, yet it is organised in a very *ad hoc* manner. The Japanese and South Korean industries appear to be much more structured and rely heavily on consortia of banks, construction companies and manufacturing businesses and can be seen to be highly structured in their approach to both business and projects. In Hong Kong, the construction industry focuses very strongly on contract and the consequences of the contract, whereas in China, the focus is on policy rather than contracts and on national advancement rather than an individual organisation's profitability. These are contextual and cultural differences that need to be considered when looking at the implementation of PPPs in different settings.

In modern society it is expected that projects will be delivered through mechanisms which make all those involved accountable for the project. Key performance areas are cost, time, quality and safety and the manner in which these different indicators are measured should be transparent, clear and auditable. This is the essence of good governance and the basis for developing trust, particularly for an out-of-country investor.

No industry can perform if its capacity is stretched beyond its capability. Currently in Hong Kong a serious labour shortage is making the task of completing projects on time and to budget virtually impossible. When embarking

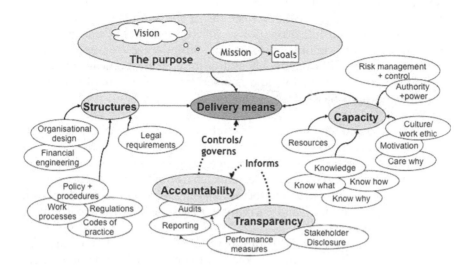

Figure 10.2 Governance issues with PPPs (Walker and Rowlinson, 2008)

on a novel delivery mode, such as PPPs are in many countries, capacity is a key element in determining whether a PPP is appropriate. Capacity includes technical and commercial knowledge and expertise as well as financial resources and the ability to plan and organise. In order to be able to make use of capacity and to ensure that accountability and transparency exist it is necessary to have structures in place that deal with the organisational, legal, financial and technical aspects of projects. Hence, capacity, delivery means and structures all come together as part of the nexus for PPPs, as shown in Figure 10.2.

In order for this to happen it is necessary that there be an overriding purpose which is clearly identified by a vision from the top. This vision must be clearly turned into a mission, a plan of action for infrastructure development in various sectors, and this should lead to clearly defined goals in terms of projects and in terms of their performance indicators.

It is this framework that will be used to analyse the four countries under consideration. It is important to bear in mind in using this model that overlying other concepts such as risk and trust are key issues in developing a vision for PPPs in a particular country. Central to this is not only the setting up of governance structures but ensuring that these structures actually work and are implemented fairly and consistently.

Infrastructure types

In dealing with infrastructure we are basically talking about a public product and its characteristics. The expression used in Korea to describe investment is 'social overhead capital'. This is a very useful way of looking at the nature of the PPP and the relationship between the two 'P's in the partnership.

Generally, we class infrastructure as soft, intangible but benefiting the whole of society or economic, tangible and stimulating economic activity. Hard infrastructure is both tangible and of long-term significance whereas soft infrastructure is intangible and is a service supplying public demand. Typically we would class infrastructure into four types:

- Hard and social infrastructure serves the community's needs and is typical of education, hospitals and water supply systems.
- Hard and economic infrastructure generally has high profitability and provides social and economic benefit such as power, roads and ports.
- Soft and social infrastructure focuses on long-term social benefits which are intangible initially and encompass issues such as the environment and energy efficiency.
- Soft and economic infrastructure may well be profitable and easy for individuals to bring to the market and these might include intellectual property and financial services.

Generally, in the early stages of development countries focus on hard infrastructure and, having partially established this, then move towards policy development to address provision of soft infrastructure. These changes in emphasis are also affected by the context in which development is taking place. Currently, China is still growing rapidly (7 per cent GDP increase per annum) and a focus on urbanisation and improving the quality of life and energy efficiency prevails. Hong Kong and Japan are relatively mature economies but with ageing populations, so there is a shift towards policy dealing with soft infrastructure issues. South Korea has reached an era of economic prosperity and is balancing the need for both soft and hard infrastructure.

Privatisation

Privatisation of public goods is a controversial topic. Since the 1980s the UK government took on board the concept of Private Finance Initiatives in providing basic physical infrastructure and social infrastructure. The reasons given for this particular approach were the need for fiscal constraints on government spending and the claim that the private sector was more efficient than the public sector in delivering these resources. Estache (2001) argues that there are three main reasons for this type of policy to develop. The first is that the sector most effective in attracting private sector investment is telecommunication followed by hard infrastructure projects. The second is that in richer countries there is more participation by the private sector. This implies that developing countries are less likely to have a well-developed Private Finance Initiative. The third point he makes is that even in high-income countries, the private sector is not as widely involved in infrastructure provision as people may believe. Indeed, in the British example, existing infrastructure was taken over and refurbished more so than new infrastructure being constructed.

In reality, one of the key issues in the Private Finance Initiative in the United Kingdom was an attempt to recoup money paid into the European Union coffers. The Conservative government had made it a policy pledge to recoup money paid into the common agricultural policy which did not appear to benefit the United Kingdom. By leveraging financing for infrastructure development projects the government was able in some ways to deliver on its policy pleasure and to use this to kick start the Private Finance Initiative.

Looking at the situation over the past 30 years, particularly in the last two decades of the twentieth century, it can be seen that infrastructure in developing countries which were previously driven by OECD countries have been replaced by sources of investment finance from within the developing countries through their own investors. For example, the increasing presence of China and India in African and Latin American countries continues to be noticed around the world. Furthermore, there are a range of mechanisms by which PPP competence develops within developing countries and is in fact exported to other developing countries. These are issues which will be discussed later with reference to Hong Kong, Japan, Korea and China.

In their paper, Estache and Fay (2007) discuss what has been learnt from 'infrastructure privatisation' so far. Their major conclusion is that the most serious problem is in addressing risk. They indicate that the most difficult time for the first generation of PPPs in East Asia was the 1997 financial crisis. On top of this crisis, which Hong Kong managed to weather relatively well, came the SARS outbreak which saw property prices plunge dramatically and GDP adversely affected. These types of risks, in 30-year concessions, must be addressed in a different way than the more technical risks in terms of the construction processes and the operational risks in terms of demand and the generation of alternative sources of supply.

Hodge and Greve (2007) discuss what they call the language game of Public-Private Partnerships. Governments often attempt to avoid using words such as privatisation and contracting out and instead talk about partnerships. This is obviously an issue of reducing opposition to policies which may be seen as encouraging the private supply of public services at the expense of the public organisations themselves. In a country like China, where until recently all industry and services were state-owned, this move towards privatisation can be seen by many to be threatening. There will, of course, be those who see this as an opportunity but, in many instances, this will be a small minority rather than a vast majority of those employed. Such issues of the promotion of PPPs have provoked strong reactions in countries such as the United Kingdom and Canada where in Quebec PPPs have been referred to as Problem, Problem, Problem. The author certainly came across such a view during the late 2000s when attending a public seminar and debate at Université de Montréal on the advantages and disadvantages of PPPs and it was obvious that in Montréal PPPs were not at all welcome.

Asian approaches to PPPs

In this section we look at the approach taken by four Asian countries in developing their PPP capability. However, we assess this by initially looking at how PPPs developed in the United Kingdom and Australia as contrasting examples. We also look at how the concept of PPP has been exported from developing countries to other developing countries as part of a policy to modernise their own industries.

The UK approach was driven by several Conservative governments during the 1980s and 1990s with this policy further developed by the Labour government over a number of years when Tony Blair came to power in 1997. It acted as a mechanism for recouping payments into the common agricultural policy and also as a mechanism for imposing financial stringency on the government and the local authorities and state businesses which were seen to be ineffective and inefficient by the government. This led to a whole series of projects which were promoted by the government as Private Finance Initiatives in both soft and hard infrastructure projects. These included rail, road, bridge and tunnel construction as well as hospitals, prisons, schools and other public services. It would be reasonable to say that there have been mixed results from this approach and that a standard approach to promoting, managing and taking over these projects has still not been developed.

In Australia, the drive for PPPs was initially generated by the need for fiscal constraints on federal, state and territory budgets. Thus, in Australia in the initial phase the two main contracting groups in the country were in a strong position to promote PPP projects and there was some concern that value for money was not being achieved. In the second phase, development projects were driven by an excess of liquidity in various funds which were searching for projects in which to invest capital and maintain a high return for their investors. Thus, the drive moved from fiscal constraints and contractor competence to an excess in liquidity and a desire to promote projects for the sake of finding an investment vehicle. In the current phase, Infrastructure Australia has taken note of an academic study by Bridge *et al.* (2015) which looked at value for money and competition in the PPP market. This study delivered recommendations as to how the market can become more competitive both internally and through the attraction of foreign investment. An initiative is currently taking place to identify the optimal number of bidders for a PPP project that would actually make it viable from both the public and the private sector viewpoints.

Thus, these two examples indicate that the drive for PPPs is both fiscal and political. The question of greater efficiency and effectiveness is one which still remains unanswered. Two countries sharing a similar heritage at opposite ends of the globe have addressed the PPP approach very differently and have developed their own unique ways of measuring the success of these projects. Consequently it should be no surprise that we find the four Asian countries, Hong Kong, Japan, China and South Korea, adopting different approaches due to different political, cultural and fiscal imperatives.

Hong Kong: entrepreneur and need driven

Walker and Rowlinson (1990) describe the 1970s as the beginning of the modern era in Hong Kong. It marked a period when significant volumes of construction took place not only in public and private housing but also in infrastructure. At a time when few countries were considering PPPs the Hong Kong government embarked on its first PPP, the construction of the Cross Harbour Tunnel. The project was constructed by a consortium of Costain International Limited, Raymond International Inc. and Paul Y Construction Company Limited. The tunnel was Hong Kong's first foray into concession contracts, the approach was very *laissez faire* and 'hands off' and was a long time in its genesis. The Hong Kong government articulated the need for a fixed crossing in the 1950s and invited the private sector to propose a scheme; government would provide the land (on a lease) and the connecting road infrastructure, for which the tunnel concessionaires had to pay. Bids were invited in 1966 but the Cultural Revolution in China brought financial and political uncertainty with the franchise finally being let in 1969 and construction completed in 1972. Within four years the tunnel was profitable and all debt was repaid by 1979 (Smith, 1999). The immersed tube tunnel is 1856 metres from portal to portal and was completed in 35 months. The 30-year private sector franchise based on the build operate transfer (BOT) model has now concluded and ownership of the tunnel has rested with the Hong Kong government since August 1999.

The Eastern Harbour Crossing was completed ahead of schedule in 1989 and is an excellent example of a construction company taking a major role in constructing and operating transport infrastructure. This project included twin road and rail tunnels and the immersed tube is 1880 metres long. This project was completed ahead of schedule, taking only 38 months out of a contracted period of 43 months. The majority shareholder of the tunnel company was Kumagai Gumi Co Ltd, which was also the main contractor, and other shareholders are China Investment Trust and Investment Corporation (CITIC), Lilley Construction Limited, Paul Y Construction Co Ltd and Maru Beni.

These two tunnel examples indicate the role of the private sector in the development of Hong Kong's infrastructure. In the case of the Cross Harbour Tunnel the impetus came from a family that had invested in Hong Kong and recognised the need for a modern transport infrastructure across the harbour. The family had investments in many areas such as hotels, the Peak Tram and agriculture. The Eastern Harbour Crossing was promoted and developed by a Japanese company which had been long established in Hong Kong and was capable of delivering the technology to provide a rail and road tunnel in exceedingly quick time. The contractors still had to tender to undertake the project but tendered and won the project and then completed and successfully ran the crossing and are still doing so. Thus, the drivers were entrepreneurs of an individual and a corporate nature recognising both the need for infrastructure and the lack of willingness or ability of the Hong Kong government to pay for it.

However, the picture in Hong Kong has not been altogether rosy. For example, the central Cross Harbour Tunnel has now been handed over to the government as part of the BOT agreement. However, there is a disparity in the prices charged at the three tunnels (there is also a Western Harbour Crossing) and the fact that the central tunnel is the cheapest has led to significant traffic congestion and overloading of parts of the road network. If all three tunnels were under the control of one organisation it would be possible to manage traffic much more effectively by using pricing mechanisms to divert traffic away from congested areas through the different tunnels. However, with the rise of the influence of the elected legislative councillors in Hong Kong the opportunities for this have diminished considerably as they are lobbied by transport operators to keep the fee within a narrow range for the central tunnel. Thus, the opportunity to manage the traffic systems by differential toll mechanisms has been lost under the current political climate.

There is also a controversial, failed attempt at PPP in Hong Kong, which is the West Kowloon Cultural Development. This has been on the drawing board for over 20 years already but little progress has been made thus far. Over 10 years ago the government called for private sector bids to construct, develop and manage this cultural centre at the end of Tsim Sha Tsui. However, despite the fact that three major developers bid for the project no decision was made to appoint any of them. The cost of developing the plans, including special purpose vehicles and finance and facilities management was of the order of HK$10 million for each bidder. As might be expected, the private sector bidders were not happy with the outcome of this failed attempt to bring PPP into the arts and culture arena.

Thus, although Hong Kong has successfully used PPPs over the years it has not put into place a robust framework for the management of the PPP process. Hong Kong has not developed a governance structure with clear and transparent planning, bidding and adjudication processes for a range of PPP projects. There appears to be no overall strategy for the use of PPPs in Hong Kong's soft and hard infrastructure development. The Hong Kong government has perhaps missed a golden opportunity to further develop Hong Kong's expertise in infrastructure and social PPPs. Hong Kong has obviously not followed the governance model as set out in Figure 10.2 by failing to articulate a clear purpose, and failing to set up appropriate structures to facilitate the development of a PPP programme. The industry certainly has the capacity to fulfil PPPs but is not directed by policy to do so. Accountability and transparency appear to be sound but with the failure in 2014 of the Mass Transit Railway Corporation (MTRC) to report fully its problems in new rail construction and the case brought against the ex-Chief Secretary and a major property developer, Hong Kong has issues to address.

Japan

Japan has ventured into the PPP markets both within Japan and abroad. This development has focused in the main around the country's big five contractors (Kajima, Obayashi, Taisei, Takenaka and Shimizu). Obayashi is probably the most experienced overseas having partnered with Australian contractors on the Melbourne City Link 20 years ago. However, the view taken by the Japanese contractors was that the venture into Australia was of the nature of a tuition rather than a full-blown commercial venture. Indeed, the Japanese contractor left the venture after completion of construction and probably forewent a good return from the successful road construction which is a 34-year franchise. Other ventures in South East Asia in terms of PPPs were often based around working with either Japanese aid projects or Japanese companies setting up their manufacturing businesses in the region. Japanese contractors have attempted to spread their PPP interest to Australia and the United States in recent years but have not been wholly successful in these efforts.

This may come as a surprise given the fact that the Japanese contractors possess impressive technologies and can produce highly successful and innovative infrastructure projects. However, much of the success has been at home and in the past may have occurred due to the cosy relationship that existed between the contractors and government for many years, based on the principle of *dangō*. *Dangō* existed in Japan for a long time until former Prime Minister Koizumi enacted a law which effectively outlawed it. *Dangō* (談合) (bid rigging) was based on the idea of a prearranged business compact in which contractors privately form an agreement in advance on bid prices, etc. It was a habitual practice of the Japanese construction industry, although it is both a violation of Japanese criminal law and the Japan Anti-Monopoly Law. In 2006, the governors of Wakayama and Miyazaki Prefectures, and former governor of Fukushima Prefecture in Japan were all arraigned or forced to resign due to nefarious connections to the Japanese construction industry involving *kansei dangō* (官製談合) or 'bid rigging instigated by government agencies'.

This relationship between the public authorities in Japan and contractors effectively almost worked as a PPP in practice. The contractor would take responsibility to complete design and undertake all measures necessary for the convenience of the public authority but the idea of a concession period did not exist. Since the outlawing of the system and with the reality of Japan's ageing population, matters have changed.

With its massive savings, which provide very little interest earnings for the savers due to the nation's persistent deflation, Japan's Post Office has a source of finance, along with banks and other finance houses, which can be sensibly tapped in order to promote and develop infrastructure development through PPPs. Japan started its road infrastructure construction through the Japan Highway Public Corporation (established in 1956) that was transformed through privatisation into East Nippon Expressway Company (E-NEXCO), Central Nippon Expressway Company (C-NEXCO) and West Nippon Expressway Company (W-NEXCO). Japan continues to develop and experiment with its approach to Private Finance Initiatives within Japan itself. A more than adequate source of finance exists within the country, there are pressing hard and soft social infrastructure needs that go along with the ageing population and the contractors have strong and specific competencies in areas such as earthquake engineering and infrastructure construction. Hence, the Japanese contractors and financiers have the potential to develop a new and profitable approach to their businesses both through internal PPPs and by exporting these to other countries. With the introduction of appropriate governance structures within Japan, the PPP markets can expand over the coming decade and fulfil the country's social and hard infrastructure needs.

In recent years, within Japan PPPs have appeared more like hire purchase agreements between the bank, the contractor consortium and different prefectures and cities. Although this approach has been successful it does not fit into a viable long-term structure. Typically, a Japanese company would expect to produce a 7 per cent return on investors' capital and so the infrastructure PPPs need further governance development in order to make them both acceptable and viable to both parties.

A different story unfolds when it comes to entering the international PPP market. This market is attractive in that it offers far higher returns than might be made on investments in Japan. However, Japanese contractors and financiers recognise the cultural and institutional barriers to exporting their expertise. For example, Japanese contractors when interviewed express concern about the need to develop trust on a longer-term basis with partners in countries such as the United States and Australia. The nature of contract is completely different in these countries compared to Japan and so a whole new set of commercial and relational skills need to be developed by the Japanese industry in relation to international PPPs.

In regard to Figure 10.2 it can be said that the Japanese industry certainly has the capacity to deliver PPPs at home but has been slow in developing governance structures to encourage more use of PPPs and to stimulate the PPP market. Accountability and transparency appear to be improving since the 'outlawing' of *dangō* but the slow development of governance structures indicates a lack of clear purpose at present.

South Korea

During Korea's rapid growth over the last two decades of the past millennium, which was based on industrialisation and exports, a commensurate investment in much-needed infrastructure did not take place. Interestingly, infrastructure in South Korea is described as social overhead capital. This is a descriptive expression and one which probably clearly defines more than most countries the nature of the contribution of hard and soft infrastructure to a nation's development.

In the mid-1990s the government introduced the Private Capital Inducement Act as a first attempt to develop a framework for PPPs. This was a structured approach to introduce a governance system to promote a relationship between private capital and the necessary infrastructure developments for the country. Like many Asian countries this particular initiative was badly hit by the Asian financial crisis in 1997 and so a need arose to review the nature of the process and the sources of finance as the new millennium approached.

The Private Participation in Infrastructure (PPI) Act was passed in 1999 and formed the basis for the current Public-Private Partnership programme which exists in South Korea. The act involves the production of an annual PPI plan and indicates how and to what extent private parties will invest in each project in the plan. It is a highly structured approach and is quite unusual compared with the approach adopted by many countries and certainly by the United Kingdom in its promulgation of its Private Finance Initiatives. Although this plan reflects a high degree of central planning and government control the act also allows for unsolicited projects, that is projects that are proposed by a private entity rather than being part of the national plan.

Unusually, the PPI Act has a minimum revenue guarantee. This provides guarantees of projected revenues from the development for the private sector party in order to encourage such ventures. Over the years, this guarantee has been fine tuned in order to limit the returns that the private party might make from projects and also reduce in some way the exposure of the public body in the agreement. The Act also contains provision for compensation for foreign exchange losses above a certain level and for the public authority to have equity shares in the project while not necessarily receiving dividends. These measures focused on attracting foreign investment into South Korea. The largest projects are controlled centrally by the ministry while smaller projects can be controlled by local governments and authorities.

The Private Infrastructure Investment Centre of Korea was established when the Act was promulgated and its role is to provide policy support for the government. For participants it is there to help appraise and develop new projects and assist in bid evaluation and the writing of concession agreements. It also has an education and training role and is an independent centre now under the Korean Development Institute. It has been renamed the Public and Private Infrastructure Investment Management Centre and reports directly to the President in order to enable it to be more independent in the views that it gives to all parties.

The focus of the PPI programme was initially on major infrastructure for transportation needs. However, since 2005 social infrastructure projects have been targeted and schools particularly have been a focus for the Centre. It should be noted that military facilities have also been a focus and these are not necessarily open to foreign investment. The percentage of investment in social overhead capital has steadily increased through the PPI programme. Although foreign banks and foreign capital form part of the PPI programme the PPI Act also set up the Korea Infrastructure Credit Guarantee Fund to facilitate the programme.

The structure of industry in South Korea is quite similar to that of Japan. It is dominated by large conglomerates (*chaebols*) that have financial, construction and service arms. As a consequence, the programme has been criticised for the dominance of the big five Korean companies in the market. This seems to be a very similar situation to that in Japan but certainly not the same as that in Australia where basically only two major companies dominate the market. It may be an inevitable consequence of the drive towards privatisation and the focus on 'efficiency and effectiveness' that large organisations form oligopolies that dominate markets. This issue was addressed in part by the UK Labour government in the new millennium by the introduction of a framework and other types of agreements that allowed consortia of smaller organisations to work on social and hard infrastructure projects and so enabled industry development at the small-and-medium-enterprise-size level. One of the more interesting PPPs in South Korea was the construction of the Incheon bridge which was led by an overseas consortium but involved all of the construction being undertaken by Korean national contractors.

In terms of the mapping on Figure 10.2 South Korea scores highly on capacity and with its PPI programme and Centre it has put structures in place to ensure projects can be established with a set of procedures and regulations that are consistent. This reflects a sense of purpose, clearly articulated goals and an overriding vision reflected in the concept of social overhead capital. Although the Centre reports directly to the President to preserve independence, the power of the *chaebols* may have an influence on transparency.

China

Although the number of projects using a PPP approach in China is large, the investment in these projects is relatively low as a proportion of all infrastructure investment. The PPP, or PuPuP (Public-Public Partnership), approach in China has, until very recently, been underdeveloped. The lack of governance structures, access to private capital and the problem of risk and transparency has held back development thus far (Asian Development Bank, 2014). However, the main drivers that make PPPs attractive in China in the contemporary context are the likes of high levels of local government debt and the daunting need to fund new infrastructure in the coming decades (Thieriot and Dominguez, 2015).

A study undertaken in the wake of the terrible Wenchuan County, Sichuan Province earthquake of May 2008 indicates that there is a place for PPPs in China's infrastructure development. An unreported study by the author showed that in the reconstruction of over 100 kindergartens in the county those undertaken using the PPP model came in at 20 per cent lower cost than those reconstructed by the public body. This is an example where the local authorities were overwhelmed with the need for reconstruction and in such a dire situation turned to the private sector for assistance. There was an immediate response leading to greater efficiency and effectiveness through careful planning and innovation, coming in part from overseas participation in the private reconstruction projects, which led to significant cost and time savings. Thus, it is worth noting that PPPs not only operate in situations where an economy is planned but are also powerful mechanisms for relief in emergencies.

As the Wenchuan example illustrates, as do the construction of many toll roads, railways and metro systems such as Beijing's subway line 4 and other infrastructure in China, China has embraced the concept of PPPs. However, we may wish to term many of them PuPuPs, Public-Public Partnerships, due to the dominance of state-owned contracting organisations and banks in the special purpose vehicles.

Any change needs a driving force in order to set it in motion and also provide the necessary governance structures to enable the change. Until very recently most of the PPP projects in China were delivered and run on an *ad hoc* rather than structured basis. China's government has recognised this and in 2013 President Xi JinPing announced that private investment in public infrastructure would become a policy issue. In 2014 the National Development and Reform Commission put forward 82 projects for private investors to provide 'social capital'. Projects included roads, harbours and the operation of railways. In order to promote this policy it was recognised that there is a need for governance structures to enable such projects to be formulated and to be managed.

This governance system was promulgated by the State Council through the publication of guidelines for the Investment of Private Capital into PPPs. In order to facilitate this process and in order to implement the policy a national PPP Centre was set up. PPP model guidelines were published by the Ministry of Finance and a total of over 200 billion ¥ worth of projects were launched. At the same time, the National Development and Reform Commission (NDRC) published guidelines on the scope and type of PPP models to be used and approved construction of the new Beijing airport, costed at over 80 billion ¥, under this model. This is the start of the building of a rational governance structure for PPPs in China which will go a long way to addressing the risk and trust issues which currently exist for local and overseas investors. The slowing down of the Chinese economy and the massive spending over the past 10 years on establishing basic infrastructure has pushed the call for more private investment in essential infrastructure in the second- and third-tier cities. Other than the President as a driving force the policy to aim for 60 per cent urbanisation in the near future makes it imperative that investment in physical and social infrastructure takes place rapidly.

Thus, Chinese businesses are in a position to seize this opportunity to develop expertise in PPPs. The natural corollary of this will be to export this expertise and develop PPPs in other developing countries. This move is already taking place with state-owned enterprises in Africa and South America. Recently the China Communication Construction Company (CCCC) purchased John Holland, a large contractor and part of the Leighton group in Australia, with an eye to exploring the Australian market which is in need of infrastructure investment. However, a recent study by Bridge *et al.* (2015) indicated that the major barrier to Chinese investment in PPPs in Australia is the cultural distance between the two nations. Additionally, Chinese contractors have relied heavily on cheap labour as their competitive advantage in overseas markets and this is obviously not an option in Australia. Thus, the future success of CCCC will be watched very closely by investors and businesses around the world. Any transformation of the company into a major PPP player will be a success story for China.

In terms of Figure 10.2 China is at the stage of just having determined its purpose and its vision for PPPs. The elements of accountability and transparency are currently a hot topic in China. President Xi is addressing these issues with gusto and it appears that a change in this respect is about to take place in China. China undoubtedly has the capacity to perform PPPs in terms of technical expertise, machinery and labour but doubts still remain over the planning and quality control systems. In terms of structures, these are slowly coming into place and over the next few years will become a much stronger element of China's delivery mechanisms. A move away from an attitude of administration towards an attitude of management will be particularly beneficial in this respect.

Chapter summary

It is clear from this chapter that there have been many approaches to the introduction, use and proliferation of PPPs in the Asia-Pacific region. At one end of the spectrum we have the Hong Kong approach that is very *laissez faire* and driven by entrepreneurial interests and society's needs. At the other end of the spectrum the approach adopted by South Korea is very structured, has a strong governance structure in place and is driven by the need for social infrastructure to be provided in a well-planned, state-directed manner. Essentially, each country has developed its own routes to providing PPPs and each has its own specific objectives and reasons for using PPP as the chosen procurement form.

These differences come about through the influence of culture and stage of development within each country, the underlying demographics and political structures and the different approaches to government funding, taxation and legal systems. The markets in each country are very different. China can be seen to be a huge market that is rapidly expanding whereas Hong Kong is a much smaller but mature market and Australia is a country in need of hard and soft infrastructure but with a relatively small population, and hence taxation base, to fund developments.

Thus, there is no one-size-fits-all solution to promoting PPPs in Asia-Pacific. The differing cultures, politics and stages of development mean that systems are developed that are appropriate for the particular country. However, this situation may well change in the near future with developments such as the funding bank in China for infrastructure development in developing regions, Asian Infrastructure Investment Bank (AIIB with 51 founding country members), and the formation of a UN-backed International Centre of Excellence on Public-Private Partnerships, to look at the future role and structure of PPPs. The Centre of Excellence will consist of a hub in Geneva and affiliated PPP specialist centres, dedicated to one sector such as roads, water, prisons, education, hosted in different countries (United Nations, 2015).

Reflections

1   The opening of the cross-harbour tunnel in Hong Kong in 1972 was the territory's first PPP and within four years all debt was paid-off and operation profitable. What were the factors that made this project so successful?

2   Many of the early PPPs, including the first Hong Kong cross-harbour tunnel, are discussed in Walker, C., and Smith, A.J. (1999). *Privatized Infrastructure: The BOT Approach*. Thomas Telford, London. This is a seminal book on the PPP form of procurement – as well as looking at the early development of this form of procurement it provides important risk, legal and case study information.

3   The practice of *dangō*, or bid rigging, in Japan can be further explored in Woodall, B. (1993). 'The Logic of Collusive Action: The Political Roots of Japan's Dangō System'. *Comparative Politics*, 25(3): 297–312.

4   How has the slowing of the Chinese economy had an impact on the country's infrastructure growth and how has this had an impact on procurement methods?

5   What are the differences between PPPs and PuPuPs and how will they influence the future of infrastructure development in China?

References

Asian Development Bank (2014). *Public-Private Partnerships in Urbanization in the Peoples Republic of China*. ADB, Manila.

Bridge, A.J., In, S.Y., and Rowlinson, S. (2015). 'Models for Engaging PPPs in Civil Infrastructure Projects: A Case of "Having Your Cake and Eating It Too"?' *12th Engineering Project Organization Conference* (EPOC), Edinburgh, June 24–26.

Estache, A., and Fay, M. (2007). *Current Debates on Infrastructure Policy*. World Bank, Washington, DC. https://openknowledge.worldbank.org/handle/10986/7651, accessed 13 January 2015.

Estache, A. (2001). 'Privatization and Regulation of Transport Infrastructure in the 1990s'. *The World Bank Research Observer*, 16(1): 85–107.

Hodge, G.A., and Greve, C. (2007). 'Public Private Partnerships: An International Performance Review'. Public Administration Review, 67(3): 545–558.

Smith, A.J. (1999). *Privatized Infrastructure: The Role of Government*, Thomas Telford, London.

Thieriot, H., and Dominguez, C. (2015). *Public-Private Partnerships in China: On 2014 as a Landmark Year, with Past and Future Challenges*. International Institute for Sustainable Development, Canada.

United Nations (2015). http://business.un.org/en/documents/9392, accessed 26 August 2015.

Walker, A., and Rowlinson, S. (1990). *The Building of Hong Kong*. Hong Kong University Press, Hong Kong.

Walker, D. H. T. and Rowlinson, S. (2008). Project Types and their Procurement Needs. *Procurement Systems – A Cross Industry Project Management Perspective*. Walker D. H. T. and S. Rowlinson. Abingdon, Oxon, Taylor & Francis: 32–69.

11 Disrupting social structure to achieve innovation on Public-Private Partnership megaprojects

A narrative analysis of instruments of power in the Singapore Sports Hub case study

Jessica Siva and Kerry London

Chapter introduction

Worldwide, infrastructure spending is projected to grow from $4 trillion per year in 2012 to more than $9 trillion per year by 2025. Overall, close to $78 trillion is expected to be spent globally between 2014 and 2025 (PWC, 2014). The Asia-Pacific market, driven by China's growth, will represent nearly 60 per cent of global infrastructure spending by 2025. In contrast, Western Europe's share will shrink to less than 10 per cent from twice as much just a few years ago. There is an anticipated $6 trillion in megaproject infrastructure spend in the South East Asia region in the next five years, hence the megaproject is a significant topic for the international construction community particularly in the Asia region.

Megaprojects have been recognised as important for rapidly developing countries seeking to build the infrastructure upon which their growth relies (Conway and Lyne, 2006). Megaprojects are significant interventions that impact at the local, regional and national level on the economy, society and environment. Megaprojects worldwide are commonly associated with excessive cost overruns and high visibility and so we often in the short term see the excessive costs and failure (Flyvbjerg *et al.*, 2003).

An important driver of infrastructure spending is financing. Dedicated infrastructure funds, pension funds and other types of investors often provide the necessary investment for megaprojects around the world (PWC, 2014). Many of the megaprojects are government capital works expenditure. The large infrastructure megaprojects are often governed through the Public-Private Partnership (PPP) instrument as it provides the opportunity to release other resources for financially restrained governments in developing areas. The complexity of these large-scale capital infrastructure projects combined with a high profile provides an environment that demands extreme stakeholder management as the projects are of significance to the region if not the national economy. The pressure to ensure that the project represents value for money can create a project environment that is striving to extend the boundaries of technological innovation. Increasingly there is an emerging understanding that infrastructure, particularly in environmentally vulnerable areas in the Asian region, needs to be resilient.

The megaproject environment thus offers a range of opportunities for both clients and project teams. However, it is also a challenging environment characterised by a high level of risk, technological and social complexity, strategic political behaviour and time and cost overruns (Priemus *et al.*, 2008). While the past decade has seen a sharp increase in the magnitude and frequency of megaprojects built it is becoming clear that many of these projects have strikingly poor performance records (Flyvbjerg *et al.*, 2003).

Architectural, engineering and construction (AEC) firms are working on megaprojects composed of multiple key partners from various countries. Past research has shown that high cost overruns and disappointing operating results can be linked to clients' political economic decisions (Flyvbjerg *et al.*, 2003). Project success is significantly impacted by the internal workings of clients and its relationship with decision-making which is often beyond the control of project teams (Crawford *et al.*, 2008). Yet the focus of past research has tended to be on the industry's role instead of the client's (Siva and London, 2012; Crawford *et al.*, 2008). In their role as project initiators and financiers, clients are considered to be the driving force on projects. Clients establish a unique culture that project team members work within by setting the boundaries within which decisions affecting budgets, design and procurement come to be made. The values that clients ascribe to their everyday practices inevitably condition how they act economically, which in turn impacts upon megaproject decision-making.

The client's centrality within a project governance network has long been recognised. Since 1944 there has been a continued trend in the quest for improved construction industry performance through client-driven strategies but with little evidence that the issues have been resolved (Simon, 1944; Egan, 1998). The various investigations and policy directions reflect a preoccupation with the development of prescriptive government standards, and best practice guidelines which assume that the decision-making is rational and can be structured and controlled with little reference to the social networks, cultural norms and undercurrents of political economy which influence project decision-making (Siva and London, 2010). We have assumed that project failures derive solely from incompetent planning and administration and that the achievement of positive change is related to the implementation of comprehensive plans, policies and procedures based on rational and democratic argument. Although recent studies have demonstrated that large, complex projects tend not to conform to the rational model, the ideal lives on (Miller and Lessard, 2008).

The overall aim of this chapter, and indeed the broader research project, is to describe the *structure* and *nature* of power relations underpinning client decision-making on megaprojects. There has been little recognition within the megaproject discourse of the power structure and social networks which affect client decision-making and the influence clients have in shaping the political economy of megaproject collaborative practice. Decision-making on construction projects is not wholly predetermined by contracts but instead often emerges from the use of power. Client decision-making on megaprojects is deeply embedded in fluid networks comprising formal and informal practices, rituals and culture whereby power is constantly exercised and exchanged. Although there is extensive literature on briefing and client participation methods seeking to guide clients as well as tools and methods to achieve successful megaprojects there is little theorising that links these issues coherently. There appears to be little research that approaches the problem from a cultural political economy perspective. This research seeks to address this gap by investigating the diverse forms of power, authority and subjectivity exercised in the client's everyday practices related to megaproject decision-making. Description of an individual detailed case study of a megaproject client in Singapore is presented using a well-known social science technique of narrative analysis. After outlining the research methods, a collection of stories are presented which reveal various types of instruments of power deployed by stakeholders in their attempt to realise their individual objectives and agendas. Prior to this an analytical model based upon cultural political economy theory and the concept of governmentality is proposed to frame the exploration of power relationships in relation to megaproject governance.

The research presented in this chapter is located within a programme of research at RMIT University in the Centre of Integrated Project Solutions and is part of an ongoing PhD study which seeks to address the following two research questions:

1   What is the nature and structure of the power relations underpinning the client's decision-making environment related to the cultural political economy of megaprojects?
2   To what extent can the merging of the concept of governmentality with narrative inquiry and social network analysis techniques assist in the description and analysis of megaproject client decision-making?

This chapter develops a typology of instruments of power to reveal the different ways stakeholders identify opportunities and constraints on megaproject decision-making and achieve various objectives through the exercise of various forms of power.

Cultural political economy

Cultural political economy (CPE) is defined as one which (Sayer, 2001, p. 688):

> emphasises the lifeworld aspects of economic processes – identities, discourses, work cultures and the social and cultural embedding of

economic activity, reversing the pattern of emphasis of conventional political economy with its concern for systems . . . [it] deals with the level of concrete and hence with firms, bureaucracies and households embedded in the relationships and meanings of the lifeworld . . . it should combine and "work up" abstractions of both system and lifeworld.

The term lifeworld encompasses the informal aspects of life which is the product of the relation between embodied actors and the cultures into which they are socialised. Systems are the formalised rationalities which have a logic and momentum of their own, going beyond the subjective experience of actors to routinise or govern specific actions through signals and rules such as prices, money, bureaucratic processes and procedures (Sayer, 2001). A key characteristic of cultural political economy is its examination of the 'embedded' nature of economic action in terms of how such actions are set within social relations and cultural contexts that impact upon those economic processes (Sayer, 2001). Cultural political economic analysis offers a way of demonstrating how the advancement of specific interests is facilitated by the political economic decisions of key players in positions of power (Anderson, 2004).

   The cultural political economy of megaproject environments is composed of power relations among a diverse range of stakeholders such as clients, project managers, architects, users, property owners, financiers, regulatory bodies, local communities – each with a stake in the project outcome and thus attributing their own value at various stages of a project. A range of 'public' activities are typically conducted, particularly on such large undertakings as megaprojects, including briefing, stakeholder management and community participation, aimed at aligning project objectives and stakeholder requirements. However, such transparent efforts may not prevent stakeholders from pursuing their self-interests. Power differentials on projects cause stakeholders to employ various strategies or tactics to place them in positions of advantage. A series of questions to this research follows:

- How do ideas get disseminated, accepted or rejected on megaprojects?
- How is power created, nurtured and employed on megaprojects?
- What forms of power are available for various stakeholders on megaprojects?
- How do responsible AEC professionals who can contribute to the quality of built environments enhance their power on projects to improve project performance?

Governmentality

The concept of *governmentality*, developed by Foucault (1979) in the 1970s through his investigations of political power offers us a useful language to frame this study. Foucault (1991a) defined government as:

1 The ensemble formed by the institutions, procedures, analyses and reflections, the calculations and tactics that allow the exercise of this very specific albeit complex form of power, which has as its target population, as its principal form of knowledge political economy, and as its essential technical means apparatuses of security.

2 The tendency which, over a long period and throughout the West, has steadily led towards the pre-eminence over all other forms (sovereignty, discipline, etc.) of this type of power which may be termed government, resulting, on the one hand, in the formation of a whole series of specific governmental apparatuses, and, on the other, in the development of a whole complex of savoirs.

3 The process, or rather the result of the process, through which the state of justice of the Middle Ages, transformed into the administrative state during the fifteenth and sixteenth centuries, gradually becomes "governmentalized".

(Foucault, 1991a)

While the word *government* may imply a strictly political meaning today, Foucault (1991b) was not restricting governmentality to state politics and placed the problem of government in a more general context embracing philosophical, religious, medical and familial sites (Lemke, 2008). By widening the context of governmentality the conception of 'governmental authorities' is extended to include families, churches, experts, professions and all the different powers engaging in 'the conduct of conduct' or the art of governing.

The concept of governmentality deepens our understanding of power by demonstrating that power not only resides at the centre of a single body but is also present in diverse locales (Garland, 1997; Rose *et al.*, 2009). The analytics of government provides us with a means to explore both the macro spaces of megaproject governance frameworks as well as the confined locales of client workplaces and the everyday practices and networks where various forms of power come to be created, distributed and exercised. Power is visible in both everyday life and institutions (Rose and Miller, 1992; Donzelot, 1979). It should not be assumed that the mere existence of a structure within a network implies an acceptance or implementation by members. Whilst such formalised structures appear highly visible, there are also other less visible relationships and dealings occurring where power is constantly exercised and exchanged.

Foucault traced a movement between the sixteenth and the eighteenth centuries and identified two distinct rationalities of governing practised by state and other agencies: the sovereign and family model, which he positioned at opposite ends of a spectrum. Whilst the former was concerned with abstract and rigid ways of thinking about power the latter model was devoted to matters to enrich the small family unit (Foucault, 1979). Distinctly, he identified a third form of rationality which took place from the mid-eighteenth century onwards – governmentality – which viewed power in terms of its populations with its own realities, characteristics and requirements; independent of government yet at the same time requiring government intervention (Rose *et al.*, 2009). These populations cannot simply be controlled by implementation of the law or programmes nor be thought of as a type of extended family. Foucault highlighted that populations have their own characteristics which need to be understood through specific knowledges and it is through these emergent understandings that the 'art of governing' is formulated.

The practices within the social realm of government are undertaken in their complex relations to the various ways in which 'truth' is conceived by the different agents (Dean, 2010). Within the context of megaproject decision-making, how clients govern themselves and others relies on what they see to be 'true' about who they are, which is in turn influenced by the rich and complex social networks, cultural norms and social obligations within which they are embedded. It is thus important to capture what rationalities of governing are implicit in the client's practices and how they relate to project team members working on megaprojects. How do clients who are at the top of the governance structure of megaproject decision-making understand their powers and the impact of their practices?

Governmentality should not be primarily viewed as a normative theory of power or governance. Instead 'it asks particular questions of the phenomena that it seeks to understand, questions amenable to precise answers through empirical inquiry' (Rose *et al.*, 2009, p. 3). The governmentality perspective seeks to pose questions relating to power without attempting to prescribe a set of principles or ideology for governing others and oneself. In doing so, we are practising a form of criticism which makes explicit the taken-for-granted character of these practices (Foucault, 1991b) in terms of how clients govern and are governed and in the ways by which they do. Through this we open up for analysis various forms of strategic games or tactics in terms of power contestations and negotiations between stakeholders on megaprojects.

Power

The word power is used often and in a variety of ways and often with assumptions of definition (Galbraith, 1983). Perhaps though it is worthwhile to consider briefly some key characteristics of power and outline a working definition for the purposes of this research. Power refers to the possibility to have one's will or decision accepted by others, the ability to influence or control the behaviour of others or more informally to 'get one's way' (Brislin, 1991, p.2). By drawing upon a common pool of diverse resources one is able to consciously realise foreseen and intended objectives or effects in a social transaction (Wrong, 1979; Ewick and Silbey, 2003). Power is, however, not an object that can be possessed whereby it is a 'probabilistic social relationship, a series of transactions whose consequences are contingent upon the contributions of all the parties' (Ewick and Silbey, 2003, p. 1333). In outlining how to study transactional processes of power Foucault offered the following conception of power:

> Power, if we do not take too distant a view of it, is not that which makes the difference between those who exclusively possess and retain it, and those who do not have it and submit to it. Power must be analysed as something which circulates; or rather as something which only functions in the form of a chain. It is never localised here or there, never in anybody's hands, never appropriated as a commodity or piece of wealth. Power is employed and exercised in a net-like organisation. And not only do individuals circulate between its threads; they are always in the position of simultaneously undergoing and exercising this power. They are not only its inert or consenting target; they are always the elements of its articulation
>
> (Foucault, 1979, p. 98)

Whilst power relations are largely asymmetrical in that those having greater access to resources can exercise greater control over the behaviour of others this does not necessarily mean that the dynamics or the givenness of situational power relations cannot be challenged or shifted. Variously referred to as secondary adjustments (Goffman, 1961), tactics (de Certeau, 1984), weapons of the weak (Scott, 1985) or acts of resistance (Ewick and Silbey, 2003), these everyday strategic games represent the ways in which those in relatively powerless positions accommodate to power while simultaneously protecting their interests (Ewick and Silbey, 2003).

Power is a part of our everyday and professional lives. Whilst power is not the only factor in decision-making on megaprojects it certainly plays a central role in how megaprojects are shaped. It is suspected that successful megaprojects have project stakeholders who recognise the critical role of power in decision-making and think of power in terms of its usefulness in relation to achieving desired project outcomes or, put simply, how to 'get things done' (Brislin, 1991). Various tactics, strategies or acts of resistance have been identified as ways in which people go about imposing their will onto others. Within the context of this research these tactics, strategies and acts of resistance will be referred to as instruments of power. Although these instruments of power have been largely drawn from other fields their underlying principles are a useful starting point for this study on megaproject decision-making to provide insights into how stakeholders 'get things done'. Some instruments of power include:

- Rule literalness: based on the premise that all transactions are governed by rules which provides opportunities to achieve both compliance and resistance (Ewick and Silbey, 2003). Any incompleteness or vagueness in systems offers opportunities for resisters to exploit (Beckert, 1999).

- Masquerade: manipulated roles used to influence transactions which tend to involve some degree of deception (Ewick and Silbey, 2003). An example of a masquerade is to present oneself as a less experienced or knowledgeable worker in order to avoid work.
- Disrupting hierarchy: this refers to those individuals who purposely ignore hierarchy of systems or processes and with it the lines of authority, respect and duty attached (Ewick and Silbey, 2003). By skipping levels of hierarchy individuals can realise their objectives.
- Friendship and social ties: this refers to an individual's emotional relationships with others. It is widely known that deals are often made because certain people had warm ties with one another (Brislin, 1991). It is based upon the premise that people are more willing to comply with the requests of friends or others with whom they have warm ties and thus individuals often find it easier to gain their support.
- Information: this refers to specialised information, usually in the form of expert advice in the context of megaprojects, which places certain individuals in positions of legitimacy or authority.
- Scarcity: this refers to individuals who make use of their access to scarce goods for example networks, expertise, etc. At times the individuals themselves may possess specific scarce skills, knowledge or expertise which can be used as an important resource whereby these individuals can pick and choose in terms of who they decide to provide their services to and in making their choices become powerful figures.

London and Cadman (2009), drawing from other social and organisational psychology literature, identified several methods of how power has been described as an influencing strategy (Cialdini, 1993; Yukl, 1998; Greene and Elfrers, 1999; Higgins *et al.*, 2003). Yukl (1998) describes influencing behaviours in relation to power and defines three sources of power – position, person and expert. First, 'position power', which is derived from legitimate power invested through statutory or organisational authority control over rewards/punishments/information; second, 'personal power', which is derived from human relationship influences or traits including friendship/loyalty and charisma; and third, 'expert power', which is a function of a leader's relatively greater knowledge about the tasks at hand when compared with subordinates who are dependent on that knowledge.

Greene and Elfrers (1999) outlined several forms of power including coercive, connection, reward, legitimate, referent, information and expert which seem to provide more detail to Yukl's descriptions. Coercive influencing tactics are based upon fear where the failure to comply results in punishment and relates to positional power. Connection refers to power resulting from connections to networks of people with influence and is both personal and political power. Reward is based upon the ability to provide rewards through incentives to comply and is typically related to position power. Legitimate is based on organisational or hierarchical position and is related to both position and political power. Referent is based on personality traits such as being likeable and admired and is personal power. Information is based on possession of or access to information perceived as valuable and can be a combination of position, personal and political power. Expert is based upon expertise, skill and knowledge and is related to an influencing power vested in a person and, if credible, the respect influences others and is considered to be personal power.

Cialdini's (1993) persuasion theory may also provide insights when considering tactics in relation to conflict resolution during decision-making. Cialdini (1993) – one of the most cited social psychologists – has identified six key principles of persuasion through a study of the psychological principles that influence the tendency of professionals to comply with requests for change. The six key principles are:

- reciprocity
- social validation
- commitment and consistency
- friendship and liking
- scarcity
- authority.

'Reciprocity' is where one should be more willing to comply with a request from someone who has previously provided a favour and concession. 'Social validation' is where one should be more willing to comply with a request or behaviour if it is consistent with what similar others are thinking or doing, i.e. we may tend to echo the behaviours of others for various reasons. 'Commitment and consistency' is where after committing to a position, a person is more willing to comply with requests for behaviours that are consistent with that position. 'Friendship and liking' refers to the situation where one should be more willing to comply with the requests of friends or other liked individuals. 'Scarcity' refers to situations where one should try to secure those opportunities that are scarce or dwindling, i.e. if people perceive they are going to miss out on something they are more likely to be persuaded to change their behaviours. 'Authority' is where one should be more willing to follow the suggestions or instructions of someone who is a legitimate authority.

This discussion on governmentality and power provides the theoretical context to critique the case study and the various stakeholders' positioning within the project decision-making as they exercise various instruments of power to enable their agendas on the project.

Research methodology

This study employs a case study strategy through the use of the narrative analysis and social network analysis methods for collecting and analysing empirical material. This chapter specifically reports early observations made from one case study of a megaproject client in Singapore. Sixteen interviews were conducted with a range of participants. In keeping with the narrative inquiry approach, the interview instrument was designed to be broad and open-ended to provide participants the opportunity to express themselves in their own words without being influenced by suggestions from the researcher as well as to invite participants to tell their own stories. Participants were asked questions in relation to three broad areas: their role in the organisation and on the megaproject, stories in relation to project issues experienced and how decisions were made to resolve issues, and their relationship with other project stakeholders.

Stories told by individual participants relating to the hidden and mundane everyday practices employed to challenge or reinforce social structures were identified and 'open coding' was conducted based on the loose association of themes and concepts. A main criteria for identifying the stories was that the acts or events described must be intentional and purposeful whereby the ultimate goal was to impose one's will by achieving compliance or a reversal of power. Following an identification of stories axial coding was conducted involving the arrangement of data according to dominant themes that emerged. A systematic examination of stories about everyday events helps to unmask how the governmentality concept works in the decision-making environment of clients on megaprojects. The stories depict first, the narrators' recognition of social structure and second, their understanding of how to operate within existing conditions towards realising their objectives. Through the analysis the complex and fluid relations of power are made visible and a typology of instruments of power employed on megaprojects can be established.

Case study background

The case study is a 1.33 billion SGD multi-use, multi-sport, leisure and entertainment complex in Singapore. The project seeks to encourage large numbers of people to adopt and pursue sports and to draw international events to its world-class facilities, which is part of the government's broader vision of creating a thriving sports, entertainment and lifestyle ecosystem (Sporting Singapore, 2001). The project involves the demolition of an existing National Stadium and the construction of new buildings including a national stadium, multi-purpose indoor arena, aquatic centre, water sports centre, Singapore information and resource centre, sports promenade and commercial space/retail mall. It also incorporates the use of an existing multi-purpose indoor arena. It is the largest sports facilities infrastructure Public-Private-Partnership (PPP) project in the world (SSC, 2012). The project is also the first PPP project in Singapore. A Special Purpose Vehicle (SPV), Sports Hub Private Limited, was formed to finance, design, build and operate the facility in partnership with the Singapore Sports Council

(SSC) over a 25-year period. The SPV invests in equity and utilises bank debt financing to build the facilities (SSC, 2012). Upon completion of the contract the facility will be handed over to the SSC.

PPP projects have multiple 'clients'. In this study the client network comprises the sports government agency responsible for setting up the project framework, SSC, and the various PPP consortium partners which have varying levels of involvement (see Figure 11.1). All the PPP partners are bound to the Project Agreement and the various protocols set out in an Interface

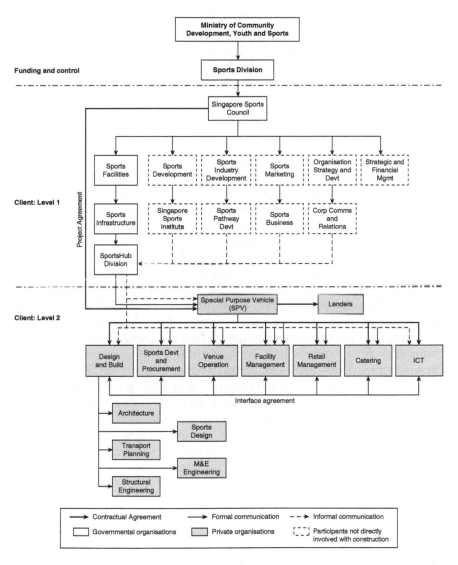

Figure 11.1 PPP structure of the Sports Hub project

Agreement. Formalised structures and protocols have clearly been established by the SSC and SPV for project coordination which are aimed at achieving control of accountability of decision-making as well as respecting the roles and boundaries of various parties. Indeed these formalised structures and protocols offer clear traceability in terms of lines of official document exchange between project stakeholders. However, what these official documents fail to record are the informal negotiations and dealings occurring in multi-level networks whereby power is constantly exchanged and exercised. Past research has identified that these informal networks and communications are equally, if not more, important than pre-established structures (De Blois *et al.*, 2011). Analysis of the Sports Hub case study supports these past findings in relation to two key areas: indirect or non-expert participants can sometimes act informally as client representatives and influence the direction of projects; and informal communication and decision-making are often made outside pre-established structures (Siva and London, 2012).

Narrating instruments of power

This section presents a collection of stories told by the interviewees – of both resistance and compliance – as the ways in which they understood and responded to various social transactions within the context of megaproject decision-making. In these stories the narrated instruments of power represent the ways in which stakeholders accommodate to or exercise power to protect their interests. Each story recounts how an aspect of the structure of social relationships, whether it's a rule, role or hierarchy, was mobilised to reinforce an existing condition or reverse a more probable outcome (Ewick and Silbey, 2003). These stories reveal the narrators' clear awareness of the ways in which opportunities and constraints are embedded in the largely taken-for-granted structure of social relationships. The stories also reveal the narrators' detailed understanding of how to respond in terms of the resources available to them in the form of a variety of instruments of power. The method for discussion is completed in either of two ways. First, a theme is articulated through a comparative discussion of the way participants described a given theme. Second, the intricacies of a theme are best exemplified through a more in-depth discussion of how one particular participant experienced the theme where only one exemplar quote may be provided although a series of quotes were considered in the analysis. In all cases the themes discussed were dominant in the interviews across participants.

Rule literalness

Rule literalness was raised as an instrument commonly used by the local government agencies to enforce project requirements. The power and authority associated with local government agencies was consistently discussed by the participants as a significant 'force to be reckoned with'. The Technical

Director of the Design and Construct Contractor firm clearly understood the hierarchy of power within the Singaporean context and that the expectations of the local agencies needed to be placed as their top priority in progressing the project.

> The authorities of Singapore have got very strong power. We got to meet their expectations first. Everybody needs to compromise a bit but usually the authorities are difficult to compromise.

Whilst in some situations or environments rules can be counted on to provide opportunities for resistance where this may involve identifying loopholes or gaps with a network of rules, this was clearly not the case in the highly rationalised setting of the Singaporean ruling government authorities. The very nature of an innovation necessitates a change from the norm whereby it is unlikely that project innovations will comply with existing codes or regulations. Given that megaprojects are characterised by innovations, code compliance and achieving regulatory approvals can often be key challenges. The project team found themselves in a difficult position of getting a specific innovation accepted by the local authorities even though the efficiency of the system far exceeded the basic requirements of the building regulations and codes. This gives rise to a situation where those responsible for providing regulatory approvals are placed in a powerful and dominating position to dictate and impose their requirements upon other stakeholders who in turn become highly vulnerable. Ultimately the drive to successfully deliver project innovations falls onto the responsibility of key stakeholders within the project team. The domination of the strict regulatory regime and the associated approval of innovations in Singapore does not appear to be a conducive or encouraging environment for innovations to be successfully delivered. As the following Senior Manager of the SSC's Sports Hub team explained, the officers working in local government agencies can be 'more tight' and demand that standard local requirements be complied with even though the project may have unique features which may contradict the regulations.

> Sports Hub has a lot of unique feature that's new . . . so that's why within the local context codes doesn't give that kind of flexibility to be applied. They may contradict to the code's requirements . . . some of the officers would be slightly more tight. They say that OK you can show me, yes I know you can meet this but you still have to meet mine.

In the case of this specific innovation, hierarchy was eventually disrupted to enable the innovation to be implemented and this will be discussed in the following section. Rule literalness can also be used as a tool to manage competing interests. As expected of any large complex project, the Sports Hub involved a variety of stakeholders during the tender evaluation phase which could have made it a potentially difficult situation to manage.

The Project Director used rule literalness as a way of controlling the stakeholders whereby the project vision and objectives was his 'biggest tool' when dealing with competing agendas or stakeholders.

> We had 60 odd people in our evaluation team. So as the stakeholder manager and project director the vision and objectives was the biggest tool that I had to control my stakeholders because in the room when we were arguing about a legal clause or how to write a specification I always brought it back to does this meet the objectives and the desired outcomes? If not we're not talking about it anymore . . . because I had very, very senior government people who had never been told what to do before but then that was my tool to control that so that not a single stakeholder was more important than the project objectives.

Disrupting hierarchy

Stories were told about how participants consciously ignored hierarchy and with it lines of authority and duty attached. In recognising the strictly regimented rules of the Building Control Authority (BCA) in Singapore, the Senior Associate from the architectural firm explained that the project team knew that there was a need to bypass hierarchy to 'bring it up to a higher level' where negotiations to achieve a compromise can be undertaken.

> The system is so regimented . . . the way that BCA is very strictly doing something and when you do something different people don't know how to deal with it. So it means you'll have to bring it up to a higher level . . . So that level of thinking is thinking about how can we reach a compromise – one that satisfies government issues as well as helps you to mitigate something that is going to cost ridiculously or become a real big issue and that compromise can only be discussed at that high level.

In dealing with such situations the project team clearly recognised that there was a limit to what they were able to achieve in terms of being able to undertake the necessary negotiations with the appropriate people at a higher level of the BCA. In this case the Design Manager from the Design and Contractor firm highlighted that there was a need to ensure that the government client, SSC, was well aware of the critical role they needed to play in assisting with the successful implementation of the innovation by working 'behind the scenes' with the BCA.

> So at one point, I mean we're confident now, but at one point we didn't have clarity that we were in violation of regulation as BCA was interpreting it and SSC was working behind the scenes to work through that. They weren't involved with the day-to-day technical aspects but we keep them very aware and they do kind of help go to government

departments and help smooth things out. So if there's something really at that level they try to help us behind the scenes.

The Senior Manager from the SSC Sports Hub team also highlighted the importance of disrupting hierarchy by going 'to the top' to achieve a degree of flexibility that the bureaucratic processes of government agencies normally stifle. He explained that as a government agency client they were able to bypass the regular channels to negotiate with the higher-ups in the BCA towards achieving a compromise.

> Sometimes when they think that they need some help and some backing and SSC being one of the government agencies to come in and maybe just stand behind there and lend them some support, we always do that . . . We know that they're having some issue with an agency we will either go together with them sometimes and even in parallel we'll go behind them and talk to the agency . . . But of course we had to pre-agree with them on some agreements on how we can operate it and things like that.

Legitimacy or position power was demonstrated to be an important form of power whereby the SSC, by virtue of being a government agency, were able to exercise their invisible rights to engage in negotiations with the BCA on the same level of authority. Approval for the innovation was only provided following SSC's private discussions with the BCA. Even though the design team had spent considerable amount of time and effort defending the scheme the BCA appeared to place much more weight in SSC's justifications in regard to the system. The process in which the bowl cooling system achieved regulatory approval demonstrates that the exercise of political power can overrule rationality in certain situations. The design team's expert power in presenting an innovative and efficient system was less effective than the SSC's legitimate power related to their position within the network of government agencies. This is of course not to say that stakeholders should ignore the environmental, social, moral and professional arguments based on rationality in support of their respective positions. Rather a strict adherence to rationality and logic alone without adequate acknowledgement and understanding of other forms of power will unlikely lead to successful outcomes. It is thus important to understand the various instruments of power stakeholders employ in megaproject decision-making and how this influences project outcomes.

Swapping deals

The social position of the local government agencies brings with it certain expectations whereby it is important for the BCA – as well as other agencies – to be seen to be 'doing the right thing' by the general public. BCA has been actively discouraging the spillage of air conditioning from indoor spaces to outdoors, both through improved regulation and enforcement, and advised that they were considering further retroactive regulation to control this practice. The project team, despite putting forward an argument in support of a proposed innovation based on a rational and technical perspective, was unable to achieve regulatory approval from the BCA as it would potentially be misinterpreted by the general public as 'throwing air con into an open space'. Providing regulatory approval for the use of the bowl cooling system may thus be seen as a precedent for the provision of air conditioning to outdoor spaces thereby undermining BCA's efforts to eliminate this unsustainable practice. Through extensive computer modelling and calculations the project team presented what they worked out to be the most efficient means of providing spectator comfort. However, it was felt that a less innovative scheme using 'four times the energy' would have been better received by the authorities simply because it would not create a public relations problem for the BCA as the proposed innovation presumably would.

> One of our arguments . . . to the BCA is . . . we could've arrived at a solution which would've used our modelling and calculations two or three times as much energy . . . but that generally wasn't accepted because to the authority they're still looking at us documenting the project where we're bringing cool air into an outside space . . . they also don't want to get into a situation that there is criticism . . . There are comparisons to other projects where semi-outdoor spaces have been cooled and the BCA hasn't allowed that and that is part of the reason why this has become a sensitive issue.

Eventually the BCA granted regulatory approval for the use of the innovation as the previous discussion in the section 'Disrupting hierarchy' has shown. The BCA advised the project team that the energy usage of the bowl cooling system must be 100 per cent offset by renewable or waste-generated energy sources. The issue of a major project associated with the Singapore government creating a public relations problem for BCA's enforcement efforts to stamp out purposeful spillage of air conditioning to the outdoors is thus resolved by the ability to advise the public that energy used for the bowl cooling system is completely offset by renewable energy. In this case there was a 'straight swap' of deals to satisfy the needs of involved parties.

> It flies in the face that something that BCA is trying to restrict . . . so how do you mitigate but the good thing about it is at least maybe the balance is that there's a balance has been struck. OK this is the value to the developer who needs to pay. OK so BCA accepts that there's a straight swap of what they can to satisfy BCA, to cover the situation.

Friendship and social ties

Even though the megaproject environment involves professional relationships between individuals within organisations, it is important to note that most of the individuals involved have emotional needs that can be satisfied only through warm ties and social relationships. People in general enjoy cordial and pleasant relationships with those they work with. Success in gaining the support of others in professional matters is often reliant upon personal friendships and networks. One way to develop social relationships in the professional environment and to generate general everyday cordiality is to take part in office social occasions such as contributions to baby showers and participation in project team lunches. For this Chief Financial Officer of the SPV, interactions at a personal level with project stakeholders at various levels was felt to be an important part of making them feel engaged.

> Even though I'm the COO I have to deal with the immediate level because all of them they are government servants and we need to engage them . . . You have to engage them informally as well and go out to lunch with them. When you go on holidays buy them some nice chocolates when someone gives birth you send them a congratulatory note and things like that. So you try to interact with them at a personal level.

Existence of prior relationships was also seen as a valuable resource by the participants. The Design Manager of the Design and Construct firm explained that prior relationships helped to build trust and provides the other party they are working with the assurance of their ability to deliver a promise, i.e. their credibility.

> Most of the authority are the same ones I worked with on the Marina Bay Sands. So the person I dealt with in the LTA is the same one I dealt with here and people in MHA, Ministry of Home Affairs, they're the same people . . . that's the I think important thing here . . . in fact green-mark to some degree "X" has a relationship with ABC and myself I think its very interpersonal with Land and Transport Authority. MHA its very much based on trust and having seen maybe how we operate and how we tend to deliver what we promise. That's a big part of it . . . They tend to be the people I know are dealing with are kind of the same age or are a little bit younger but you kind of maybe have similar levels of interests.

Furthermore understanding the dynamics of social relationships can also be beneficial in helping one realise their aims. Brislin (1991) classified this as a knowledge base and this was certainly the case for the Senior Associate in the Sports Architecture firm who knew specific individuals with different areas of expertise or in positions of power who could help his cause. The Senior Associate indicated that conversations were often 'held quietly' to 'attune' others to provide him with the information he required.

> Some conversations are held quietly to try and prepare to get some-
> one else ready to be bringing the information out in the form which
> is going to work for everybody . . . but the people who are doing the
> work . . . they're not necessarily attuned to precisely the message that
> we want to hand over to the authority. So we do a little bit of work
> in making sure that they become attuned to that . . . Sometimes we
> have to use other people to help us get the message across and that's
> where someone like "X" has a great degree of influence in holding it
> together . . . We can use people like "X" . . . to help our cause and we
> do and he is a very effective operator in that position.

The use of certain relationships and networks can thus be more effective
than others depending on situational needs. This demonstrates that there
are ways in which those in weaker positions of power can achieve stronger
positions through smart linkages. This also demonstrates that despite the
importance of organisational and project structures in formalising communication flows and coordination, how work is carried out on a daily
basis tends to have more to do with the informal relationships and interactions between members within and across organisations as they strategically
assess and 'work out deals' to better achieve what they require.

Information

The analysis highlighted the consistent use of information in the form
of technical rationality by stakeholders in gaining the support of others
towards realising their aims. For the project team, using technical rationality to support their argument was often seen as a more 'black and white'
way to demonstrate the characteristics of a proposal solution.

> What we have at our disposal are our technical ways of showing some-
> thing. So technical solutions are although they're not real – they're
> theoretically proven so that's quite good in some ways . . . to a certain
> degree it's more black and white to be able to show that something
> works or doesn't work.

When aligned well, technical rationality can also be used to help support
an argument towards achieving one's hidden agenda. The Senior Associate
from the architecture firm told a story about how they were experienc-
ing difficulty convincing the client to accept the design team's proposal
to disconnect an existing train station from a new walkway. The client
had requested for a large canopy to be built between the station and the
walkway so that people would remain under cover as they exited the station. The design team's proposal was largely driven by an aesthetic desire
in that maintaining the integrity and outlook of the existing station was
seen as important whereby connecting it to the walkway would destroy

the architectural language of the structure. The client, on the other hand, was more concerned about the functional use of the canopy to protect the public from inclement weather. In the end additional canopies were proposed to the side of the station rather than to have the one large canopy at the main entrance which was accepted by the client. The argument put forward though was one centred around the importance of crowd control whereby it was necessary to disperse people from multiple exits and hence side canopies rather than the single large canopy even though the underlying aim was an architectural reason to maintain the 'fantastic gesture view' of the station.

> In the end we managed it technically by saying that the big opening to the station, the big door that big mouth of the station that's too big to allow that many people to come in . . . So the whole system is actually of a limited capacity . . . but from a functional point of view we're saying we'll put canopies to the side which is what we did. We had to demonstrate why it was not unreasonable to not have this huge canopy sitting over the main station . . . So meeting with SSC with LTA with URA as well and doing technical studies like pedestrian modelling to show where people move, to look at, the rate at which people are taken away from the station, on the trains to show that you don't need that, to show that you can't deal with that capacity and we can't force people in, they don't know where to go. So therefore you allow that fantastic gesture view to still be there. If we put a roof on it you take away a lot of that. So the idea is that when you stand as you come out of the station you see the window which is framed by that big view and from that window you see west plaza so that's your window and you frame that view so that was the architectural reason for doing it . . . aesthetics . . . it is a better solution architecturally.

Whilst the use of technical rationality may not work on every occasion, as demonstrated in the earlier discussion in 'Disrupting hierarchy', it can sometimes work in favour of specialists on projects. In this case, specialist knowledge relating to crowd control became particularly powerful as it offered the design team a way to justify their aesthetic desire.

> Using crowd control as a diversion to an aesthetic which has been to some degree is true . . . we probably didn't like it . . . we couldn't bear the look of this thing. And the argument of crowd control it just happened to align with what we wanted aesthetically. I mean really with crowd control we've done a lot of crowd control work. You wouldn't put a station at that spot but since it's pre-existing. It's a machine to get people coming out quickly and safely. Architects, we have an aesthetic desire or like doctors have credibility.

Chapter summary

This chapter described an analytical model which was developed based upon cultural political economy theory and the concept of governmentality to examine megaproject client governance. Various instruments of power were drawn from other fields such as business, sociology and social psychology to inform the analysis of how stakeholders go about imposing their will onto others on megaprojects.

The observations of the case study of the Singapore Sports Hub in relation to the use of various instruments of power confirmed initial assumptions made that although formalised protocols were established for project communication and coordination, decisions were often made outside the pre-established structures. Furthermore this decision-making was influenced by informal communication embedded in multiple levels of social networks comprising various stakeholders in positions of power. These observations highlight the significant influence of the structural and behavioural characteristics of networks on decision-making.

Project stakeholders typically draw from a common pool of sociocultural resources including organisational, disciplinary, symbolic and professional. Some stakeholders, however, can be advantaged by having greater access to the resources which can be effectively employed in transactions towards realising their aims. Therefore even though power can never be possessed categorically, some individuals enjoy a greater probability of realising their goals or objectives simply by virtue of their location within a social structure. Megaproject decision-making is thus a network problem requiring an understanding of social structures. Different types and forms of social networks may be essential for achieving different project outcomes at various stages of project decision-making. The structure of social networks embedded in the multi-level environment in which client decision-making is undertaken may contribute towards understanding the way decisions and actions occurring at the confined locales of client workplaces can impact on project outcomes at higher levels. However, to date there is still little known in terms of the nature and structure of power relations in megaproject client decision-making where various forms of power come to be created, distributed and exercised.

The stories presented in this chapter are evidence of the narrators' consciousness of structure and of how social structure works and further to that their understanding of how at times it can be exploited to serve specific purposes. Within the stories the narrators expressed how power was reversed or compliance was achieved through the use of various instruments of power. A typology of instruments of power was developed to make explicit what had largely been taken for granted previously. Through the telling of stories of the narrators' success with or against power an understanding of how to disrupt social structure has been achieved which makes clearer how instruments of power can be used in other situations to achieve better project outcomes. Examples of *rule literalness, disrupting hierarchy, swapping deals, friendship and social ties, information and scarcity* were found in this case study as stakeholders sought to disrupt the formal organisational designed social structure of the PPP contract ties to create, transform and negotiate positions to influence decision-making and achieve individual objectives.

Further work is required though to map explicitly the social ties between stakeholders to demonstrate how the structure of power relations influences decision-making. Therefore the next stage of analysis involves a social network mapping of the informal links between stakeholders through the stories collected.

Reflections

1 The European Union (EU) has an excellent Web-based resource that focuses on how megaprojects can be designed and delivered more effectively. Archival news, case studies, research etc. can be found at: http://www.mega-project.eu.

2 A handbook that presents state-of-the-art research on the decision-making processes in the deliverance of megaprojects has been produced by Priemus, H. and van Wee, B. (2015). *International Handbook on Mega-Projects*, Cheltenham, UK: Edward Elgar. The book provides coverage on political, technical, managerial, social, economic and environmental issues affecting megaprojects.

3 Singapore's Building and Construction Authority (BCA) is an agency under the Ministry of National Development, who champion the development of an excellent built environment. For further details, please visit: http://www.bca.gov.sg.

4 The Singapore Sports Hub consists of a number of sporting and leisure facilities, including the 55,000 capacity National Stadium. More information on these venues can be found at: http://www.sportshub.com.sg.

5 The Singapore Government's Ministry of Finance produced a *PPP Handbook* in 2012. It puts a Singapore perspective on PPPs and as well as providing background information on the topic it also focuses on structuring deals, the procurement process and managing relationships. It can be viewed at: http://www.mof.gov.sg/Portals/0/Policies/ProcurementProcess/PPPHandbook2012.pdf.

References

Anderson, T. (2004). Some thoughts on method in political economy, *Journal of Australian Political Economy*, 54: 135–145.

Beckert, J. (1999). Agency, entrepreneurs and institutional change: the role of strategic choice and institutionalised practices in organizations, *Organisation Studies*, 20: 777–799.

Brislin, R. (1991). *The Art of Getting Things Done: A Practical Guide to the Use of Power*, New York: Praeger Publishers.

Cialdini, R. (1993). *Influence: The Psychology of Persuasion*, New York: Quill.

Conway, M. and Lyne, L. (2006). The world's top superprojects: the best of the big, *The Futurist*, 40: 32–39.

Crawford, L., Cooke-Davies, T., Hobbs, B., Labuschagne, L., Remington, K. and Chen, P. (2008). Governance and support in the sponsoring of projects and programs, *Project Management Journal*, 39: 43–55.

De Blois, M., Herazo-Cueto, B., Latunova, I and Lizzarralde, G. (2011). Relationships between construction clients and participants of the building industry: structures and mechanisms of coordination and communication, *Architectural Engineering and Design Management*, 7: 3–22.

de Certeau, M. (1984). *The Practice of Everyday Life*, Berkeley, CA: University of California Press.

Dean, M. (2010). *Governmentality: Power and Rule in Modern Society*, London: Sage.

Donzelot, J. (1979). *The Policing of Families*, New York: Pantheon Books.

Egan, J. (1998). *Rethinking Construction*, London: Department of the Environment, Transport and the Regions.

Ewick, P. and Silbey, S. (2003). Narrating social structure: stories of resistance to legal authority, *American Journal of Sociology*, 108 (6): 1328–1372.

Flyvbjerg, B., Bruzelius, N. and Rothengatter, W. (2003). *Megaprojects and Risk: An Anatomy of Ambition*, Cambridge: Cambridge University Press.

Foucault, M. (1979). Governmentality, *Ideology and Consciousness*, 6: 5–21.

Foucault, M. (1991a). Questions of method, in (eds) Burchell, G. and Miller, P., *The Foucault Effect: Studies in Governmentality*, Chicago: Chicago University Press, pp.73–86.

Foucault, M. (1991b). Governmentality, in (eds) Burchell, G. and Miller, P., *The Foucault Effect: Studies in Governmentality*, Chicago: Chicago University Press, pp.87–104.

Galbraith, J. (1983). *The Anatomy of Power*, London: Corgi Books.

Garland, D. (1997). Governmentality and the problem of crime: Foucault, criminology and sociology, *Theoretical Criminology*, 1 (2): 173–214.

Goffman, E. (1961). *Asylums*, New York: Anchor Books.

Greene, R. and Elfrers, J. (1999). *Power the 48 Laws*, London: Profile Books.

Higgins, C., Judge, T. and Ferris, G. (2003). Influence tactics and work outcomes: a meta-analysis, *Journal of Organisational Behaviour*, 24: 89–106.

Lemke, D. (2008). Power politics and wars without states, *American Journal of Political Science*, 52(4): 774–786.

London, K. and Cadman, K. (2009). Impact of a fragmented regulatory environment on sustainable urban development design management, *Journal of Architectural Engineering and Design Management*, 5: 5–23.

Miller, R. and Lessard, D. (2008). Evolving strategy: risk management and the shaping of mega projects, in (eds) Priemus, H., Flyvbjerg, B. and van Wee, B., *Decision-making on Mega Projects: Cost-Benefit Analysis, Planning and Innovation*, Cheltenham, UK: Edward Elgar, pp. 145–172.

Priemus, H., Flyvbjerg, B. and van Wee, B. (2008). *Decision-making on Mega Projects: Cost-Benefit Analysis, Planning and Innovation*, Cheltenham, UK: Edward Elgar.

PWC (Price Waterhouse Coopers) (2014). *Capital Project and Infrastructure Spending Outlook to 2025*, Research by Oxford Economics.

Rose, N. and Miller, P. (1992). Political power beyond the state: problematics of government, *British Journal of Sociology*, 43: 173–205.

Rose, N., O'Malley, P. and Valverde, M. (2009). *Governmentality*, Legal Studies Research Paper No. 09/94, The University of Sydney.

Sayer, A. (2001). For a critical cultural political economy, *Antipode*, 33 (4): 687–708.

Scott, J. (1985). *Weapons of the Weak*, New Haven, CT: Yale University Press.

Simon, E. (1944). *The Placing and Management of Contracts*, London: HMSO.

Siva, J. and London, K. (2010). Client management on international mega projects: investigating the client's complex decision-making environment, *1st International Conference on Sustainable Urbanisation*, Hong Kong, 15–17 December.

Siva, J. and London, K. (2012). Client decision-making to support innovations on megaprojects, *7th International Conference on Innovation in Architecture, Engineering and Construction*, Sao Paolo, 15–17 August.

Sporting Singapore (2001). *Report of the Committee on Sporting Singapore*, July 2001.

SSC (Singapore Sports Council) (2012). Singapore Sports Hub – information sheet.

Wrong, D. (1979). *Power: Its Forms, Bases and Uses*, New York: Harper & Row.

Yukl, G. (1998). *Leadership in Organisations*, Sydney: Prentice-Hall.

12 Public-Private Partnerships (PPPs) in China

The past, present and future

Jing Xie, ShouQing Wang, Marcus Jefferies and Yongjian Ke

Chapter introduction

According to Ke *et al.* (2014) the continuous economic growth in China has resulted in an immense demand for infrastructure and in order to meet the country's development needs the Chinese government has been active in encouraging and supporting the participation of private participation in the provision of public infrastructure which has led to a huge investment opportunity for Public-Private Partnerships (PPPs). However, China has often been criticised for having immature regulatory and institutional frameworks for PPPs (Chan *et al.* 2010) and therefore, given the country's growth in infrastructure, it is worthwhile examining both the historical and the current application of PPPs in China in order to derive lessons for the future.

The encouragement of private capital in the financing of public service facilities in China originated in the urbanisation process, and the tremendous economic growth created huge demand for infrastructure such as roads, water supply, waste treatment and power generation facilities. Simultaneously, the supplementary investment from the private sector was highly promoted due to the government's budget concerns (Wang *et al.* 2012). World economic growth weakened considerably during 2012 and it is expected to remain subdued in the subsequent years, according to the United Nations in its latest issue of the World Economic Situation and Prospects (2013). China has faced weakened investment demand because of financing constraints and excess production capacity elsewhere. The economic woes in Europe, Japan and the United States have adversely spilled over to developing countries through weaker demand for their exports and heightened volatility in capital flows and commodity prices. To overcome these challenges the Chinese government has prioritised the urbanisation process, expedited industrialisation and raised rigid domestic demand. This has helped with China's urbanisation rate increasing from 36.2 per cent to 51.3 per cent from 2000 to 2011. However, the rate of urbanisation in small and medium-sized cities is far below the national average of 33.9 per cent, ensuring that urban infrastructure and public service delivery still has room for improvement. China needs better planning to improve the quality of urbanisation, particularly when addressing inadequate services provided to new urban dwellers from rural areas, and an imbalanced distribution of resources between megacities and small and medium-sized cities (China Daily 2013).

This chapter reviews the application and development of Chinese PPPs by focusing on their background, evolution, frameworks, government organisations and policies, implementation, and risk. It then summaries the drivers and barriers of PPP application, existing problems and challenges, critical successful factors and lessons learnt from previous projects. Finally, a simple comparison of PPP practices in China and Western countries is provided, followed by suggestions on how to improve the PPP process in China with an outlook on possible application and development in the near future. This chapter updates the findings of Liu and Yamamoto (2009) who undertook similar research and also builds on the findings of Chapter 10, where Rowlinson explored the emergence of PPPs across the broader continent of Asia, by providing a focused investigation into the development of PPPs within the context of the People's Republic of China.

Evolution

The PPP approach was adopted in China to relieve the pressure on the government's budget for infrastructure development. It was first introduced in 1984 in the Shajiao B power plant in Guangdong province, and developed by local governments around the same time as the privatisation process began (Wang *et al.* 2012). The first booming phase of private participation in infrastructure development in China began in the 1990s and after 1996, several state-approved pilot Build-Operate-Transfer (BOT) projects were awarded in order to promote this form of PPP on a broader scale. These projects included the Laibin B power project in Guangxi province and the Chengdu No. 6 Water Project in Sichuan province. Thereafter the involvement of private investors in infrastructure development has shown dynamic growth (Wang *et al.* 2012).

At the end of the 1990s, in response to the adverse influence of the Asian financial crisis, the central government invested large amounts of treasury bonds in infrastructure projects and was determined to clean up the unregulated or illegal projects that had led to the end of the first phase of the private investment boom (Shen *et al.* 2005). Foreign investors were the major players during this period, usually charging higher fees and preferring to operate projects in the more developed regions of China. Therefore, even in the state-approved pilot BOT projects such as the Laibin B power plant, the Chinese government took far too many risks due to their lack of experience (Wang and Ke 2009). Most foreign investors would request a guarantee of a fixed or even minimum return which would reduce their motivation to improve the operational efficiency.

To solve these issues, the General Office of the State Council issued in 2002 a 'Notice on Relevant Issues Concerning the Appropriate Handling of the Existing Projects Guaranteeing the Fixed Return of Investments by Foreign Parties'. Thereafter, existing projects with an agreed fixed rate of return from local government were managed by modifying relevant contracts, selling shares to local government, transferring projects back to local government or even terminating contracts (Wang 2006). The Chinese government was accused of breaching its written promise in the concession agreements and this directly led to a lack of confidence from foreign investors.

At the turn of the twenty-first century, infrastructure shortages limited economic growth and at the same time imposed great budgetary pressure on the Chinese government. Despite state-owned enterprises becoming the principal players in PPPs, major changes in the evolution of PPP from the first to the second phase occurred due to the improvement of the legal framework. The Chinese government has produced several documents for promoting and guiding private investment through the PPP process. These reflect several improvements with regard to PPP implementation, such as extending the group of key PPP players from solely foreigners to all investors, widening the PPP implementation concession from the previously used BOT model, and providing more operational procedures. In particular, the 'Several Opinions of the State Council on Encouraging and Guiding the Healthy Development of Private Investment' issued in May 2010 recommended widening the scope for private investment in PPPs by including railway, water, petroleum, gas, telecommunication, land control, exploration and development of mineral resources, national defence, technology industries, housing, medical and health industries, and education and welfare services projects. Private enterprises are also allowed to establish financial institutions and participate in the reform of state-owned enterprises (Wang *et al.* 2012).

The Chinese government's current view is that it should encourage and support private investors to participate in infrastructure development and public service supply, to progressively promote a number of regulations for private investment in public utilities, to adopt international contractual practices and to work out an equitable risk-sharing scheme (Ke *et al.* 2014). At the Communist Party of China's (CPC) 18th National Congress in 2013 reform initiatives were expressed in order to 'Allow private capital investment in urban infrastructure and operations through franchising', and the Chinese Ministry of Finance subsequently launched a comprehensive plan for the promotion of the PPP model by establishing its PPP Steering Group in 2014. Simultaneously, the National Development and Reform Commission launched 80 demonstration projects in May 2014 to encourage private capital to participate in the construction and operation of infrastructure. The (draft) Infrastructure and Public Utilities Concession Law (i.e. PPP Law) was formulated by the National Development and Reform Commission Regulation Department and solicited for comments to the broader industry. This institutionalisation of the PPP model in 2014 has therefore been seen as the catalyst for forging into the future and cultivating the Chinese-style PPP model (Liu 2014).

Current status of PPP ventures in China

General debate

Generally speaking, PPP is identified as an innovative procurement approach for delivering infrastructure projects and has enormous potential in developing countries such as China. PPP models that attract foreign and private capital for infrastructure investment and development are welcomed by the Chinese government, especially at a local level. PPPs are not, however, a panacea or a 'quick fix' solution for delivering project finance and realisation. The debate lies in whether further adoption of PPPs would allow private investors to charge a premium for providing a public service that was previously 'free' under traditional procurement methods (Wang 2006).

PPP projects often require extensive expert input, have high preliminary costs, and require lengthy negotiation periods, which may not offer a good return to all parties (Chan *et al.* 2010). Private sector consortia may not be experienced enough and financially capable of taking up the infrastructure projects owing to a lack of relevant skills and experience (Ke *et al.* 2009). More importantly, the adoption of PPPs has put the Chinese legislative framework under pressure. In China, the PPP regulatory system, including cost auditing, tariff regulation and definition of scope of responsibility and remit of relevant regulatory agencies, has not been fully established. This has led to a great deal of resistance from the conservative wing of the government and has slowed down some of the key reform projects (Zhang 2009).

Macro-economic, political, legal and governance environment

The Chinese government has been making progressive efforts to promote private capital and improve the financial market. The public-private capital collaboration programme aims to stimulate the adoption of a PPP model and is an important means to support the construction of new urban development. Urbanisation is regarded as the main driver for modernisation in China, building harmony and a prosperous society. China's urbanisation is expected to reach 60 per cent in 2020, resulting in investment demand of approximately 42 trillion RMB (Wang 2014). PPPs are an effective solution to financing needs, they favourably attract social capital, broaden the urbanisation financing channels, and achieve diversified and sustainable funding mechanisms. They appear to provide a great opportunity for private investment as most local governments still suffer from severe budget pressure, especially as the central government is now trying to cool down the overheated real estate sector, which will further reduce the local government's revenue from selling land.

Typical frameworks and/or models of PPPs in China

Economic infrastructure projects

BOT is still the most popular model for toll roads, where private investors collect the tariff from end-users directly. The user pay mechanism means that operators of toll roads in China carry the risk of traffic volume and toll revenue. The tariff is regulated by the provincial authority for road transportation and should be adjusted according to the operation cost. However, in most cases there has been a failure to adjust the tariff according to the concession agreement (Zhang 2009). There are also many projects in which the returns of investment are unreasonably high due to unfair concession agreements signed by local governments. As a result, the central government decided in June 2011 to investigate all toll road projects and if necessary regulate unreasonable concession terms. In addition, joint ventures between public and private sectors are rarely observed in toll road projects.

The most critical issues in the railway sector are the lower project financial self-liquidating ratios with unclear subsidiary and profit mechanisms. These make private participation much more hesitant, except in the case of some dedicated lines with independent tariff settlements. Being the most conservative and cautious ministry, the previous Ministry of Railways had avoided the popular reform and the railway sector was the least open to private investors. The same difficulties are observed in urban transit development. Huge investment requirements and the imposition of low fares owing to public welfare concerns greatly reduce the possibility for private investors to obtain a reasonable financial return from the construction and operation of railway and transit projects. Some pilot projects, such as the Beijing Metro Line 4 and Shenzhen Metro Line 4, have encountered financial difficulties and the city governments have had to subsidise them with a premium, even though successful lessons from Hong Kong's Mass Transit Railway Corporation's (MTRC) PPP expertise has been introduced. Under the existing Chinese land law, local governments are not allowed to grant the land around subway stations to investors without competitive tendering. Hence, the integration of land use and transportation development, which should have provided large space to support a higher intensity of urban activities and in turn increase the ridership of the transit railways, could not be realised. More importantly, the integration could also have allowed the investors to gain profit from real estate development so as to compensate partially for construction and operation costs (Wang *et al.* 2012).

In light of the current legal framework, two common business models have been adopted, i.e. Subsidise-Build-Operate-Transfer (SBOT) and Build-Subsidise-Operate-Transfer (BSOT). In a SBOT project, the government is responsible for subsidising the construction, while in a BSOT project, the government would provide a subsidy during the operation period. The key to these two solutions is the improvement of the financial self-liquidating ratio by the government undertaking part of the cost or providing part of the revenue (Wang *et al.* 2012).

Social infrastructure projects

In the water sector, the difference between sewage treatment and water supply is again the financial self-liquidating ratio of projects. The total cost of a sewage treatment plant cannot be covered by the collected waste-water tariff, but the tariff for water supply would normally offer an appreciable return to the investors. Therefore, most water supply plants in China adopt the model of so-called 'plant-pipeline bundle' which means the water companies collect a tariff from end-users. Local governments usually procure the sewage treatment plants by means of BOT or Transfer-Operate-Transfer (TOT), and leave investment and operation of the waste-water pipeline network to the government. Sewage treatment companies generally collect fees from local governments according to their treatment volume, irrespective of how much the government charges the end-users for waste-water treatment (Wang *et al.* 2012).

The revenue of a city gas project consists of both connection and commodity charges, where the connection charge is applicable when a user applies for access to the pipeline network, and the commodity tariff refers to the price of gas throughput on the transmission network. The connection charge was first introduced in Guangdong at the end of 2006 (Zhang 2009) but its legality is controversial. Nonetheless, given the lack of governmental fiscal support for the construction of a pipeline network, it may be reasonable and necessary to charge for connection.

PPP implementation process

Pre-transaction preparation

Several PPP guidance documents have been issued by both the Ministry of Housing and Urban-Rural Development (MoHURD) and various local governments as shown in Table 12.1. In accordance with these guidelines, such as Clause 8 in 'Administrative Measures on the Concession of Municipal Public Utilities' by MoHURD, a public bidding process is required for selection of a private investor. A typical procurement procedure is shown in Figure 12.1 (Wang *et al.* 2012).

The current Tendering and Bidding Law in China is incompatible with the nature of PPP procurement for infrastructure. Procurement of PPP projects differs from that of traditionally procured engineering and construction works as the former in most cases comprises financing and long-term operation in addition to engineering and construction, while the latter is confined to engineering and construction only. Several contradictions therefore

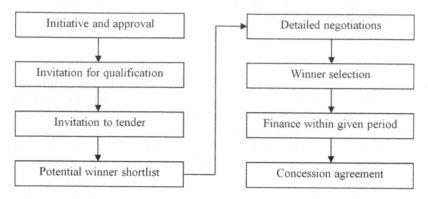

Figure 12.1 Typical bidding procedure for selection of PPP investors

exist between the bidding procedure for PPP projects and the Tendering and Bidding Law, and these contradictions constitute potential obstacles to PPPs. For instance, according to the Tendering and Bidding Law, investors are not allowed to negotiate with the government on key issues such as the tender price. There is also a requirement for at least three potential bidders, which may not be appropriate for large and complex infrastructure sectors such as urban railways. Infrastructure PPP projects in China should adopt more appropriate competitive procedures for the selection of private investors (Zhang 2009). For PPP models with a high degree of standardisation, such as 'Build-Transfer' (BT) and 'Operate & Maintenance' (O&M) schemes, which are adopted in projects with high financial attractiveness, an open competitive bidding would be appropriate. For PPP models with a lower degree of standardisation, such as the joint venture arrangements used in projects with lower financial attractiveness, a direct bid offer or a competitive negotiation would be more suitable (Wang *et al.* 2012).

At present, there are two types of administrative arrangement commonly used for PPP bidding. The first one involves establishing a separate tendering office formed of officials and professionals. Rules and regulations related to the tender process are formulated jointly by relevant government departments, including the Construction Commission, Planning Commission, Fiscal and Auditing Bureau, Industrial and Commercial Administration Bureau, with the Construction Commission representative taking the lead. The second one is to employ a qualified (registered) tendering agency and entrust it with the whole process. Evaluation criteria normally fall into pricing and non-pricing factors as pre-set in the invitation to tenderers. In the Laibin B power plant project (Wang and Ke 2009), a bipartition system of evaluation criteria was used: an electricity tariff (60 per cent weight in the evaluation) and financial proposal, technical proposal and operation, maintenance and transferral (OMT) proposal (40 per cent weight in the evaluation). Of the latter 40 per cent weighting, the financing proposal accounts for 60 per cent of it, while the technical proposal and OMT proposal account for only 20 per cent each (Wang *et al.* 2012).

Legal and regulatory provision for PPPs

The State Development and Reform Commission (SDRC) and MoHURD (formerly the Ministry of Construction) have issued several regulations and related documents on PPPs. Provincial governments have also issued relevant guidance. There is no organisation at national level in China specifically responsible for PPP projects, such as those in the UK, the United States and Australia. Most PPP projects are usually managed at provincial or municipal level, providing they follow sector ministry guidelines or regulations. The following project, Chengdu No. 6 Water Plant, is used as an example to illustrate the approval and registration process (Wang *et al.* 2012). The Chengdu No. 6 Water Plant is a pilot BOT project promoted by the central government and supported by the SDRC and the State Administrative Bureau of Foreign Exchange. The Chengdu Municipal Government (CMG) strongly supported the project and took on the responsibility for obtaining the required approvals from the central government before the bidding stage. A BOT Coordinating Committee, made up of officials from different bureaus and departments from the CMG, was established to assist with coordinating the acquisition of the approvals as presented in Table 12.1. It took the CMG one year to facilitate the invitation-to-tender stage, which is considered very quick compared to the standard Chinese benchmark for this type of project (Chen 2009). The complexity of the approvals systems in China for a PPP project could be significantly simplified and accelerated if a 'single-window' system, comprising project identification, evaluation, submission and approval, was adopted. This was reinforced almost 20 years ago by the United Nations Industrial Development Organization (1996).

Selection of concessionaire

Local governments are proactively promoting PPPs to relax budget constraints and this tends to push the burden of private capital onto the somewhat incomplete Chinese economic system. Nonetheless the lack of expertise, or a centre of excellence on PPP projects, as well as knowledge accumulation and cross-boundary transfer are all hurdles faced by investors in the selection process. The lack of in-depth project feasibility and the necessity to assess the adoption of PPP terms and conditions, advantages and disadvantages, a risk-sharing mechanism, supervision and regulation are widely identified in the Chinese market, in particular the application of VfM (Value for Money) principles and benchmarking tools such as a Public Sector Comparator (PSC). The existing selection procedures are often accused of

Table 12.1 Main approval processes for the Chengdu No. 6 Water Plant Project

No.	Approval	Approval authority
1	**Approvals for establishment and operation of the project company**	
1.1	Approval for project company establishment	Chengdu Foreign Trade and Economic Cooperation
1.2	Pre-registration of project company	State Administration of Industries and Commerce
1.3	Business-opening registration and operating licence	State Administration of Industries and Commerce
1.4	Foreign exchange registration	Chengdu Branch of State Administration of Foreign Exchange (SAFE)
1.5	Taxation registration	Local Taxation Administration
1.6	Fiscal registration	Local Fiscal Administration
1.7	Customs registration	Customs
1.8	Approvals on labour administration	Labour Administration of Chengdu
2	**Approvals for project financing**	
2.1	Approvals of financial agreements	State Development and Reform Commission, SAFE and Ministry of Foreign Trade and Economic Cooperation
2.2	Foreign debt registration	Chengdu Branch of SAFE
2.3	Registration for foreign security	Chengdu Branch of SAFE
2.4	Registration of mortgage raised on water plant facilities	Chengdu Administration of Property, Chengdu Land Use Authority, and Chengdu Administration of Industries and Commerce
2.5	Audition and approval of loan repayment	Chengdu Branch of SAFE

Source: Chen (2009)

being insufficient, with public participation mechanisms, transparency and a serious shortage of information disclosure all cited. With respect to government and corporate interests, the focus appears to be comparatively less on public interest (Wang *et al.* 2012).

Any limitations of the private investor are often overlooked in Chinese PPPs. The fuzzy definition of the public and private sectors, government and corporate roles and responsibilities, as well as the status of government-linked/state-owned enterprises are worth paying attention to. Investors prefer profit-driven projects in developed regions so incentive policies that strike the balance of regional/projects need to be developed. There are also unbalanced market perceptions for state-owned enterprises vs private companies. State-owned enterprises have powerful relationships and strong lines of communication with government, and therefore hold significant bargaining power and have a great advantage in securing competitive bank loans for the longer duration of PPP projects. This makes the involvement of private capital of less interest to the public sector, furthermore the local government is also more inclined to partner with state-owned enterprises. Pricing issues (such as electricity, water supply, transportation, etc.) of the project are still under government control, thus unprejudiced and reasonable pricing and dynamic adjustment mechanisms are the downstream work of the Chinese PPP industry in order to motivate greater private sector capital involvement.

Financial closure

Under the current monopoly of the Chinese domestic financial system, banks prefer higher-margin projects, are reluctant to share risk, and are unwilling to strike limited recourse project financing. Moreover, due to the banks' inexperience in project financing they require the project owners to provide full guarantees or mortgage and expand the scope of recourse. There is an obvious scarcity of financing channels, with limited applications of international syndicated loans, an immature corporate bond market and direct financing difficulties. PPP financing is still highly reliant on bank loans and insurance institutions. With strict project finance warranties, commonly used international project financing guarantees in China do not carry enough legal or policy support, which makes financial closure much more difficult to achieve. Addressing these issues along with developing a clear definition of the roles and responsibilities for administering PPPs are finally on the agenda of the Chinese government to achieve easier financial closure.

Improving PPPs in China

Evaluating and revising PPP policies

Figure 12.2 illustrates the number of PPP projects each year in China according to the World Bank (2015) PFI database and clearly shows the two booming phases of private investment in Chinese infrastructure projects. Most of China's infrastructure market has opened up to private investors, except for some special sectors such as key rail, port and airport projects which are the responsibility of central government. The openness to private investment has been specifically seen in projects such as toll roads and municipal utilities such as water, power and gas, which are overseen by local government. The most important incentive for local government to open up the infrastructure market to private capital is the pressure of inadequate fiscal resources (Chan *et al.* 2009).

China's legislation often demands direct involvement from both the community and market. Both of these issues have led to comprehensive blueprints which can be applied to PPP practice and provide for further

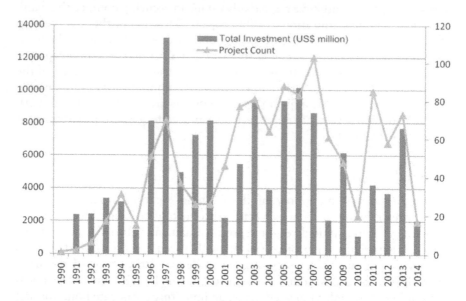

Figure 12.2 Private investment in infrastructure development in China (Source: World Bank, 2015)

development of the PPP law and innovation model. PPP legislation should be established on the basis of the fundamental concept of this procurement method and a thorough understanding of the key elements to suit the specific macro-economic environment of the project. The PPP model is a multi-functional approach to provide public facilities and services following a 'partnership, risk-sharing, benefit-sharing' principle, complying with the concession contract, and thereafter offering a long-term stable collaborative agreement. Chinese PPP projects are anticipating a reasonably standardised and orderly approach without incurring additional costs and risks (Liu 2014).

PPP regulatory system and institutional framework

PPP legal and regulatory systems form the basis of the PPP institutional framework and this establishes a clear, long-term value orientation. PPP policy reflects the objectives of the government's tenure, including any midterm policy schemes for private sector participation in infrastructure and public services. It can be affected by the adjustment of government tenure or the changes in periodical policy emphasis; however, PPP project processes, institutional responsibilities, governance and public financial management all constitute the sustainable implementation of PPP projects in the Chinese public management system.

The Chinese finance sector and the Development and Reform Committee are closely involved in PPP projects, both having incentives to promote and coordinate the management of PPP initiatives. Improving efficiency and revitalising the existing public assets is within the scope of administrative responsibilities for State-Owned Assets Management sectors. Sectoral ministries, including the building and construction authorities (municipal utilities, affordable housing), transportation (roads, ports, airports, railways), energy (thermal power, hydropower, new energy, energy transport and storage), environmental protection (water, solid waste, environmental protection, environmental governance, environmental remediation), irrigation (water resources and hydraulic engineering), as well as education, health, sports, culture and social affairs are the responsibility of the industrial jurisdiction who manage the planning and regulatory responsibilities of the projects. Other than railway and major power plants, the ownership for the majority of the urban infrastructures and public services are still with the city government.

When considering the PPP reforming characteristics and ministry responsibilities, a newly created inter-ministerial PPP administration is the best solution for governance, however, this violates institutional reform. The alternative option is to adjust and optimise the functions of one of the existing ministries outlined above in order to establish a PPP administrative centre, for example, the finance sector or the Development and Reform Commission departments seem an appropriate choice.

Relationship contracting

Relationships between multiple project team members have been interpreted and analysed in some detail over the past few decades (see also Chapters 2, 8 and 9 of this book). In a formal contract, parties usually act in an atomised manner, with more concern for their own personal interests (Williamson 1973), whereas in relationship contracting (RC) approaches, more collaborative working arrangements are involved (Davis and Walker 2008; Love *et al.* 2011; Jefferies *et al.* 2014; Walker and Lloyd-Walker 2015).

International architectural, engineering and construction (AEC) firms have shown significant interest in the Chinese market and its construction industry (Ke 2014). It is well understood when contracting parties adopt RC, as opposed to formal contracting methods, construction projects may achieve favourable outcomes. However, public projects in China usually face more constraints in adopting RC, especially in the centrally planned economy, as close relationships may lead to allegations of corruption. Therefore undertaking comprehensive studies in adopting RC practices in public construction projects in China is becoming increasingly important.

In the centrally planned economy, the government controls the project implementers and participants are hesitant to adopt RC due to the immaturity of the Chinese PPP administrative system. Public sector clients are usually restrained from offering future relationship incentives because of the need to go through competitive bidding processes in the procurement of contractors' and consultants' services. Furthermore, the use of taxpayers' money in public projects makes it politically expedient to award contracts to the lowest bidder. Ke (2014) investigated PPP practice in China and identified the motivating factors which led to RC adoption, such as better cost/time/quality outcomes and competitiveness due to the significant involvement of the private sector which led to increased client satisfaction. However, when compared to their international counterparts, Chinese AEC firms were found to be less innovative, as a result of a low contribution rate of technology in the construction industry (Lu *et al.* 2008).

Barriers to RC implementation are also reviewed and classified, as lack of experience/knowledge in RC, misalignment among project participants, adversarial environment, cost and time to conduct RC, uniqueness of public projects. In the Chinese philosophical and collectivistic culture connection is the way of life. It is believed that building trust and maintaining good relationships with business counterparts will bring many long-term benefits. Therefore the Chinese project counterparties are not fully aware and more reluctant to structure the RC practices (Ke 2014). The Chinese public sector continues to take a short-term philosophy in many aspects of governance and is also conservative in terms of developing RC, therefore resulting in a distinct status quo in RC initiatives due to the lack of interest and incentives in sustainable long-term relationships (Ke *et al.* 2011). Being self-protective of their position in the traditional authorisation hierarchy, Chinese government clients also show an unwillingness to empower their respective representatives. Various hurdles are experienced, especially when upcoming project opportunities are open for tender, and RC development is less appreciated, which demands that highly competent professionals are engaged to overcome the difficulties.

Ke (2014) has recommended that the Chinese government makes use of its dominant position in order to encourage wider adoption of RC by introducing supporting public policies and encouraging firms to work collaboratively. It is suggested that contracting parties share knowledge from organisations that have successfully implemented RC practices. It is further recommended that Chinese public clients appoint highly competent representatives to manage their projects so that they can be empowered to make informed decisions. Instead of initiating a strictly open tender process, China could learn from Australia by adopting Alliance contracting so as to extend relationships from project to project. While the leadership of the new generation of central Chinese government has entered an unprecedented era of anti-corruption and open policy decisions this has meant less of a focus on the likes of RC and other procedures for innovative procurement, such as the early involvement of various project stakeholders, especially contractors. However, given the international advances in this area, such as those discussed in Chapter 7 of this book, it is worthwhile for the Chinese government to explore these issues in order for the construction industry, and broader economy, to continue to develop and evolve.

Brief outlook for future PPPs in China

PPP capacity at central, sub-regional and municipal levels

Investment has significantly driven economic growth in China over the past two decades. Chinese gross fixed capital formation, the measure of investment reported in the national accounts, has grown at an average annual rate of 11 per cent in real terms since 2000 (Wilkins and Zurawski 2014). As discussed previously in this chapter, a significant proportion of investment in China has been directed towards increasing the amount of and improving the quality of infrastructure over the last 20 years. In the 10 years since 2004, investment in infrastructure has accounted for between 25 to 35 per cent of total 'fixed asset investment' in China, and has grown in nominal terms by an average annual rate of 20 per cent. Much of this has been driven by urbanisation, as cities require substantial infrastructure development to support a growing population which has subsequently resulted in improvements to a range of social and economic indicators.

However, China still has a long way to go compared with the levels of development in the majority of developed economies, particularly with urban rail/transit infrastructure which is relatively underdeveloped in many large Chinese cities. The forecast is for infrastructure growth in China to remain strong, due to demand and investment, and this could also be mitigated by reforms proposed by the authorities such as those that will increase the private sector's participation in the allocation, execution and financing of infrastructure development (Wilkins and Zurawski 2014).

Possible applications and development of PPPs

In addition to converging with the standards of developed economies, the ongoing process of rural-urban migration in China will continue to increase the stress on the demand for improvements in national infrastructure. The government's recent urbanisation plan targets an increase in the urbanisation rate to 60 per cent by 2020, a 6 per cent increase from the current level. This implies an additional 100 million people migrating from agricultural and rural areas to the cities, and an estimated $US7 trillion of investment (74 per cent of one year's worth of GDP) spread over the next six years (Wang 2014). To accommodate this larger urban population, substantial investment in municipal infrastructure and transport infrastructure is required. China is currently facing demographic challenges, as a result of its population control policies and the nature of its economic development process. Investment in the social sector (e.g. age care) is likely to receive increasing attention from policymakers (Wilkins and Zurawski 2014).

There are, however, significant risks, mainly related to the dominant role of the government in allocating and financing infrastructure investment, out of which around 85 per cent is undertaken by the state, a much higher percentage than most other countries. In the absence of a strong framework of project prioritisation and transparent cost-benefit analysis, the reliance on government-directed investment creates the potential for misallocation of resources through inefficient investment (Wilkins and Zurawski 2014). The challenges of procuring the infrastructure facilities and services are not unique to China. The sustainable financing of infrastructure investment has recently been on the agenda of the G20, which continues to discuss ways to facilitate the efficient allocation of global savings to long-term infrastructure financing. Actions that can improve project preparation, planning and funding of infrastructure are a key priority in G20 countries (Chong and Poole 2013). China had also invited the World Bank to collaborate and strategise PPP projects and the related social capital planning.

According to the *China Daily* (2013), the economic planner, in the form of the National Development and Reform Commission (NDRC), has invited social capital to invest in a list of 80 projects and this is seen as the most aggressive step in bringing private funds to infrastructure investment. The list of projects covers construction and operation of railways, roads, harbours, wind power stations and oil pipelines. Opening these sectors to social capital will speed up changes in investment and financing regimes and diversify investment sources. Most of these industries were previously dominated by state capital and were off-limits to private and foreign investors. These 80 projects will be trials in attracting social capital, used to map out laws and regulations, and also help with overcapacity in industries like steel, cement and glass. As emphasised by NDRC, government departments should simplify review and approval procedures and step up supervision to create a fair environment for competition.

The implementation of these demonstration projects is an important measure to optimise investment structure to stimulate market activity and promote sustained and healthy economic development. Targets remain as to improvements in policy, innovation management, simplifying procedures that comply with the PPP law, strengthening effective regulation, creating a fair and competitive environment, and to further expand the openness to social capital.

Chapter summary

Infrastructure investment in China has continued to proceed at a swift pace, in turn significantly contributing to economic growth and improved living standards. In addition to driving the longer-term development of the Chinese economy, infrastructure investment has also been used as a countercyclical policy tool to stimulate economic activity. This was most evident during 2008 and 2009, when the government rapidly implemented a stimulus programme targeting infrastructure growth in response to the global financial crisis. Rail investment has increased significantly and is an example of how the construction of long-term infrastructure projects in China has been quickly mobilised to provide at least short-term economic stimulus.

Chinese reform proposals have highlighted the need for private capital in infrastructure investment in order to allow market forces to play a more important role in allocating resources. Proposed reforms include supportive measures for PPPs in order to attract more private capital. Alongside these reforms and following pilot programmes launched in Harbin and Luoyang, the State Council recently announced plans to launch 80 PPP infrastructure projects. Efforts to simplify and decentralise the investment approval process should help to attract more PPPs, but, as international experience of this model has shown, there are a lot of preconditions required in order for these partnerships to be implemented successfully. In addition to broader institutional settings that mitigate private investors' concerns over political risk, a key aspect in China, PPPs require careful risk assessment and appropriate risk sharing.

Past experiences echo those in the present; although the PPP approach has been successfully adopted in various projects across diversified infrastructure sectors, there is still enormous scope for improvement in order to encourage stakeholders to bring their comprehensive skills to the project. This is especially evident in the project assessment and tender evaluation stages, public accountability process, national financial market and the allocation of fair risk-sharing mechanisms in the concession agreement (Wang *et al.* 2012). At the 21st APEC Finance Ministers' Meeting convened in Beijing on the 22 October 2014, finance ministers of the APEC economies highlighted that PPPs have provided an innovative approach to facilitate feasible infrastructure development as opposed to the traditional public-funded models of procurement. Promotion of PPPs is not the ultimate goal in China, but the approach is an important means to infrastructure development in this rapidly growing economy. The successful delivery of Chinese PPP projects can effectively expand the supply of public infrastructure, strengthen the quality and efficiency of public services, improve the use of public funds, share the risks more appropriately and maximise the use of private sector funding and technology.

Reflections

1   The Asian Development Bank (2014) produced *PPPs in Urbanization in the People's Republic of China*, which is an excellent summary document of a workshop held in Beijing in 2013, which they co-hosted with China's Ministry of Finance and their National Development and Reform Commission. Of particular interest are the various PPP case

studies that are very well documented in this publication. It can be accessed at: http://www.adb.org/sites/default/files/publication/42860/public-private-partnerships-urbanization-prc.pdf.

2   This chapter mentions that 80 'demonstration' PPP projects were launched by China's National Development and Reform Commission in 2014. What is the current extent of these projects and are there any early indicators of *success*?

3   What are the major causes of the need for reform in the provision of public infrastructure in China and how has the implementation of PPPs contributed to this evolution?

4   One hundred million more residents will inhabit Chinese cities by the year 2020. This creates and enormous challenge when providing adequate infrastructure for this increased population. The following paper discusses these challenges: Thieriot, H. and Dominguez, C. (2015). *Public-Private Partnerships in China: On 2014 as a Landmark Year, with Past and Future Challenges*. International Institute for Sustainable Development, Winnipeg. The paper can be accessed directly at: http://www.iisd.org/sites/default/files/publications/public-private-partnerships-china.pdf.

5   While not specifically PPP-focused Zou (2014) presents a very useful overview of construction project tendering in China and also proposes strategies for improvement. This is crucial background reading for any construction student, professional or academic who has any interest in the Chinese construction industry. Full citation details are: Zou, Patrick X.W. (2014). 'An overview of China's construction project tendering'. *International Journal of Construction Management*, 7(2): 23–39.

References

Chan, A.P.C, Lam, P.T.I., Chan, D.W.M., Cheung, E. and Ke, Y.J. (2009). 'Drivers for Adopting Public Private Partnerships: Empirical Comparison between China and Hong Kong Special Administrative Region'. *Journal of Construction Engineering and Management*, **135**(11): 115–124.

Chan, A.P.C., Lam, P.T.I., Chan, D.W.M., Cheung, E. and Ke, Y.J. (2010). 'Potential Obstacles to Successful Implementation of Public-Private Partnerships in Beijing and the Hong Kong Special Administrative Region'. *Journal of Management in Engineering*, **26**(1): 30–40.

Chen, C. (2009). 'Can the Pilot BOT Project Provide a Template for Future Projects? A Case Study of the Chengdu No. 6 Water Plant B Project'. *International Journal of Project Management*, **27**(6): 573–83.

China Daily (2013). 'China to Issue Urbanization Layout in 2013'. *China Daily*, 3 June 2013.

Chong, S. and Poole, E. (2014). 'Financing Infrastructure: A Spectrum of Country Approaches'. *Bulletin*, September Quarter 2014: 65–76, Reserve Bank of Australia.

Davis, P.R. and Walker, D.H.T. (2008). 'Case Study: Trust, Commitment and Mutual Goals in Alliances'. In (eds) Walker, D.H.T. and Rowlinson, S. *Procurement Systems: A Cross Industry Project Management Perspective*, London, Taylor & Francis, pp.378–399.

Jefferies, M.C., Brewer, G. and Gajendran, T. (2014). 'Using a Case Study Approach to Identify Critical Success Factors in Alliance Contracting'. *Engineering, Construction and Architectural Management*, 21(5): 465–480.

Ke, Y. (2014). 'Is Public-Private Partnership a Panacea for Infrastructure Development? The Case of Beijing National Stadium'. *International Journal of Construction Management*, 14(2): 90–100.

Ke, Y.J., Wang, S.Q. and Chan, A.P.C. (2009). 'Public-Private Partnerships in China's Infrastructure Development: Lessons Learnt'. *Proceedings, Changing Roles: New Roles, New Challenges*, Noordwijk ann Zee, Netherlands, 5–9 October, pp.177–188.

Ke, Y., Ling, F.Y.Y., Kumaraswamy, M., Wang, S.Q., Zou, P.X.W. and Ning, Y. (2011). 'Are Relational Contracting Principles Applicable to Public Construction Projects?'. *COBRA – Proceedings of RICS Construction and Property Conference*, Salford, UK, September 12–13, pp.1364–1374.

Ke, Y., Jefferies, M., Shrestha, A. and Jin, X. (2014). 'Public Private Partnership in China: Where to from Here'. *Organization, Technology & Management in Construction: An International Journal*, 6: 1156–1162.

Liu, S.J. (2014). 'China PPP Model Development and Implementation'. *Jun He Legal Review*.

Liu, Z. and Yamamoto, H. (2009). 'Public Private Partnerships (PPPs) in China: Present Condition, Trends and Future Challenges'. *Interdisciplinary Information Sciences*, 15(2): 223–230.

Love, P.E.D., Davis, P.R., Chevis, R. and Edwards, D.J. (2011). 'Risk/Reward Compensation Model for Civil Engineering Infrastructure Alliance Projects'. *ASCE Journal of Construction Engineering and Management*, 137(2): 127–136.

Lu, W., Shen, L. and Yam, M. (2008). 'Critical Success Factors for Competitiveness of Contractors: China Study'. *Journal of Construction Engineering and Management*, 134(12): 972–982.

Shen, J.Y., Wang, S.Q. and Qiang, M.S. (2005). 'Political Risks and Sovereign Risks in Chinese BOT/PPP Projects: A Case Study' (in Chinese). *Chinese Businessman: Investment and Finance*, 1: 50–55.

United Nations Industrial Development Organization (1996). *Guidelines for Infrastructure Development through Build-Operate-Transfer (BOT) Projects*, Vienna, UNIDO.

Walker, D.H.T. and Lloyd-Walker, B.M. (2015). *Collaborative Project Procurement Arrangements*, Newtown Square, PA, Project Management Institute.

Wang, B.A. (2014). 'Speech for the Government and Social Capital Collaboration (PPP) Training Program', Government and Social Capital Cooperation Seminar, 17 March.

Wang, S.Q. (2006). 'Lessons Learnt from the PPP Practices in China', Keynote Speech, *Asian Infrastructure Congress*, Hong Kong.

Wang, S.Q. and Ke, Y.J. (2009). 'Laibin B Power Project: the First State-Approved BOT Project in China'. In *Public Private Partnerships in Infrastructure Development: Case Studies from Asia and Europe*, Weimar, Bauhaus-Universität Weimar, pp.101–129.

Wang, S.Q., Ke, Y.J. and Xie, J. (2012). 'Public Private Partnership Implementation in China'. In (eds) Winch, G.M., Onishi, M. and Schmidt, S. *Taking Stock of PPP and PFI Around the World*, Association of Chartered Certified Accountants (ACCA), pp.29–36.

Wilkins, K. and Zurawski, A. (2014). 'Infrastructure Investment in China'. *Bulletin*, June Quarter 2014: 27–35, Reserve Bank of Australia.

Williamson, O.E. (1973). 'Markets and Hierarchies: Some Elementary Considerations'. *American Economic Review*, 63(2): 316–325.

World Bank (2015). *Private Participation in Infrastructure (PPI) Project Database*, http://ppi.worldbank.org/features/Archive/ppi_Archive.aspx?SectionID=3, accessed 12 April 2015.

Zhang, L. (2009). 'Development of Infrastructure Sectors in China: Status Quo and Trends', seminar presentation, The University of Hong Kong, 28 February.

13 A new method for minimizing financial deficit for private investment in infrastructure projects

Sang Hyuk Lee and Myungsik Do

Chapter introduction

During the last three decades, the Korean economy has grown rapidly with a great amount of capital invested in constructing infrastructure facilities. As the infrastructure capacity has grown, maintenance costs for facilities have constantly increased. However, as concerns for the environment and welfare have also increased, the total budget for infrastructure facilities has been continuously reduced. Therefore since the 1990s, the invigoration of private investment has been treated as an important issue by the Korean government as it provides an effective solution to the financial constraints it has faced. The scope of private investment projects is expanding from existing ones in road and transportation facilities to social infrastructure facilities such as schools, hospitals, and residential accommodation (PIMAC, 2010). As in developed countries, private participation in infrastructure provision in Korea is a concept which involves the public and the private sectors working in cooperation and partnership to provide infrastructure and public services.

The infrastructure provides essential social capital and facilities in Korean society. However, the government's revenue and budget has been reduced for infrastructure investment, while at the same time the budget for other sectors such as welfare has increased dramatically. Therefore, the role of private investment offers an alternative solution for expanding infrastructure capacity. As a supplement to government investment, private investment has not only supported the constant provision of economic infrastructure, such as highways, bridges, airports, and railways, but also expanded since the 1990s to investment in social infrastructure like educational, cultural and welfare facilities provided by the government (Do and Kwon, 2009).

The involvement of the private sector has led to more effective, efficient and innovative contracting and procurement methods resulting in project cost savings, improved serviceability and more appropriate risk allocation. In contrast, typical public-sector-led projects involving a traditional single contract typically dictates that responsibilities of design, construction, finance, and operation are all passed onto the private sector with very little equitable risk analysis, therefore making it very unappealing for private sector involvement (Suhr and Kim, 2002).

In order to revitalize private sector investment into infrastructure markets, the Korean government has supported private investment by creating new law: for example, the Minimum Revenue Guarantee (MRG). After establishing the MRG policy, the government also developed a new method for solving the lack of private investment, which is referred to as Standard Cost Support (SCS). These two policies play a key role in stimulating private investment for the development of Korean infrastructures.

This chapter builds on the Asian-inspired theme of innovative procurement discussed in Chapters 10, 11, 12, 14 and 15 by outlining the background and current status of private investment and policy stimulation for infrastructure development in Korea. Policies such as MRG and issues such as overestimated demand forecasting are examined along with the recent solution of SCS, which was initiated to further encourage private investment in public infrastructure.

Current status of private investment in Korea

Infrastructure can be defined as facilities and policies for a base of economic and welfare activity such as energy facilities, highways, railways, airports, seaports, and educational and medical facilities. Among these, the biggest portion in Korea is transportation-related infrastructures. Furthermore, due to urban sprawl and for some political reasons the capacity of these infrastructures has been expanded for 30 years. However, since the Asian Financial Crisis the government has faced budget constraints. Private investment has consequently been considered a new method for implementing infrastructure facilities in Korea (Ahn and Kim, 2006).

Private investment has been continuously increasing since the introduction of the Act on Private Partnerships in Infrastructure (PPI) and has taken a key role in providing infrastructure in a timely manner, making up for lack of public investment. The number of private investments in infrastructures was initially only one project after The Enactment of the Promotion of

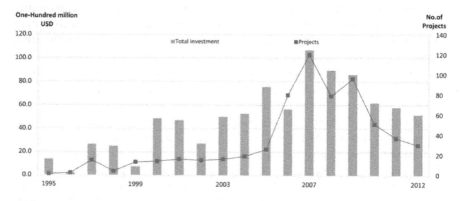

Figure 13.1 Trends in private investment (Source: KDI, 2013)

Private Capital into Social Overhead Capital Investment Act was legislated in 1994. Moreover, after a few years, private investment in financial infrastructure investment slowed down due to the Asian Financial Crisis in 1997. In 1999, the PPI Act was amended in order to stimulate investment in private infrastructure, and the number of projects and amount of investment increased to 120 projects and US$10.69 billion in 2007 (see Figure 13.1). However, private investment has shrunk since 2009 as the benefits of the PPI Act were reduced (Joo *et al.*, 2010).

In 2012, private investment contracts for 487 projects were operating using BTO (Build-Transfer-Operate), BOT (Build-Operate-Transfer), BOO (Build-Own-Operate) and BTL (Build-Transfer-Lease) methods and 118 projects were under construction. Furthermore, 27 projects were awarded as private investment contracts, as shown in Table 13.1.

Private investment in the implementation of infrastructure projects in Korea has employed BTO, BOT, BOO, and BTL methods (Table 13.2). BTO is when a private sector investor builds an infrastructure facility, transfers the facility to the government, and then operates the facility on a contractual basis for a specific period of time. The BTO method has been widely used in the construction of transportation facilities including highways, railways, airports, and seaports (Back, 2007). Environmental facilities dominate the list of the highest number of projects, while costing the least per project, but highway projects account for 28.0 percent (US$37.4 billion) of all investment (KDI, 2013).

The BTL method, introduced in 2005, is where the private sector designs, finances, and builds an infrastructure facility on leased public land, and operates the facility for a specific period of time and then transfers ownership to the government. BTL has been particularly utilized in the building and reconstruction of old educational facilities, for example elementary and middle schools and university dormitories, and environmental facilities.

Table 13.1 Status of private investments in 2012

	No. of Projects	Proportion (%)	Amount of Investment (Billion USD)	Proportion (%)	Amount of Investment per Project (100 Million USD)
Under Operation	487	76.8	53.5	60.3	1.1
Under Construction	118	18.6	25.3	28.6	2.1
Contract Awarded	27	4.3	9.7	10.9	3.6
Operation Complete	2	0.3	0.2	0.2	1.0
Total	**634**	**100.0**	**88.7**	**100.0**	**1.4**

Source: KDI (2013)

Table 13.2 Proportion of private investment by type of investment

	Methods	No. of Projects	Proportion (%)	Amount of Investment (Billion USD)	Proportion (%)	Amount of Investment per Project (100 Million USD)
Managed by Central Government	BTO	60	36.4	42.0	77.8	7.0
	BOT	1	0.6	0.3	0.6	2.9
	BOO	4	2.4	0.8	1.5	2.0
	BTL	100	60.6	10.9	20.1	1.1
	Total	165	100.0	54.0	100.0	3.3
Managed by Local Government with Central Government Support	BTO	31	19.3	14.3	66.5	4.6
	BTL	130	80.7	7.2	33.5	0.6
	Total	161	100.0	21.5	100.0	1.3
Managed by Local Government	BTO	109	35.4	5.4	40.6	0.5
	BOT	2	0.6	0.3	2.2	1.4
	BOO	4	1.3	0.5	3.8	1.2
	BTL	193	62.7	7.1	53.4	0.4
	Total	308	100.0	13.3	100.0	0.4

Source: KDI (2013)

Moreover BTL has contributed to the expansion and improvement of military facilities and also cultural/touring facilities.

Stimulation of private investment

Private investment for developing infrastructure was introduced in Korea after the legislation in 1994 of The Enactment of the Promotion of Private Capital into Social Overhead Capital Investment Act. This act legislated for the construction of a highway from Seoul to Incheon International Airport and was initiated by the Korean government using the BTO method. Incheon International Airport was originally planned as a government-financed project; however, it was turned into a BTO in order to release pressure on the government's budget. This private investment performed an important role in supporting the operation of Incheon International Airport because of the early completion of the Airport Expressway (Kim *et al.*, 2004). Subsequently, in 1998 the act was amended to the Act on Private Partnerships in Infrastructure (PPI). This amended act included MRG policy in order to overcome financing difficulties for constructing and operating infrastructure facilities by private investment after the onset of the 1997 Asian Financial Crisis.

MRG policy was initially an effective method making up for operational deficits in private investment projects in accordance with the agreement between the government and private investors in regard to project demand forecasting. The main contents of MRG policy were guarantee periods and guarantee levels. The guarantee periods were not specified in the policy but were decided in each project by negotiation. For instance, the Incheon International Airport Expressway project has a 20-year guarantee period for MRG. Furthermore the guarantee levels were within 90 percent of demand forecasting for government proposed projects and within 80 percent of demand forecasting for the privately proposed projects (Ko, 2008).

At this early stage in private investments in Korea, MRG policy could play a key role of releasing the financial risk of construction and operation and stimulating private investment for infrastructure. However, MRG policy also brought about negative effects such as the overestimation of traffic demand forecasting. The private sector could receive revenue for deficits of facility operation by traffic demand forecasting each year and overestimation is a huge burden on the government's financial management. The typical example of overestimation of traffic volume is Incheon International Airport Expressway. Table 13.3 shows the gap between estimated traffic volume and observed volume over 12 years.

Consequently, even though MRG policy has advantages for private investment, the benefits of MRG were reduced in government-proposed projects and abolished in privately-proposed projects due to erroneous traffic demand forecasting and increased government budget deficits. Table 13.4 illustrates how total paid revenue of central government expenses increased significantly from 2001 to 2011. The guaranteed rate of return of each project shows MRG percentages in the concession agreement. The numbers in brackets are the ratios of the estimated volume to observed volume. Most of the observed traffic volumes are not more than 60 percent of estimated volume.

Table 13.3 Example of overestimation of traffic volume (Incheon International Airport Expressway)

Year	Estimation (vol./day) (A)	Observation (vol./day) (B)	Difference (A-B)	Ratio (A/B)
2001	110,622	51,939	58,683	2.13
2002	121,496	54,244	67,252	2.24
2003	133,438	55,323	78,115	2.41
2004	146,554	59,780	86,774	2.45
2005	119,026	62,831	56,195	1.89
2006	125,322	65,751	59,571	1.91
2007	131,954	68,711	63,243	1.92
2008	138,930	64,956	73,974	2.14
2009	146,282	62,165	84,117	2.35
2010	93,094	53,490	39,604	1.74
2011	95,920	51,815	44,105	1.85
2012	98,936	52,970	45,966	1.87

Source: Macquarie Korea Infra Fund (2013); Kim *et al.* (2004)

Table 13.4 Annual paid revenue and traffic demand accuracy (in million US$)

Projects	2001	2002	2003	2004	2005	2006	2007	2008	2009	2010	2011	Total	Guaranteed rate (%)
Incheon Airport Expressway	53.7 (47%)	62.1 (45%)	86.6 (42%)	91.7 (41%)	60.0 (53%)	64.5 (52%)	69.3 (52%)	81.8 (47%)	87.9 (43%)	70.5 (58%)	80.7 (54%)	808.9	80
Cheonan-Nonsan Expressway	–		36.7 (47%)	35.1 (52%)	35.4 (55%)	36.7 (56%)	35.4 (58%)	42.9 (55%)	44.0 (57%)	51.1 (58%)	45.2 (61%)	362.6	82
Daegu-Busan Expressway			–	–	–	30.6 (56%)	30.1 (61%)	43.4 (56%)	62.4 (55%)	53.8 (56%)	52.9 (55%)	273.3	77
Seoul Beltway			–	–	–	64.5 (52%)	69.3 (52%)	6.0 (82%)	13.4 (93%)	39.0 (97%)	38.8 (92%)	97.3	90
Incheon Airport Railway			–	–	–	–	99.9 (6%)	145.6 (6%)	153.6 (6%)	110.9 (8%)	123.4 (7%)	633.0	65
Woomyeonsan Tunnel				9.5 (39%)	8.7 (45%)	7.9 (49%)	6.9 (52%)	4.5 (56%)	4.7 (61%)	3.5 (67%)	2.5 (67%)	48.3	79
Sub-total	53.7	62.1	123.3	137	104.2	142.1	244.6	329.9	366.0	328.8	343.7	2235.7	–

Note: all calculation errors appear in source material
Source: Ministry of Strategy and Finance (2010); Macquarie Korea Infra Fund (2011, 2012)

In 2006 MRG policy was eventually abolished in all private investment projects to prevent an inappropriate estimate of demands and to release the financial burden on the government. At this point in time, despite the shrinkage in benefits arising from MRG for infrastructure projects, private investments have gradually been increasing.

New solution for existing problems

More recently, circumstances for private investment have changed due to the long-term issue of low-interest-rate project financing. This has led to the abolishment of MRG policy to relieve the burden on budget management of both central and local governments. Subsequently, a process referred to as Standard Cost Support (SCS) has been applied in order to make up for the operational deficits of private investors. SCS is an agreement to a financial reward when actual operating revenue cannot reach the estimated standard revenue.

In the past, infrastructure facilities have been widely implemented by the central government to promote dramatic economic growth. However, infrastructure projects led by the central government could not be effectively managed due to poor financial management as well as inefficient human and business resources management. Therefore, private investment has emerged as an alternative method to carry out infrastructure projects. Initially, private investment was recognized as a good solution to the financial burden on the government. In contrast, however, the financial solvency of central and local government has worsened considerably due to impractical MRG provisions and higher fund-raising expenses.

In order to address this damaging effect of private investment, investors have tried to refinance using existing refinancing methods. Nevertheless, such methods are not effective enough to resolve the financial burden on the government because the private sector has to refinance with a low interest rate and then any benefits from refinancing are shared in a 50–50 arrangement with the government. If not, an MRG agreement is renegotiated.

SCS can be defined as an arrangement whereby the government repays the total private investment costs with a premium to private investors with whom it also renegotiates operating rights and profit distribution. First the government estimates standard operation costs (SOC) (operational costs plus financial costs) and then it subsidizes any shortfall when actual operational revenue (AOR) does not reach the SOC figure. However, when AOR has the potential to exceed the SOC, the government can redeem certain levels of profit from the private investors. Figure 13.2 compares the MRG and SCS methods for private investment.

The government can anticipate two positive effects through the SCS method. First, the fiscal sustainability of central and local governments can be improved. The government needs to pay back a debt for project costs at an interest rate of 4–5 percent instead of at least 9–15 percent guaranteed

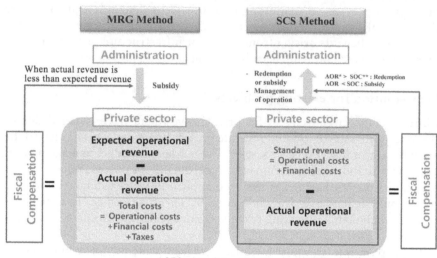

-AOR* : Actual operational revenue -SOC** : Standard operation cost

Figure 13.2 Comparison of MRG and SCS methods

revenue per year through the MRG agreement. For example, the Busan-Geoje Fixed Link (an 8-km bridge-tunnel that connects the South Korean city of Busan to Geoje Island) has been supported by around US$56.4 million per year from an MRG agreement because the actual operational revenue was less than the expected revenue due to overestimated traffic demand. After settlement of the SCS agreement between the government and private sector, the government expects a maximum US$5.0 billion savings during the contract period. Similarly, the Seoul Subway Line 9 project (an extension to the broader subway network) received government subsidies of US$13.4, US$30.3, and US$20.0 million in each year due to the MRG agreement. At present, a refinancing plan is being renegotiated with an SCS after existing private investors withdrew from the Seoul Subway Line 9 project; the government expects a US$3.7 billion saving from the refinancing plan. Table 13.5 shows the status of private investment projects with changing revenue supporting systems.

The Korean government can further improve the financial stability of infrastructure projects and enhance the decision-making process in both business and operational aspects, particularly when deciding upon the service fare of the facility. For instance, the private sector stakeholders of the Seoul Subway Line 9 project tried to increase service fares even though the Seoul Metropolitan City (government) denied them from doing so. This case led to obvious conflict between the private sector's business decision in favor of profits and the government's decision in the public interest. Under the SCS agreement the government can retrieve the primary right of decision

Table 13.5 Status of private investment projects with refinancing plans using SCS (in billion US$)

Projects	Present	Revision	Expected Budget Saving	Current status
Busan-Geoje Fixed Link	MRG	SCS	5.0	Tentative settlement
Yongin Light Rail	MRG	SCS	0.9	Settlement
Daegu Beltway East	MRG	SCS	0.2	
Kwangju 2nd Beltway Section 1	MRG	Repurchase	0.3	Under discussion
Seoul Subway Line 9	MRG	SCS	3.7	
Machang Bridge	MRG	SCS	0.3	
Busan-Gimhae Light Rail	MRG	SCS	0.3	

Source: the authors

from the private sector in an attempt to minimize any operational conflict. The SCS can solve many of the government-"owned" problems in using private investment for infrastructure development through improving fiscal sustainability and enhancing the decision-making process.

Chapter summary

During the last decade infrastructure projects in Korea have faced several critical issues, particularly with decreasing government budgets for infrastructures and increasing government budgets for the likes of welfare. Thus, the government encouraged private investment in infrastructure facilities not only to address constraints on the public purse but also to enhance efficiency and competition through the private sector. Therefore, private sector involvement is recognized as playing an important role in expanding and improving Korean infrastructure.

In order to stimulate private investment, the government enacted laws to support investment such as the MRG policy. However, the MRG policy had some negative effects, such as overestimations in traffic demand forecasting, and put a financial burden on the public sector as the government had to subsidize the private sector according to unfair aspects of the MRG agreement. Eventually, the benefits to the private sector of the MRG policy were shrunk and the policy was abolished in 2009.

Recently, the government realized that an alternative to the aborted MRG policy was needed to support, adequately and fairly, private investment for the continued development of infrastructure. The SCS method appears to be a better option to encourage private sector investment. It is an agreement for financial reward when actual operating revenue cannot reach the estimated standard (target) revenue. This method has advantages for the government such as improving fiscal sustainability and enhancing decision-making rights. Some private-sector-financed infrastructure projects have tried to change compensation method from MRG to SCS, for example, the Busan-Geoje Fixed Link and Seoul Subway Line 9 projects.

Reflections

1   The Republic of Korea has been implementing PPP projects for almost a decade. A two-volume report prepared by the Korea Development Institute (KDI) presents an in-depth assessment of the different components of PPP frameworks and looks at success factors, regulatory reform, risk, and case study projects. It can be accessed at: http://www.adb.org/publications/public-private-partnership-infrastructure-projects-case-studies-republic-korea.

2   The Macquarie Korea Infra Fund (MKIF) has numerous infrastructure assets, including the Incheon Airport Expressway and the Machang Bridge and is listed on both the Korean and London Stock Exchanges. The MKIF website contains significant information on how this type of private investment is becoming more common for the development and operation of infrastructure projects. It can be accessed at: http://www.macquarie.com/mgl/mkif/en.

3   What have been the positive aspects of the Korean government's Standard Cost Support (SCS) policy and is this something that exists or could be successful in other Asian countries?

4   Park *et al.* (2015) present a practical tool for articulating best value criteria during the procurement of public sector building projects in Korea. Data is obtained from sampling 180 stakeholders drawn mainly from a pool of government construction and project management experts in the Republic of Korea. The results of the study identified eight criteria for best value in Korea's construction projects, including that of "operational cost." See Park, J., Ojiako, U., Williams, T., Chipulu, M., and Marshall, A. (2015). Practical Tool for Assessing Best Value at the Procurement Stage of Public Building Projects in Korea. *Journal of Management in Engineering*, **31**(5): 06014005.

5   The Korean construction market was opened up to overseas investment and contractors in 1997. The Construction Bureau is a multi-discipline government organization with the resources and capability to provide a one-stop service in architecture, contract management, engineering, and project management for medium- to large-scale construction projects. Further details can be accessed at: https://www.pps.go.kr/eng/jsp/wedo/contract/overview.eng.

References

Ahn, H. and Kim, M. (2006). *Regional Allocation of Transportation Infrastructure Investment and Development of Local Economy*, Korea Research Institute for Human Settlements (KRIHS), Anyang, Republic of Korea.

Back, S. (2007). A Study on the Evaluation of Value for Money in Private Provided Infrastructure with a Focus on BTO Projects, *Journal of Korean Society of Transportation*, 25(1): 49–59.

Do, M. and Kwon, S. (2009). Current Status and Perspectives of Asset Management in South Korea, *Proceedings of International Workshop on Asset Management Implementation in Asian Countries*, International Islamic University Malaysia, Kuala-Lumpur, pp.17–35.

Joo, J., Ha, H. and Park, D. (2010). Analysis of Factors of IRRs and Spread on Korea's BTO Projects, *Journal of Korean Transportation*, 28(2): 135–150.

Kim, S., Oh, J., and Kang, S. (2004). An Experience of Construction and Operation of the Incheon Airport Exclusive Expressway by Private Capital, *Journal of Korea Society of Civil Engineers*, 24(2): 143–148.

Ko, C. S. (2008). *A Study of Realization of Government Subsides by Abolition of Minimum Revenue Guarantee (MRG) for Private Transportation Infrastructure Investment Project*, Unpublished Master's Thesis, University of Seoul.

Korea Development Institute (KDI) (2013). *Public-Private Partnership in Infrastructure in Korea Brochure*, Asian Development Bank, Manilla.

Macquarie (2011). Macquarie Korea Infra Fund (MKIF), http://www.macquarie.com/mgl/mkif/en (accessed September 8, 2015).

Macquarie (2012). Macquarie Korea Infra Fund (MKIF), http://www.macquarie.com/mgl/mkif/en (accessed September 8, 2015).

Macquarie (2013). Macquarie Korea Infra Fund (MKIF), http://www.macquarie.com/mgl/mkif/en (accessed September 8, 2015).

Ministry of Strategy and Finance (2010). Economic Indicators, http://english.mosf.go.kr (accessed September 8, 2015).

Public and Private Infrastructure Investment Management Center (PIMAC) (2010). *Act on Private Participation in Infrastructure*, http://www.ksp.go.kr/ (accessed September 8, 2015).

Suhr, M. and Kim, M. (2002). A Study on the Effective Operation Model of Joint-Venture Contract with a Focus on the Example of Domestic Construction Industry, *Korea Journal of Construction Engineering and Management*, 3(3): 103–111.

14 Economic infrastructure projects and PPP framework implementation in Indonesia

Policy and processes

Wishnu Bagoes Oka and Pradono

Chapter introduction

As evidenced across various jurisdictions, infrastructure development plays an important role in supporting economic activity and often becomes the highlight of government budget and expenditure plans. The importance of infrastructure development in an urban planning context can be seen in its ability to direct growth and deal with issues such as urban sprawl, while from an economic point of view, infrastructure development can serve as an effective tool to implement government fiscal policies. Under the public-private partnership (PPP) framework, policymakers and industry participants often view infrastructure development from an investment perspective to understand its economic value and the financial returns that can be generated from its operation. In this chapter, discussion focuses on the implementation of a PPP framework in Indonesia, the importance of economic infrastructure projects and their implication, project information aspects and their relation to developing a strong PPP market, as well as the role of the private sector. This chapter provides an excellent addition to the section of the book that focuses on South East Asia by investigating the implementation of PPP frameworks and policy in Indonesia.

PPP investment framework in Indonesia: the case of economic infrastructure

Infrastructure development practice around the world recognizes the distinction between economic and social infrastructure projects. From an economics standpoint, economic infrastructure is defined as "large capital intensive natural monopolies such as highways, other transport facilities, water and sewer lines, and communication" (Gramlich, 1994). The economic infrastructure project category consists of facilities or assets that directly affect economic activities, which include projects from the transportation, utilities, communications and energy sectors. In contrast, the social infrastructure project category consists of schools, hospitals, defense buildings, prisons, and stadiums (Inderst and Stewart, 2014). PPP framework implementation worldwide has seen wide application in both economic and social infrastructure projects. In Europe, for example, the UK's experience with its Private Finance Initiative (PFI) is reflected in its vast project portfolio including various infrastructure sectors such as the National Health System, the Nottingham Express Transit, the M6 Toll Road, and Skynet. The wide PPP investment framework implementation across various infrastructure sectors can also be seen in less advanced PPP markets such as Brazil, which delivered two major projects under its current PPP investment framework—São Paolo Transit Development and the Minas Gerais Prison Complex—and the Philippines, which is currently undertaking a wide array of both social and economic infrastructure projects under their PPP investment framework.

Under the current PPP investment framework introduced in 2009, Indonesia's project portfolio consists of 27 projects in the planning stage (e.g., potential and prospective projects), 20 in procurement, and one

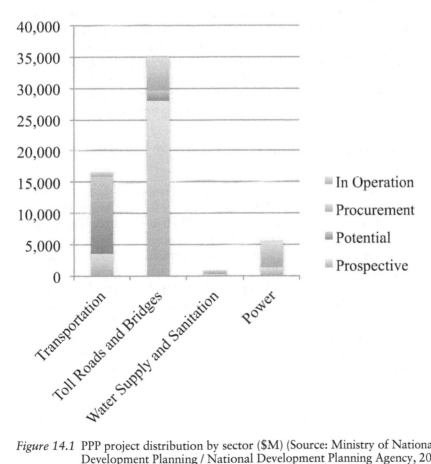

Figure 14.1 PPP project distribution by sector ($M) (Source: Ministry of National Development Planning / National Development Planning Agency, 2013)

currently in operation. As seen in Figure 14.1, Indonesia's PPP investment framework categorizes projects under four different sector types: (1) transportation, (2) toll roads and bridges, (3) water supply and sanitation, and (4) power, all of which fall under the category of economic infrastructure projects. Of all these 48 economic infrastructure projects, the transportation and toll roads and bridges sectors are the two most common infrastructure sectors, contributing 67 percent of the total project count or 89 percent ($51 billion) of the total PPP investment portfolio.

Project categorization under Indonesia's current PPP investment framework recognizes the distinction between the transportation and toll roads and bridges sector, with airport, seaport and railway projects included under the transportation sector category. This sector distinction is the result of the historical large amount of investments made in the toll roads and bridges sector that has established the sector as the most developed infrastructure sector among all categories. Historical infrastructure investment shows that although Indonesia has experienced major changes, including policy regimes and economic trends, there seems to be a consistent trend that favors the development of economic infrastructure projects over social infrastructure projects. This raises an important policy question for the government and by using an investment point of view and applying various different analysis lenses, we will approach the current PPP investment framework in Indonesia and try to identify the trends and causes that have led to the dominance of economic infrastructure projects.

Lens 1: Indonesia's fiscal realities and current PPP investment framework

For the period between 2010 and 2014, the government of Indonesia set up an economic growth target of 7 percent and in order to achieve this target, infrastructure investment in the region of $213B is required. The funding source for this substantial infrastructure investment will include multiple levels of government for an expected total of approximately $139B, leaving a funding gap of $74B to be filled through non-government funding sources (Ministry of National Development Planning/National Development Planning Agency, 2011a). This government fiscal constraint means that any infrastructure expenditure to fill the funding gap needs to be captured as an off-balance-sheet expenditure. Under this approach, asset acquisitions and other long-term liabilities resulting from infrastructure investment will not appear as part of the government's expenditure and fiscal plans. In order for the assets and liabilities to not be recorded under the government's balance sheet, this approach requires that the majority of project responsibility be allocated to parties outside of the government (Pricewaterhouse Coopers, 2005). This may include responsibilities over the project operational phase, including revenue collection responsibilities that form the basis of investment profitability in private sector business cases.

Under Indonesia's current PPP investment framework, responsibility over revenue collection is a common feature that can be found in concession agreements between the public and private sectors. This allows project construction and maintenance costs to be deferred to infrastructure users through fees collected by the private sector. In other words, the revenue component of the project plays an important role in project decision-making and also determines the feasibility of the project. The revenue component of an infrastructure project can more often be found in economic infrastructure projects than in social infrastructure projects. This can be seen in various types of user fees commonly applied on utility and road infrastructure projects worldwide.

It is therefore well understood that economic infrastructure projects have emerged as a favorable investment due to their ability to generate revenues through user fees. While this off-balance-sheet approach may be sufficient to fill the gap in infrastructure funding, it may not be sufficient enough to explain the government's policy preference on economic infrastructure projects under Indonesia's current PPP investment framework.

Lens 2: Investment returns of economic infrastructure projects

In minimizing the funding gap in public infrastructure investment, bonds can be identified as a financing source that are effective to spread the cost of infrastructure an project throughout its life cycle, although its interest on debt implication may result in increased government long-term liabilities. For investors, Indonesian bonds are considered less popular and contribute to only 8 percent of total corporate borrowing, with long-term bonds (e.g., 10-year maturity or longer) equivalent to just 5 percent of Gross Domestic Product (GDP). Furthermore, Indonesia's local currency bond market is considered amongst the lowest in terms of size, compared to other Asia-Pacific countries as seen in Table 14.1.

Although recent efforts have been made by the government, including through the introduction of dollar-denominated Islamic bonds in 2009, the lack of appetite for bond financing puts Indonesia in a more difficult situation compared to other countries given the dire need to address the infrastructure funding gap. Arguably this has, to a certain extent, led to the current government approach to pursue other types of funding sources, such as through direct investment under the PPP investment framework. In attracting direct investment, Indonesia has also been facing fierce competition from other countries. Moreover, PPP investment opportunities in economic infrastructure projects offer an advantage over social infrastructure projects and thus have the ability to attract potential investors. Investments in core economic infrastructure projects can offer interesting opportunities for investors for various reasons. From an investment point of view, economic infrastructure projects have several advantages compared to social infrastructure projects, primarily due to their ability

Table 14.1 Asia-Pacific countries bond market size

Country	Size of Local Currency Bond Market ($B)	Government Bond (%)
Japan	11,663	92
China	3,811	73
Australia	1,918	25
Korea	1,471	39
India	835	73
Malaysia	327	60
Thailand	279	79
Taiwan	250	70
Singapore	241	59
Hong Kong	178	53
Indonesia	*111*	*83*
Philippines	100	87
Vietnam	25	96

Source: Citi Group (2013)

to generate revenue. Revenues generated from the operation of economic infrastructure facilities can offer high and steady returns that could potentially allow Indonesia's PPP investment framework to compete with other types of investment.

Middle-income economies are often characterized by urbanization, motorization, and industrialization, all of which leads to a rapid growing demand for infrastructure. As a result, infrastructure investments in middle-income countries have higher returns and are therefore more attractive to investors (Japan International Cooperation Agency, 2004). As a middle-income country, Indonesia's growing demand for roads and utilities infrastructure is a reflection of the rapid growth in economic infrastructure projects. This means that the revenues generated from operating these economic infrastructure facilities can offer a lucrative profit margin for investors. In addition to that, investments under concession agreement will allow the private sector to collect revenues throughout the concession period, normally around 35 years for toll roads in Indonesia, and offer stable returns. Again, from this point of view, economic infrastructure projects emerged as a more favorable sector under the PPP investment framework compared to social infrastructure projects.

Lens 3: Economic infrastructure and government policy alignment

Indonesia's economy has been growing rapidly over the last decade and has been successful in sustaining GDP growth at an average of 6.2 percent within the last four years (World Bank, 2014). This rapid economic growth has been a result of government's current medium- and longer-term economic plans and has been the central piece of government policy that has so far positioned Indonesia as one of the front-runners among developing countries. Furthermore, the government policy emphasis on economic growth, through the *Masterplan for Acceleration and Expansion of Indonesia Economic Development* unveiled in 2011, sets out a multi-billion-dollar infrastructure investment strategy to achieve an economic growth target of 7–8 percent per year after 2013 (Ministry of National Development Planning / National Development Planning Agency, 2011b).

Economic growth, as suggested by many empirical studies, relies on the development of economic infrastructure projects. Esfahani and Ramirez (2003), for example, identified a strong linkage between infrastructure investment in the power and telecommunication sectors with GDP growth. A report issued by Standard Chartered (2011) shows that roads and bridges infrastructure plays an important role in reducing logistical bottlenecks that hinder economic growth, and the recent development of Suramadu Bridge connecting Madura Island and Surabaya is a perfect example of how economic infrastructure can boost economic growth and manage the spatial distribution of welfare in the medium-term (Suhono, n.d.). In addition, the completion of Indonesia's first private railway between Muara Wahau and Bengalon in 2011 has also demonstrated how economic infrastructures can connect resource-rich regions and help expedite economic growth (Middle East Coal, n.d.). All these exemplify the importance of investments in economic infrastructure projects to support Indonesia's medium-to-long-term economic growth, and given the government's current policy emphasis on economic development, it is well understood that investments in economic infrastructure projects has been more favorable as opposed to investments in social infrastructure projects.

Through these different lenses, we can understand why economic infrastructure projects are an important central piece of the PPP investment framework, and going forward, government investments in the economic infrastructure sector will remain a priority as a result of government policy emphasis on economic development. Furthermore, investment in economic infrastructure projects becomes more important when taking into consideration that Indonesia's current economic infrastructure quality is ranked below that of most of its neighboring countries in the South East Asian region, including Singapore, Malaysia, Brunei Darussalam, and Thailand according to the Global Competitiveness Report (Schwab, 2014). Therefore, to increase the country's competitiveness at a regional and global scale, government investment is required to improve the quality of economic infrastructures.

Economic infrastructure projects and their potential implications

Despite the importance of investing in economic infrastructure projects, the government will also need to be aware of the potential issues that may arise from conflicting interests between the public and private sectors, which tend to be more evident in economic infrastructure projects. For example, government concerns include ensuring that public investments made, such as infrastructure investments, provide the greatest benefit to society. This is reflected through the common use of analytical tools (e.g., cost-benefit analysis, net present value analysis) to take into account social costs and benefits, which serve as an important assessment tool for public sector decision-making (HM Treasury, 2003). In addition, infrastructure development also has urban planning implications, which often reference growth plans and planning legislations developed by various levels of government. These examples can show that government's interest in infrastructure development is mainly driven by the macroeconomic approach that views infrastructure as a public good that serves a greater purpose for the society.

On the other hand, private sector is more concerned with the microeconomics of infrastructure investment, which rely on project financial assessment (European Union, 2008). It is well understood that the very basis of private sector business operations is to gain profit and this can be reflected through the use of financial, risk assessment tools and criteria to assess the feasibility of the infrastructure project. The financial assessment tools often used include financial net present value (NPV) and internal rate of return (IRR) assessments, along with the use of indicators including profitability index and debt-equity ratio. From these two different interests of the public and private sectors can emerge potential implications, including increased government financial responsibility, risk liability and longer-term project liabilities.

Viability gap funding and increased government financial responsibility

One of the implications of the conflicting interests between the public and private sectors is increased government financial responsibility in the form of additional financial incentive. One of the central issues in project financial feasibility assessment that informs the private sector decision-making process is the project rate of return. Infrastructure projects often vary in demand depending on factors such as population growth, geographical location, local economic growth, and the existence of other competing facilities. There are cases where an infrastructure project is expected to generate insufficient returns, although significant economic benefits associated with the project can be delivered to the greater public. As seen in Figure 14.2, projects under the current Indonesian PPP investment framework, for example, have an average financial IRR of 14.44 percent, with projects under the toll roads and bridges sector as the primary contributor with 22 projects and an average IRR of 13.79 percent. As an illustrative comparison, institutional investors such as the Canada Pension Plan Investment Board booked 8.8 percent return in 2013 on its infrastructure asset portfolio, while the

Australian Infrastructure Fund has delivered an annual average return of 10.0 percent (Canada Pension Plan Investment Board, 2013; Australian Infrastructure Fund, 2013). This means that there may be a number of PPP projects under the transportation and toll roads and bridges sectors that might fall below these preferred rates of return, some of which may be projects of significant economic importance. Since government tends to view the importance of infrastructure as a public good, while the private sector values infrastructure projects for their rates of return, conflicting interests may arise and require a proactive approach by government.

In the context of PPP implementation in Indonesia, a public policy agenda that pushes strongly towards the generation of economic growth has allowed the government to step in and assume additional responsibility to improve the financial feasibility of economic infrastructure projects. This can be seen through the PPP investment framework where a certain amount of government contribution can be provided for projects of significant economic importance that have marginal financial feasibility. This government financial contribution, often referred to as "viability gap funding," mostly in the form of construction support, will allow the project to achieve a certain degree of financial feasibility that is sufficient

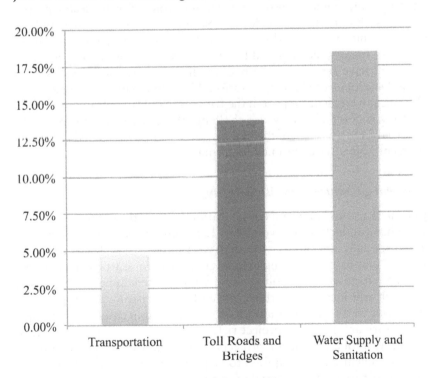

Figure 14.2 Project average financial IRR (Source: Ministry of National Development Planning / National Development Planning Agency, 2011c and 2012)

to attract potential investors. PPP projects in Indonesia often experience marginal financial feasibility and therefore receive viability gap funding incentives to improve their financial feasibility. The Medan-Kualanamu-Tebing Tinggi toll road, for example, received this viability gap funding in December 2012 for part of its construction. Subsequent stages of this project did not receive any funding of this nature as the project was now deemed to be economically feasible.

In total, 20 of all 48 PPP projects under the current investment framework are planning to obtain, are in the process of applying for, or have been approved for government financial support. The breakdown for each sector can be seen in Figure 14.3. Although only one project has been approved for government financial support so far, the number of projects that are planning to request government financial support based on their preliminary business case reflects the potentially large amount of financial responsibility the government may assume in the future.

Under the current PPP investment framework in Indonesia, the Ministry of Finance in coordination with the National Planning Agency will provide the necessary financial contribution towards a PPP project after assessing requests from the government contracting agencies (e.g., ministry, provincial government, municipal government) and project private sector proponents. The PPP Central Unit under the National Planning Agency and the Fiscal Risk Unit under the Ministry of Finance oversee the assessment of viability gap funding as requested by the government contracting agencies, and as such, have access to fiscal budget allocation decisions (Ministry of National Development Planning / National Development Planning Agency, 2013). This government support in the form of viability gap funding may be deemed as a necessity, but may also challenge the underlying principle of the off-balance-sheet approach of the PPP investment framework and will affect government's already-constrained fiscal plan.

Government guarantees and risk liability

Project risk is also an important consideration for both the public and private sectors when conducting an investment. Large investments in infrastructure projects especially are often associated with larger risks. These risks can come in various forms including inaccurate project forecasts, natural disasters, changes in government policy, or even government approval processes, all of which will increase the risk profile of an infrastructure project. The private sector views returns as a tradeoff to risks and in cases where project rates of return are marginal, project risk profile plays an important role in attracting investors. Again, for projects with significant economic importance, government may be required to step in and improve the project risk profile in order to attract potential investors.

In the case of PPP implementation in Indonesia, government intervenes in this condition by assuming a certain amount of risk through government

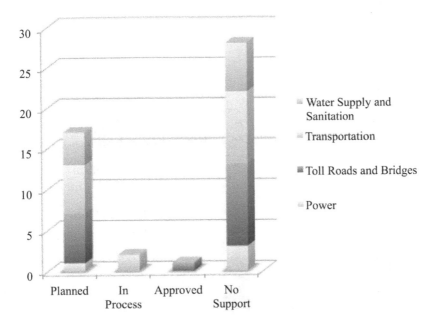

Figure 14.3 PPP projects and government financial support (Source: Ministry of National Development Planning / National Development Planning Agency, 2013)

guarantees which offer protection to an infrastructure project in case a certain risk occurs. Basically, the principle of government guarantees is similar to that of insurance products, in which the project will have to pay a certain risk premium. Under the current PPP investment framework, the government has set up a state-owned enterprise in the form of Indonesia Infrastructure Guarantee Fund (IIGF) to deal with the provision of government guarantees. IIGF, together with the PPP Central Unit of the National Planning Agency, conducts assessments on PPP projects, and with the approval of the Risk Management Unit of the Ministry of Finance determines the government guarantee requirements for these projects prior to the project procurement phase. This means that the winning bidder would then need to establish a separate guarantee agreement with IIGF, while related government contracting agencies would need to develop a recourse agreement with IIGF (Organisation for Economic Co-operation and Development, 2012).

As seen in Figure 14.4, 27 out of the total 48 projects in the 2013 government-issued annual PPP project publication are either planning to request, are in the process of requesting, or have received government guarantees. So far, only one project has been approved for government guarantee and as many as 26 projects are planning to request government guarantees based

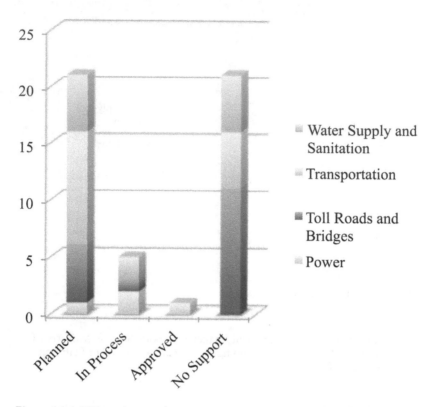

Figure 14.4 PPP projects and government guarantees (Source: Ministry of National
Development Planning / National Development Planning Agency, 2013)

on their respective business cases. This indicates that government risk liability will increase significantly if these projects proceed with government
guarantees. In particular, the transportation sector has a relatively large
proportion of projects that may require government guarantees compared
to other infrastructure sectors. This may suggest that larger risks are more
prevalent in the transportation sector. On the other hand, the roads and
bridges sector has a large number of projects that may not require government guarantees, which may reflect the sector's maturity and experience in
dealing with project risks.

The risks that are insured by IIGF are different on a case-by-case basis,
but may include political risks, project performance risks, and demand risks
(Pricewaterhouse Coopers, 2011). In addition, the government through,
Indonesia Infrastructure Financing (IIF), may also provide financing guaran-
tees through products such as credit enhancements and performance bonds
(Indonesia Infrastructure Finance, 2012). This shows that the government,

through the various products of IIGF and IIF, may have an increased risk liability, and in cases where the risk materializes, the guarantees provided to the project will result in financial implications for the government.

Government long-term liabilities: equity and debt financing

Project financing structure often involves various financial instruments, with debt and equity as the most commonly used in infrastructure projects. These two instruments form an integral part of project financing structure and are often used as indicators for project financial assessment through the common use of debt-equity ratio. The application of debt-equity ratio for infrastructure projects can vary depending on the type of infrastructure project and the geographical location of the project. In Europe for example, the average composition between equity and debt for PPP projects is 80 percent and 14 percent respectively, with the remaining 6 percent funded using bonds. For economic infrastructure projects in particular, the equity portion is much higher compared to that in social infrastructure projects between 2006 and 2009 (European Investment Bank, 2010). Using available information on 14 Indonesian PPP projects from various sectors that are currently in the planning stage, we have discovered that the projected debt-equity ratio is 70:30 across all three sectors (see Figure 14.5).

In certain cases, the procurement process for PPP projects may involve project proponents that have developed strong business cases but lack equity and/or debt components in their financing structure. In cases where the projects procured are of important economic significance, these proponents may progress as winning bidders with equity and/or debt support provided by the government. Under the Indonesian PPP investment framework, equity financing is available from the government through one of its state-owned enterprises, IIF. IIF operates as a commercially oriented non-bank financial intermediary for infrastructure projects, through the development of several products, including long-term financing (e.g., 10-years and beyond) and minority equity stakes (Pricewaterhouse Coopers, 2011).

Viability gap funding, government security, equity and debt financing form important features of PPP concession agreements, and these tools also exist in other developing countries with newer PPP framework implementation. More established and advanced PPP markets such as Canada do not recognize these kinds of government support and guarantees. Furthermore, these established and advanced PPP markets are often supported by strong construction and investment sectors that can alleviate some of the government financial responsibilities and risk liabilities. In this sense, it is important for Indonesia to assess its own progress in developing and establishing a strong PPP market.

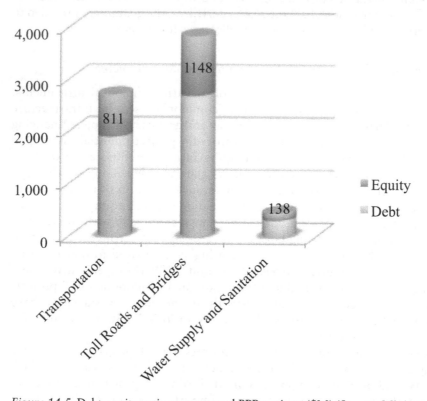

Figure 14.5 Debt-equity ratio on proposed PPP projects ($M) (Source: Ministry of National Development Planning / National Development Planning Agency, 2013)

Developing a strong PPP market

Infrastructure projects are often characterized by their immense risk profile and complexity, and these characteristics are more often found in economic infrastructure projects. This may be due to the fact that economic infrastructure projects are often large in size, stretch over a large area (e.g., in the road, water and electricity sectors), and often have a revenue component attached to them, which adds another layer of project complexity. The dominance of economic infrastructure projects in the current Indonesian PPP investment framework therefore requires a strong and mature PPP market to deal with these project risks and complexities. In other words, sufficient capacity needs to exist in the construction and investment sector, while a strong confidence in the PPP market is also needed to attract potential investors. All of these require strong partnership between the public and private sectors, and government in particular can play an important role in fostering this strong relationship.

Historically, Indonesia's construction sector has been an important contributor to the GDP, with 2013 GDP contribution at 9.99 percent and averaging 8.79 percent over the last 10 years (Statistics Indonesia, 2004–2013). In comparison, Lowe (2003) found that the construction sector's GDP contribution varies between developed and developing countries, at around 7–10 percent and 3–6 percent respectively (Lowe, 2003). Under this categorization, Indonesia's construction sector size is comparable to that of developed countries relative to the whole economy and may reflect sufficiency in terms of the capacity of its construction sector.

Developing confidence in the PPP market on the other hand remains a challenge for the Indonesian government as private sector interest saw no improvement since the implementation of the current PPP investment framework in 2009. Over its five years of implementation, the current PPP investment framework only managed to allow seven projects to progress towards an advanced stage of procurement. One project that is currently in operation is the 12.7 km Nusa Dua-Tanjung Benoa Toll Road that provides important access to Bali's international airport. Despite being delivered ahead of schedule, the outcome of this project's procurement process sent a strong message on the lack of private sector interest. This can be seen in the winning consortium, which consists of seven state-owned enterprises that will deliver the project under a Build-Operate-Transfer concession.

Since PPP publications ensure transparency, equal access to project information, and help attract a wide range of attention, including from potential investors (Farqurharson, *et al.*, 2011), a great deal of interest remains to assess trends in project selection, schedule, and cost consistency in Indonesia's annual PPP publications to progress towards a stronger PPP market.

Trends in PPP project selection

Project selection is an integral component of PPP processes and plays an important role in filtering projects that may be of interest to potential investors. Governments around the world generally have their own periodical publication to inform the general public on projects that will be offered through their PPP investment framework. The UK for example, acknowledge the importance of a clear and consistent PPP project pipeline to provide the business sector in general, and potential investors in particular, with a sense of confidence (CBI, 2011). This will provide potential investors with options to choose their preferred investments that may suit their interest from a pool of potential projects. As one of the world's advanced and established PPP markets, the UK has been successful not only in offering different varieties of projects (e.g., by type, size, PPP model) in their annual PPP project list, but also in consistent project selection.

Table 14.2 PPP project consistency

Sector	Consistent	Inconsistent
Toll Roads and Bridges	16	20
Water Supply and Sanitation	11	45
Transportation	8	39
Power	4	12
Total	39	116

Source: Ministry of National Development Planning / National Development Planning Agency (2009–2013)

Through its annual PPP publication between 2009 and 2013, Indonesia has offered various kinds of potential PPP projects. Some that were included were eliminated or replaced in subsequent publications. The fact that these year-on-year project list inconsistencies occur frequently can be a worrying sign for potential investors. Through five annual PPP project list publications, only 39 projects offered to potential investors have been selected consistently year-on-year and in comparison, 116 projects were eliminated from the published annual PPP project lists. This inconsistency, portrayed in Table 14.2, is especially more prevalent in water supply and sanitation projects as well as transportation projects. For example, there are several railway projects under the transportation sector that were introduced in various PPP project list publications, but were eliminated later in subsequent publications; these include projects of significant size and importance such as Kuala Kurun-Palangkaraya-Pulau Pisang, Puruk Cahu-Kuala Pembuang, and Tumbang Samba-Nanga Bulik railway projects, with an estimated project value of roughly $2B each.

Toll road project selection in five annual PPP publications, for example, indicates that increased project size aligns with improved project selection consistency, with toll road projects larger than $600M showing the most significant consistency, as shown in Figure 14.6. This finding could provide an input for the future selection process of potential PPP toll road projects. On the other hand, certain sectors such as water supply and sanitation projects do not indicate any strong publication consistency trend relative to project size, as shown in Figure 14.7. This improved consistency on larger potential toll road projects can be better explained from the perspective of potential investors.

When looking for investment opportunities, there are criteria that potential investors may have in mind; these could include project type, rate of return, risk profile, and project size. Certain investors might look for projects within a smaller size range due to their financial capabilities, while larger institutional investors such as hedge funds and pension funds often look for projects that require larger capital investments that will allow a constant revenue stream over a longer investment horizon to align with the cash requirement that suits their business operations (Sharma, 2013).

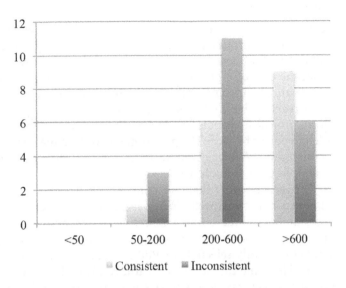

Figure 14.6 Consistency in toll road projects by value ($M) (Source: Ministry of National Development Planning / National Development Planning Agency, 2009–2013)

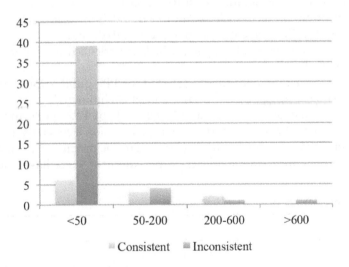

Figure 14.7 Consistency in water supply and sanitation projects by value ($M) (Source: Ministry of National Development Planning / National Development Planning Agency, 2009–2013)

Project size is therefore crucial for potential investors, and the consistent selection of larger projects and elimination of smaller projects in a government-published annual PPP project list may indicate a government policy agenda that targets larger institutional investors.

Project cost variance

As mentioned previously, project size is an important criteria for potential investors, and consistency in project size throughout the planning process can provide assurance and confidence for potential investors. It is, however, well understood that the cost of a project often varies during the planning phase as more information and business case findings become available, and once the project reaches the final procurement decision-making phase or commercial close, project financials are locked through a binding contract or concession agreement between the government contracting agencies and the winning consortium. Average cost variance for each project varies according to its sector, as seen in Table 14.3.

The distinction between projects that are consistently selected for PPP publications and projects that are cancelled/eliminated is highlighted to provide a better understanding of the cost variances between these two categories. The cost for energy infrastructure projects has a bigger variance for projects that are consistently selected for PPP publication compared to the ones that are not. The same case applies for toll roads and bridges and transportation projects, which saw a lower average cost variance for projects that are cancelled/eliminated in government annual PPP publications. In the case of transportation sector projects, the large average cost variance amount is attributed to the significant outliers that largely affect the sector average. This includes projects such as the Integrated Gedebage Terminal, which experienced a 484 percent cost increase in 2010, and the development of New Bali International Airport, which experienced a significant cost increase in 2011.

Overall, projects that are consistently selected for PPP publications have bigger cost variance than projects that are cancelled/eliminated. This may be due to the fact that projects consistently selected for PPP publications experience more due diligence as part of their progression between stages. For example, projects that progress from planning to the procurement stage or from early procurement to the advanced procurement stage have gone

Table 14.3 Project average cost variance by sector

Summary		
Sector	Average Cost Variance	
	Consistent Projects	Cancelled Projects
Toll Roads and Bridges	113%	105%
Water Supply and Sanitation	98%	120%
Transportation	420%	92%
Power	106%	100%

Source: Ministry of National Development Planning / National Development Planning Agency (2009–2013)

through more due diligence processes and business case studies than projects that do not progress between stages. As part of the due diligence processes and business case studies, project size and features are often refined, which generally lead to cost adjustments. These project cost variances, to a certain degree, can therefore be acceptable; however, improvements toward a better project preliminary forecast may be necessary to minimize cost variances and improve investors' confidence to develop a stronger PPP market.

Project schedule

As part of government best practices, competitive and open procurement policy provides not only transparency, but also value for money for goods and services provided by the government. Especially for developing countries which tend to be associated with political instability and uncertainty, open and competitive procurement provides a response to government public scrutiny as well as sending a strong message to the general public, including investors, on business case-based government decision-making processes. The procurement processes for PPP infrastructure projects are nonetheless similar and will also serve as communication opportunities between the public and private sector to scope the private sector's interest and construction sector's capacity. This shows that a reliable procurement process can play an important role in developing a stronger PPP market.

Procurement processes for PPP projects around the world are generally based on the same model and Indonesia, like most countries, implements a staged procurement process starting from Request for Qualification (RFQ) and Request for Proposal (RFP), followed by commercial close once the preferred bidder has been determined. Countries with a mature PPP market often adopt a strict procurement timeline that enables potential PPP projects to progress through various planning and procurement stages consistently. The UK government, for example, announced their commitment to implement a more efficient procurement timeline for their PF2 model by making project funding approval contingent on the project being procured in less than 18 months (HM Treasury, 2012). Australian PPP projects have an average procurement time of around 18 months, while Spain implements a more aggressive procurement standard of five to eight months (KPMG, 2010). Although the standard or average procurement duration serves only as a performance benchmark due to its varying implementation on a case-by-case basis (e.g., depending on project size, type, and complexity), this provides assurance for potential investors and allows them to plan their business operations in advance. In this context, a procurement standard provides a sense of comfort and reliability for potential investors to invest with confidence.

The current PPP investment framework for Indonesia does not provide a standard planning and procurement duration, as it differs on a project-by-project basis. The schedule outlined for each proposed project indicates

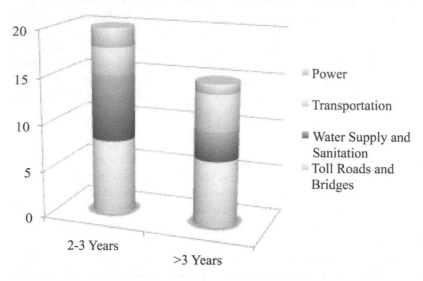

Figure 14.8 Projects that have stayed in the planning stage for more than one
year (Source: Ministry of National Development Planning / National
Development Planning Agency, 2009–2013)

government commitment to implement the project procurement schedule
in around one-and-a-half to two years, starting from project business case
finalization, the procurement process and up to commercial closure. In
practice, however, the planning and procurement schedule for Indonesian
PPP projects often varies from the agreed timelines. Using the government-
published annual PPP project list, we have identified potential PPP projects
that have stayed in the planning stage for more than a year, aggregated at a
sector level. Figure 14.8 reveals that a large number of projects have stayed
in the planning stage for more than one year and some even staying as long
as five years in the planning stage. The project breakdown on a sector basis
shows us that most of the projects that have stayed or have been staying
over a long period in the planning stage comes from the toll road sector.
Understandably, planning for toll road projects require more due diligence
due to its large size and coverage over a large amount of area, which may
require inter-jurisdictional studies and approvals. This may be different
with other transportation projects such as airports and seaports, in which
the asset construction will be conducted at a specific limited site, although
its operations may encompass different regions.

With regard to schedule implementation, the PPP procurement process
implementation in Indonesia does not differ much compared to project
planning implementation. Figure 14.9 shows that toll road projects again
are the largest contributor to this list, with a large number of projects that
have stayed in the procurement stage for around two to three years. In other

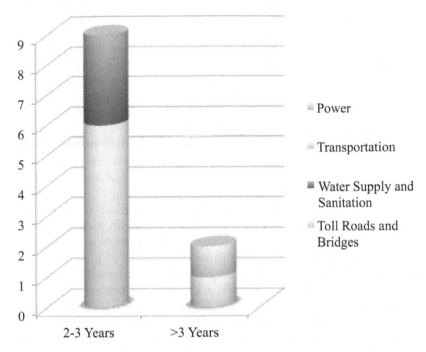

Figure 14.9 Projects that have stayed in the procurement stage for more than one year (Source: Ministry of National Development Planning / National Development Planning Agency, 2009–2013)

words, both project planning and the procurement process combined can take up to roughly four to six years, or even more. These schedule delays can indicate either the government does not have sufficient expertise and experience to progress projects forward within the projected timeline, or the lack of interest from potential investors has stopped potential projects from progressing. Furthermore, inconsistencies in PPP project selection, variances in project cost development, and project schedule delays show the importance of conducting thorough and accurate project planning. To ensure improvement towards better PPP project planning can be achieved, government effort is required, especially when taking into account that in some projects the private sector plays a big role in the project planning process.

Increased private sector role and project performance

Under the generic PPP framework applied worldwide, projects are identified and initiated by government contracting agencies (e.g., central, provincial and municipal governments) which take full responsibility in conducting the project planning by developing studies to determine project financial and economic viability. This then leaves the responsibility of the remaining project stages, such as detailed design, project construction, financing, operating and/or the maintenance of the infrastructure facility, depending on the PPP model, in the hands of the procurement winning bidder. Transportation projects under the UK's PF2 model, for example, require the government local authority to develop business cases for potential PPP projects and seek approval from the Department for Transport (UK Government, 2014).

In Indonesia's current PPP framework, however, the private sector is allowed to identify projects' potential for PPP delivery and initiate an unsolicited proposal. Through the unsolicited PPP proposal stream, the private sector project proponent will be involved early in the project planning stage by developing their own proposal, including project feasibility study and financing plan. After government approval has been sought and the project advances to the procurement stage, the initial private sector project proponent may enter a competitive procurement process with certain privileges, including added value to a maximum 10 percent of the proponent's tender during the competitive procurement process, or compensation by the government through the purchase of the proposal's intellectual property rights (Hodges and Dellacha, 2007). This unsolicited PPP proposal stream is a departure from the generic PPP framework, which allows an even bigger role for the private sector to improve public-private collaboration, but may also have important policy implications that require government attention.

Studies have suggested that inaccurate forecasts used by the private sector have resulted in poor project performance. Demand forecast in particular is directly associated with project revenue and is an important component of economic infrastructure PPP projects. JP Morgan's assessment on 14 toll road projects in the United States showed that only one project exceeded its traffic forecast and half of the projects demonstrated significant inaccurate forecasts (JP Morgan, 1997). A study by Flyvbjerg (2008) analyzed optimism bias in cost forecasting for rail, bridges, and tunnels and road projects and found average inaccuracies of 44.7 percent, 33.8 percent, and 20.4 percent respectively. These could indicate that economic infrastructure may be more prone to project inaccuracies, and project planning conducted by the private sector may very well increase the chances of these inaccuracies occurring. This can be seen in many cases, including the Croydon Tramlink transit project in the UK, which experienced ridership shortfall as a result of competition against other modes of transportation.

The use of overly optimistic ridership/revenue assumptions are more obvious in the design, construction, operation, and maintenance phases of PPP projects (Taylor, 2005) and can become an important issue because all of these project phases form the basis of generic Build-Operate-Transfer concession agreements used in the Indonesian PPP investment framework. Furthermore, by not taking into account government planning of other competing transportation modes, the private sector has failed to integrate a macro transportation planning perspective in order to develop an accurate ridership forecast. In this context, government objectivity and a prudential approach when viewing risks related to project costs and financial returns can be viewed as playing important roles to safeguard the project planning process from overly optimistic forecasts. The current PPP investment framework in Indonesia had only two unsolicited projects as of 2013. These two projects were currently in planning at that time and came from the toll roads and bridges and water supply and sanitation sectors. Using cost variance as an indicator, we can see that one of the two projects, Southern Bali Water Supply, has been performing poorly in the planning stage with a cost variance of -23 percent between 2012 and 2013. In addition, several unsolicited projects were eliminated from the annual PPP project publication, such as Batang Toru Hydroelectric Power Plant and Jambi Coal Power Plant. The lengthy negotiation processes have been identified as the main cause as to why these projects never progressed to advanced procurement stages (Hodges and Dellacha, 2007). Going forward, it remains to be seen whether more unsolicited projects can progress through the procurement phase and perform better in terms of project planning accuracy and consistency. In this case, government's role in safeguarding the project planning process through planning approvals should ensure minimum room for overly optimistic planning assumptions.

Chapter summary

One of the key aspects of Indonesia's current PPP investment framework implementation is the dominance of economic infrastructure projects. Although viewed as a lesser government priority, social infrastructure projects can also play an important role in the society by providing vital services such as health and education that will allow better quality of living. Hence, awareness of the importance of social infrastructure projects should be escalated into proportion, especially given the fact that the condition of these social infrastructure assets is deteriorating. The challenge is now evident for the Indonesian government to develop further its PPP investment framework to enable investment in social infrastructure projects. This includes exploring other financing models that may be applicable for social infrastructure projects such as social impact bonds. This challenge becomes more apparent given that social infrastructures are often associated with sustainable longer-term economic growth. For example, education infrastructure facilities can provide quality human resources that may attract investments, as well as improvements to human development index indicators often associated with sustainable and longer-term economic growth.

Improvements on PPP information processes are also important in both the short-term and long-term. In the short-term, a reliable project publication can improve potential investor's confidence to help project implementation progress in between stages, while in the longer-term, the development towards an established and advanced PPP market will allow the government to move away from assuming financial responsibility and risk liability arising from PPP projects. To achieve this, government needs to ensure that sufficient mechanisms are in place to implement better PPP policies and processes.

Reflections

1   Previous PPPs in this part of Asia have been predominantly economic projects. What social infrastructure models are applicable to the infrastructure development in Indonesia?
2   The Indonesian government's intention is to become one of the 10 major world economies by 2025. An important part of this is the "Masterplan for Acceleration and Expansion of Indonesia's Economic Development." This and other key issues can be accessed at: http://www.indonesia-investments.com/projects/public-private-partnerships/item70.
3   In recent years, the Indonesian capital, Jakarta, has been plagued by floods, therefore a joint venture between the government and the Netherlands has resulted in a 32-km giant sea wall being developed that will house an airport, harbor, toll road, and residential and industrial areas on a 4,000-hectare site. What is an appropriate procurement method for this mega-project?
4   The UK Foreign and Commonwealth Office commissioned a study by Strategic Asia in 2012 to explore opportunities for PPPs in Indonesia. The report provides an excellent analysis of the "Master Plan for the Acceleration and Expansion of Indonesia's Economic Development (MP3EI)" and can be accessed at: http://www.strategic-asia.com/pdf/PPP%20%28Public-Private%20Partnerships%29%20in%20Indonesia%20Paper.pdf.
5   In 2013 the Indonesian government produced an infrastructure projects plan based on PPPs. The report is very much written as a guide as it includes policy frameworks, analyzes current PPP projects, and identifies potential projects for this form of procurement. It can be accessed at: http://www3.bkpm.go.id/img/file/PPP%20BOOK%202013-compact.pdf.

References

Australian Infrastructure Fund. (2013). *ASX Announcement*. Hastings Funds Management.
Canada Pension Plan Investment Board. (2013). *Annual Report 2013: People, Purpose, Performance*. Canada Pension Plan Investment Board.
CBI. (2011, November 15). *CBI Comments on Government's Review of PFI*. Retrieved August 24, 2014, from CBI website: http://www.cbi.org.uk/media-centre/press-releases/2011/11/cbi-comments-on-governments-review-of-pfi/.
Citi Group. (2013). *Market Infrastructure Developments Impacting Asian Bond Markets*. Citi Group, Securities and Fund Services.
Esfahani, H. S., and Ramirez, M. T. (2003). Institutions, Infrastructure, and Economic Growth. *Journal of Development Economics* (70), 443–477.

European Investment Bank. (2010). Infrastructure Finance in Europe: Composition, Evolution and Crisis Impact. *EIB Papers*, 15 (1), 16–39.

European Union. (2008). *Guide to Cost Benefit Analysis of Investment Project*. European Union, Directorate General Regional Policy.

Farqurharson, E., de Mastle, C., Yescombe, E., and Encinas, J. (2011). *How to Engage with the Private Sector in Public-Private Partnerships in Emerging Markets*. World Bank, Public-Private Infrastructure Advisory Facility.

Flyvbjerg, B. (2008). Curbing Optimism Bias and Strategic Misrepresentation in Planning: Reference Class Forecasting in Practice. *European Planning Studies*, 16 (1), 3–21.

Gramlich, E. (1994). Infrastructure Investment: A Review Essay. *Journal of Economic Literature*, 32 (September), 1176–1196.

HM Treasury. (2003). *The Green Book: Appraisal and Evaluation in Central Government*. HM Treasury.

HM Treasury. (2012). *A New Approach to Public Private Partnerships*. HM Treasury. The National Archives.

Hodges, J., and Dellacha, G. (2007). *Unsolicited Infrastructure Proposals: How Some Countries Introduce Competition and Transparency*. World Bank, Public-Private Infrastructure Advisory Facility.

Inderst, G., and Stewart, F. (2014). *Institutional Investment in Infrastructure in Developing Countries*. World Bank, Financial and Private Sector Development.

Indonesia Infrastructure Finance. (2012). *Annual Report 2012: Fulfilling Aspirations*. Indonesia Infrastructure Finance.

Japan International Cooperation Agency. (2004). *A New Dimension of Infrastructure: Realizing People's Potential*. Japan International Cooperation Agency.

JP Morgan. (1997). Examining Toll Road Feasibility Studies. *Municipal Finance Journal*, 18 (1).

KPMG. (2010). *PPP Procurement: Review of Barriers to Competition and Efficiency in the Procurement of PPP Projects*. KPMG Corporate Finance.

Lowe, J. (2003). *Construction Economics* (2nd Revised Edition). Wiley-Blackwell.

Middle East Coal. (n.d.). *Fact Sheet*. Retrieved August 11, 2014, from Middle East Coal website: http://www.mec-holdings.com/download/mec_factsheet.pdf.

Ministry of National Development Planning / National Development Planning Agency. (2011a, January 13). *Directorate for Public Governance and Territorial Development*. Retrieved August 4, 2014, from Organisation for Economic Co-operation and Development website: http://www.oecd.org/gov/regulatory-policy/47377678.pdf.

Ministry of National Development Planning / National Development Planning Agency. (2011b). *Masterplan Percepatan dan Perluasan Pembangunan Ekonomi Indonesia 2011–2025*. Ministry of National Development Planning / National Development Planning Agency.

Ministry of National Development Planning / National Development Planning Agency. (2011c). *Public Private Partnerships: Infrastructure Projects Plan in Indonesia*. Ministry of National Development Planning / National Development Planning Agency.

Ministry of National Development Planning / National Development Planning Agency. (2012). *Public Private Partnerships: Infrastructure Projects Plan in Indonesia*. Ministry of National Development Planning / National Development Planning Agency.

Ministry of National Development Planning / National Development Planning Agency. (2013). *Public Private Partnerships: Infrastructure Projects Plan in Indonesia.* Republic of Indonesia, Ministry of National Development Planning / National Development Planning Agency.

Organisation for Economic Co-operation and Development. (2012). *OECD Reviews of Regulatory Reform: Indonesia.* Organisation for Economic Co-operation and Development.

Pricewaterhouse Coopers. (2005). *Delivering the PPP Promise: A Review of PPP Issues and Activity.* Pricewaterhouse Coopers, Advisory Services in Infrastructure, Government and Utilities.

Pricewaterhouse Coopers. (2011). *Electricity in Indonesia: Investment and Taxation Guide.* Pricewaterhouse Coopers.

Schwab, K. (2014). *The Global Competitiveness Report.* World Economic Forum.

Sharma, R. (2013). *The Potential of Private Institutional Investors for the Financing of Transport Infrastructure.* OECD, International Transport Forum.

Standard Chartered. (2011). *Indonesia: Infrastructure Bottlenecks.* Standard Chartered, Global Research.

Statistics Indonesia. (2004–2013). *Gross Domestic Product at Current Market Prices by Industrial Origin 2004–2013.* Retrieved August 22, 2014, from Statistics Indonesia website: http://www.bps.go.id/eng/tab_sub/view.php?kat=2&tabel=1&daftar=1&id_subyek=11&notab=1.

Suhono, A. (n.d.). *Integrated Regional Development Planning: Indonesia's Experience.* Retrieved August 11, 2014, from United Nations Centre for Regional Development website: http://www.uncrd.or.jp/content/documents/1073IRDP%20EGM%202013%20-%20P14_Indonesia.pdf.

Taylor, J. H. (2005, February). *UK House of Commons.* Retrieved August 30, 2014, from Select Committee on Transport: Memorandum by Serco Integrated Transport (LR 68): http://www.publications.parliament.uk/pa/cm200405/cmselect/cmtran/378/378we50.htm.

UK Government. (2014, July 7). *PFI/PF2 Tracker: Department for Transport.* Retrieved August 24, 2014, from UK Government website: https://www.gov.uk/government/publications/pfipf2-tracker-department-for-transport.

World Bank. (2014). *Data: Indonesia.* Retrieved August 12, 2014, from the World Bank website: http://data.worldbank.org/country/indonesia.

15 Relationally Integrated Value Networks (RIVANS) for Public-Private Partnerships (PPPs)

Mohan Kumaraswamy, Jacky Chung, Weiwu Zou and Kelwin Wong

Chapter introduction

This chapter makes a strong case for prioritising relationship management in PPPs and builds upon the discussion in Chapter 8, which focused on adding value through relationships, and also Chapter 9, which identified several innovative examples of relationship management in a PPP case study. The perennial quest for improvements in the procurement and provision of public infrastructure and services has traversed many partnership and collaborative models and modalities. This chapter draws together two streams of research, the first being on Relationally Integrated Value Networks (RIVANS), which proposes deeper 'integration' and hence super-charged teamworking, through a focus on clearly identified and hence well-defined common 'best value' objectives of the whole network of stakeholders (Kumaraswamy *et al.*, 2003; CICID, 2008) in built infrastructure projects in general. The second stream of research drawn upon here is that on the imperative for improved relationships and teamworking in Public-Private Partnership (PPP) projects given their greater complexities, many stakeholders and uncertainties, as well as longer timelines (Kumaraswamy *et al.*, 2007; Zou and Kumaraswamy, 2009).

The synergies between the above two research streams have been previously demonstrated by Kumaraswamy *et al.* (2008) in terms of developing integrated 'value networks' to target better performance in PPPs through the envisaged much better relationships and common value focus. However, now that both research initiatives have drilled down even further ([i] Kumaraswamy *et al.*, 2010; Anvuur *et al.*, 2011; and [ii] Zou, 2012; Zou *et al.*, 2014), it is now timely to revisit the potential benefits and bonuses in applying RIVANS to PPPs. Furthermore, while the initial research initiatives were launched from Hong Kong, the relational approaches envisaged in the RIVANS frameworks have been tested elsewhere too, e.g. in mainland China (Ling *et al.*, 2014a) and Singapore (Ling *et al.*, 2014b). Moreover, the latter reports on an extension of RIVANS to the 'total asset management' of built assets, i.e. through their full life-cycle. Indeed, this Singapore exercise was one of four parallel studies conducted (the others being done in the UK,

Sri Lanka and Hong Kong), with the parent study having been launched in Hong Kong in 2011 (CICID, 2012; Wong *et al.*, 2014). This set of studies was specifically on RIVANS for TAM (total asset management). This extension of RIVANS to the operational phase brings it even closer to PPP scenarios, hence their juxtaposition within this chapter.

The next section in this chapter applies the enhanced value focus of RIVANS to PPPs, followed by a section examining the importance of integrative relationships in PPPs. This sets the stage for the section examining the potential for enhancing PPP performance through RIVANS-type strategies and approaches. The two subsequent sections, while drawing on the study on improving relationships in PPPs, move from probing critical success factors in their relationship management to developing specific RIVANS-enhanced relationship management frameworks for PPPs. The next two sections highlight the added value in proactively enlisting asset management team representatives upfront, as co-creators of value, and in mobilising their knowledge at the planning and design stages to better address end-user needs. The last two sections draw the foregoing streams together in integrating relationship management with value focus frameworks to provide a solid base from which to launch RIVANS for PPPs.

Importance of value delivery in PPPs

The concept of Public-Private Partnership (PPP) formalised collaboration between the public and private sectors in providing public services. This can even be traced to the development of the postal station network in the Roman Empire two thousand years ago. Postal stations were constructed and managed by a private partner under a five-year contract, which were awarded by municipalities under competitive bidding (PPIAF, 2009). More recently, PPPs have been used for the delivery of public facilities and services for over 200 years, for example in water supply in Paris. Even more recently, the private Water Works Company of Boston provided drinking water to residents in 1962 (HDR, 2005).

There are many definitions of PPPs. In general, PPP is described as 'a long term partnering relationship between the public and private sectors to deliver services', which aims to increase private sector involvement in the delivery of public services (MOF, 2004). Specifically, the contractual agreement between a public authority and a private partner enables both to share their resources including skills and assets as well as their risks and rewards in the delivery of public services (NCPPP, 2014).

In summary, PPPs combine the best resources and strengths of the public and private sectors to deliver the targeted public services. Some essential characteristics are listed below (Lee, 2005; Infrastructure Australia, 2008; Yong, 2010):

- Long-term contract: PPPs usually involve a long-term PPP contract (normally between 10 and 30 years) between a public contracting authority and a private partner based on the procurement of services, not of assets.
- Partnership agreement: the above long-term contract is viewed as a 'partnership' between the public and private sectors, drawing on the relative strengths of each party to establish a complementary relationship.
- Risk transfer: PPPs also involve the transfer of risk from the public to the private sector that can best manage it in contract. For example, the private partner is usually responsible for delivering the required services and the authority is involved in regulation and procurement of such services.

PPPs can be applied to many types of infrastructure projects, such as wastewater treatment works, public motorways, toll roads, power plants, telecommunications infrastructure, tunnels, school buildings, airport facilities, toll bridges, government offices, prisons, light rail systems, railways, parking stations, subways, museum buildings, harbours, pipelines, road upgrading and maintenance, health services, waste management, etc. (HDR, 2005; Grimsey and Lewis, 2004). Moreover, PPPs can be applied in many different contractual arrangements, such as Build-Finance, Build-Operate-Transfer (BOT), Build-Own-Operate (BOO), Build-Own-Operate-Transfer (BOOT), concession, cooperative arrangements, Design-Build-Finance (DBF), Design-Build-Finance-Maintain (DBFM), Design-Build-Finance-Maintain-Operate (DBFMO), Design-Build-Finance-Operate (DBFO), Design-Build-Operate (DBO), Finance Only, joint ventures (JV), leasing, Operation & Maintenance Contract (O&M), Private Finance Initiative (PFI), etc. (Grimsey and Lewis, 2004; CCPPP, 2014). Furthermore, PPPs can be applied in many different ways in terms of project objectives. It may be postulated that the evolution of PPPs can be divided into three generations, based on dominant objectives, as in the pictorial summary in Figure 15.1. It should be noted that although only the 'dominant objective' is visible 'upfront' in each generation 'box', the dominant objectives of any preceding generation may still remain important. Indeed their relative priorities depend on each specific project, so what is indicated is a general progression of priorities over time.

First generation: project financing

The first generation of PPPs focused on project financing, i.e. on mechanisms for mobilising private capital for developing and delivering public facilities and services to overcome government fiscal constraints. In this framework, the private partners finance the project upfront and receive their return on investment by collecting charges or payments made either by users (e.g. motorway tolls), by the authority (e.g. periodic lump sum payments) or by a combination (e.g. low user charges with government subsidies) during the

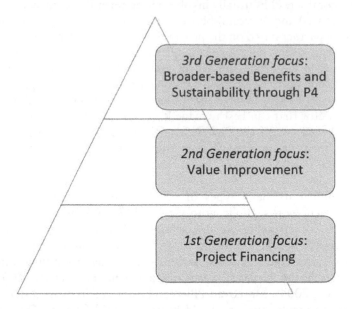

Figure 15.1 Classifying PPP evolution in terms of dominant objectives

contract period. The authority may contribute through land supply, financial commitments to project investors, purchase of the agreed services, etc. Based on this model, in 1992 the UK government introduced the Private

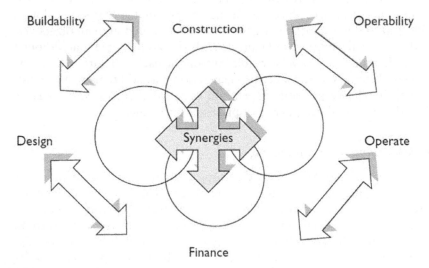

Figure 15.2 PPP synergies (Source: Cartlidge, 2006)

Finance Initiative (PFI) which transformed its role from being facility financiers, owners and operators into purchasers of services from the private sector (Construction Industry Council, 1998).

Second generation: value improvement

The second generation of PPPs focused on value improvement rather than only on mobilising private finance (Zhang and Kumaraswamy, 2001; Clifton and Duffield, 2006). Private sector efficiencies are mobilised and project performance is improved through the releasing of synergies in the delivery process (Aziz, 2007). For example, Figure 15.2 shows that PPP improves the buildability by strengthening the synergy between the design process and the construction process. It also improves the operability by strengthening the synergy between the construction phase and the operation phase (Cartlidge, 2006). Life-cycle costs and value are also improved by linking design to operational needs and priorities.

PPPs can inject additional value to projects in comparison with the traditionally fragmented delivery and operation processes. Some examples are listed below (MOF, 2004):

- Promoting innovation in design. PPPs allow the private sector to bring in innovative solutions.
- Optimising the whole life-cycle costs. PPPs allow government to outsource design, build, maintenance and operation to a private partner who is in a good position to optimise the design for economy in construction, operations and maintenance.
- Achieving better asset utilisation. PPP allows government to share its facilities with a private partner who is more knowledgeable in determining the optimal way of sharing these facilities with end-users.
- Optimising the sharing of risks between the public and private sectors. PPPs allow government to transfer capital risks to the private sector through private finance. Moreover, PPPs also allow government to transfer design and operation risks to the private sector, which can manage them better.

In summary, PPP contributes to pool the resources and skills of both public and private sectors to deliver better values in public services.

Third generation: Public-Private-People Partnerships

There is a growing trend towards including more stakeholders and even general public representatives or 'people' in Public-Private-People Partnerships (P4) involving the engagement of public stakeholders like non-profit-making and non-governmental organisations (NGOs) to capture the social concerns for PPP projects (Ng *et al.*, 2013). For example, a modified PPP scheme named Revitalising Historic Buildings through Partnership Scheme was introduced in Hong Kong in 2007, aiming to promote the adaptive re-use of government-owned historic buildings for the benefit of the general public. This scheme encouraged inputs from NGOs as well. Under this scheme, government provided a start-up fund to help 'build' (renovate) and 'operate' the building throughout a fixed concession period. NGOs re-invested the revenue generated from their primary business running the building to subsidise their secondary business – the provision of selected public services to the community (Chung, 2012).

Mobilising the fourth 'P' (people) at the planning and implementation stages spreads the potential benefits to a broader base. Proactively involving 'people' also draws attention to the social and environmental aspects of the sustainability tripod. Another related strand of research has recently focused on this advantage in post-disaster reconstruction in particular (Zhang and Kumaraswamy, 2013).

Importance of relationships in PPPs

PPP projects are usually complex in nature because they are usually large in scale. Moreover, they have longer contract periods, usually with a wider scope of services and more risks involved. More importantly, PPP projects involve numerous stakeholders from the public and private sectors compared to non-PPP projects. Figure 15.3 provides an example of a special purpose vehicle for a DBFO Project.

Relationship is generally described as the way in which two groups feel and behave towards each other (Sinclair, 2001). Figure 15.3 shows that there are multi-level relationships among stakeholders in PPP projects and we can categorise their relationships into two main types. For example, the primary relationship is built on the PPP contract between the public sector authority and the consortium. The secondary relationships are built among the consortium members including equity financers, construction contractors and facilities management operators (Smyth and Edkins, 2007). Stakeholder relationships can thus become very complicated in PPP projects.

Moreover, PPP projects usually have very long contract periods, hence stakeholder representatives and relationships will keep changing through planning, development and operational stages too. For example, a consortium will move from an 'internal' bidding team to a project delivery, then operational team, while maintaining relationships with the public authority (Zou et al., 2014) over a long period. Therefore, it is important to build and maintain good working relationships among stakeholders in PPP projects.

Need for RIVANS in PPPs

The above discussion shows PPP moving from the first generation that targeted private finance mobilisation to the second generation focusing

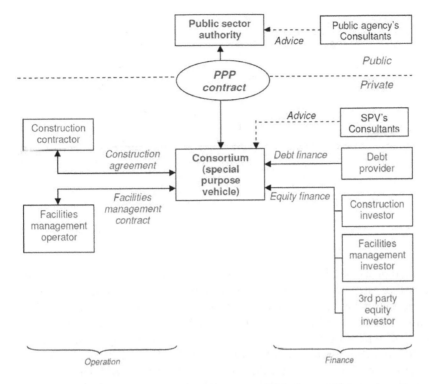

Figure 15.3 An example of a special purpose vehicle for a DBFO project (Source: MOF, 2004)

on value improvement through the enhanced synergies highlighted in Figure 15.2.

It is expected that the stronger integration of project stakeholders will yield superior synergies and bring much greater value to the outcomes. It is important to engage these stakeholders effectively and efficiently. However, PPP involves numerous stakeholders. It is therefore very difficult to transform them into truly integrated teams. Indeed this difficulty arises even in many large non-PPP infrastructure development projects. Consequently, there is a need to develop better ways to integrate stakeholders. It was proposed that this can be achieved by identifying, prioritising and focusing on their common best value objectives.

The concept of Relationally Integrated Value Networks (RIVANS) was introduced for the above purpose in 2003 (Kumaraswamy *et al.*, 2003). RIVANS was conceptualised as a holistic framework for 'relational' integration of hitherto mutually suspicious project stakeholders into cross-linked 'value networks'. The identification and focus on 'network values' (common value objectives) is expected to empower the optimal integration of project teams, and activate dormant synergies through relationally

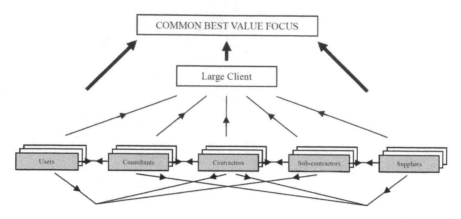

Figure 15.4 Strategic network of a large client (Source: Anvuur *et al.*, 2011)

integrated processes. Specifically, RIVANS proceeds well beyond the 'organisational' or 'structural' integration of 'functions' such as design, construct, finance, operate, etc., e.g. as in design-build or PPP procurement types. The 'relational integration' levels targeted in RIVANS also transcend the collaborative expectations in 'partnering' or even 'alliancing'. Furthermore, the integrative common (overall 'network') value focus should also help to channel potentially divergent value streams towards a confluence of consolidated high performance levels (CICID, 2008; CICID, 2009). A basic RIVANS initiated by a large construction client, with a portfolio of ongoing projects is visualised in Figure 15.4.

As shown in Figure 15.4, RIVANS can help diverse stakeholders such as users, consultants, contractors, sub-contractors and suppliers to identify common value elements and work together towards best network value. This serves as a useful vehicle to drive integration across a supply-value network. RIVANS is particularly valuable for large clients, as well as on megaprojects with many stakeholders and longer time-frames as in PPP projects, since the 'investment' in good relationships should yield long-term benefits and even better value. But such benefits can only accrue if there is excellent 'relationship management' in the crucial teamworking, i.e. after setting up the RIVANS strategies, structures and mechanisms aimed at common value objectives.

Critical success factors of relationship management in PPPs

Relationship management (RM) is not a new concept. Indeed, it has taken on many forms, addressing specific organisational constituencies (customers, partners, special service providers, employees, suppliers, etc.). It borrowed concepts and tools from relationship marketing, and became a formal approach to understanding, defining and supporting a broad spectrum of inter-business activities related to providing and consuming knowledge and services via networks. Particularly in a coopetitive business environment, good relationships are becoming even more important and fundamental to an organisation's success, hence the relationship-based management philosophy has expanded its field of application to construction management and PPPs (Cheung and Rowlinson, 2011; Jefferies and McGeorge, 2009; Jefferies *et al.*, 2014; Smyth and Pryke, 2008; Zineldin, 2004; Zou *et al.*, 2008; Zou and Kumaraswamy, 2009).

In PPPs, RM can be defined as a set of comprehensive strategies and processes of partnering with selected counterparties, including the project stakeholders, to create superior value for the PPP project through developing sustainable relationships. In order to successfully formulate and implement RM strategies, a set of critical success factors (CSFs) has been proposed and investigated in a research study (Zou, 2012; Zou *et al.*, 2014). The findings presented in Table 15.1 are extracted from this study.

Based on 42 responses from a cross-section of experienced PPP participants, the summary indicates that: 'Commitment and participation of senior executives' (mean value = 3.75) is considered to be the most influential factor impacting the success of RM; 'Defining the objectives to be achieved with the implementation of the RM strategy/exercises' and 'Integration of the different divisions of the organisation so as to meet the general RM objectives of the company and of each of the divisions/groups' (mean value = 3.44) were both ranked the second most influential factors. In order to improve the effectiveness of RM strategies, however, greater efforts should be made to improve commitment and participation of senior executives, define common objectives, increase effective communication approaches/channels between the PPP main parties and stimulate staff's commitment, as they are perceived as the most difficult factors to improve, according to the research findings presented in Table 15.1. These factors have also been addressed by RIVANS for their significance in improving the PPP network value.

In a PPP contract between the public sector and private sector, both parties cross over their traditional boundaries and also become more inter-related, interactive and integrated, thereby also spreading, hence diluting some extent of direct control, both internally and externally. RM in PPP cannot be legislated or become purely contractual, but its development depends on a solid underpinning in the contract, as well as the commitment of all involved parties, especially the senior executives. The identified CSFs in this study are important elements for developing a systematic framework for RM in potential RIVANS in PPP. This framework can provide general guidelines for developing organisational and operational mechanisms to foster better relationships, and hence much better, if not best value from PPP.

Table 15.1 Ranking of the eight identified critical success factors (CSFs)

ID	CSF	A	B	C	D
		*(A) Relative importance that should be assigned**	*(B) Relative importance assigned at present**	*Potential to improve (A-B)*	*Difficulty to improve (1 to 3) Difficult, moderate, easy#*
CSF 1	Commitment and participation of senior executives	**3.75**	2.44	**1.31**	**1.63**
CSF 2	A multidisciplinary team responsible for implementation of the RM	3.40	2.25	1.15	1.82
CSF 3	Defining the objectives to be achieved with the implementation of the RM strategy/exercises	**3.44**	**2.34**	**1.10**	**2.09**
CSF 4	Integration of the different divisions of the organisation so as to meet the general RM objectives of the company and of each of the divisions/groups	**3.44**	**2.53**	**0.91**	**1.81**
CSF 5	Publishing/disseminating the objectives, benefits and implications of the project to all the staff	3.25	2.22	1.03	1.94
CSF 6	Staff's commitment to the RM strategy	3.16	2.38	0.78	1.72
CSF 7	Integrating Information Systems (IS) for consistency and availability of information related to RM in the organisation	3.12	2.41	0.71	1.75
CSF 8	Effective communication approaches/channels between the PPP main parties	3.33	2.30	1.03	1.67

* For 'Mean scores': 1 (extremely unimportant) to 5 (extremely important)

For 'Mean scores': 1 (difficult) to 3 (easy)

Bold values highlight the three most important CSFs identified in this study (CSF1, CSF 3 and CSF4)

Developing relationship management and RIVANS frameworks for PPPs

PPP is increasingly advocated for boosting 'value' in selected physical and social infrastructure projects (Kumaraswamy *et al.*, 2007). However, evidence of 'additional value' is as elusive as the truly integrated teams that have been targeted for many years in infrastructure projects in general. Both 'better value' and 'greater integration' are even more critical in PPP scenarios for both RM and RIVANS (Kumaraswamy *et al.*, 2007; Kumaraswamy

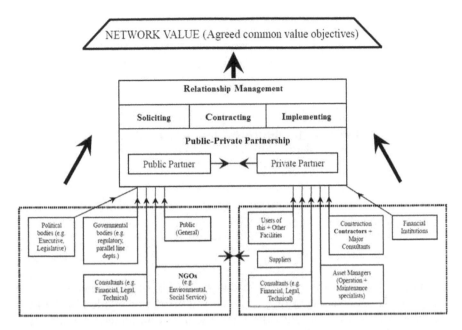

Figure 15.5 Integrating PPP stakeholders via relationship management to achieve common 'network value' (Source: adapted and developed from Kumaraswamy *et al.*, 2008)

et al., 2010; Zou *et al.*, 2014). This dual focus on long-term relationships and overall value for 'true integration' is particularly useful in the context of PPPs. A basic framework focusing on both longer-term relationships and network value is presented in this section. It provides a strategic approach for incorporating RM and win-win solutions among public and private sectors through integrating various PPP stakeholders as shown in Figure 15.5.

The framework illustrates how typically multitudinous and multifarious stakeholders (usually more in number and variety than in a non-PPP project) must be integrated and focused on continuously managing relationships between public and private partners through several quite distinct stages – namely soliciting, contracting and implementing (Schaeffer and Loveridge, 2002) – and targeting common value objectives, i.e. network value.

Defining common value objectives becomes even more critical, and ironically more difficult, in a PPP scenario. However, an effective relational approach can create a foundation for developing common objectives among the main players with different institutional missions and strengthen the forces that pull them together to decrease their 'distance', in moving from the equilibrium shown in Figure 15.6, to becoming 'closer' (Figure 15.7(b)) rather than the further 'pushing apart' (Figure 15.7(a)) scenario.

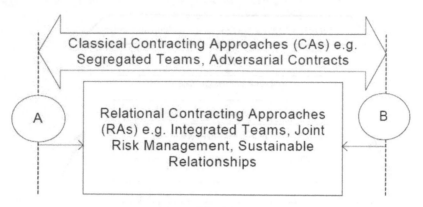

Figure 15.6 Equilibrium of push-pull forces between any two team members
(Source: adopted from Kumaraswamy *et al.*, 2007)

In order to target the scenario in Figure 15.7(b) in preference to that in Figure 15.7(a), the concept of 'partnership' must be embedded in the governance and processes of PPP, otherwise it will become a mere formality and the partnership ethos will not be embraced.

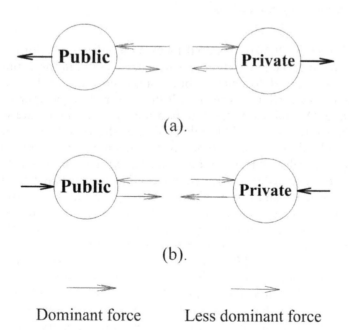

Figure 15.7 Potential relational effects of the proposed framework to help
move to (b) (Source: adopted from Kumaraswamy *et al.*, 2007)

A set of innovative procedures is proposed with a guideline checklist. Each of these proposed procedures has been assigned an ID number, from 'a' to 'i', pertaining to the soliciting, contracting and implementing stages. The corresponding checklists for each of the procedures (presented in Table 15.2) provide suggestions on 'what' needs to be done for better relationship building and maintenance. While Table 15.2 was developed for this chapter, it is based on a basic checklist compiled by Zou (2012) for better RM in PPPs. Developments include a structuring according to phases, as well as specific additions on RIVANS. When adopting these guidelines, the listed items should be considered in the context of the specific project characteristics, as well as the wider environmental and contextual conditions.

The framework for integrating RM and RIVANS engenders proactive project delivery modalities by fostering cooperation among public and private parties with a long-term mind-set, and can therefore focus on whole-life-cycle performance and value creation in infrastructure provision. The framework also provides mechanisms to integrate diverse partners both organisationally and relationally (Kumaraswamy *et al.*, 2010); it involves the entire PPP supply chain, which includes contractors, subcontractors and financiers, etc., in order to build up their long-term relationships and maximise the values and benefits from synergising diverse stakeholders. Truly 'relational integration' and 'overall value' may then be achieved in PPPs based on the proposed framework. The government is in a better position to implement the framework by taking the lead, say on some demonstration / pilot projects and to create a conducive environment, thereby also defusing apprehensions of the private sector on various risks, including expropriation/nationalisation and allegations of collusion/corruption. Another important aspect when applying RM and RIVANS to PPP is to involve the whole network team as early as possible, i.e. including the operational and maintenance groups, indeed the whole asset management team.

Recognising asset management teams as co-creators of value in addressing end-user needs

PPP projects are typically complex in nature with significant time and resources dedicated to the procurement process. The projects usually span long periods as well, adding to the uncertainties. Therefore, adequate consideration of long-term end-user needs and any impacts on the general public is of paramount importance. However, involving additional stakeholders in the design process, conducting more thorough consultation/engagement with stakeholders and mobilising asset management (AM) teams (or at least their representatives) earlier, are generally viewed as too costly and cumbersome. But lower initial project costs do not necessarily lead to better value, while being usually inversely related to life-cycle costs.

While stakeholder consultation or engagement processes are often conducted before and during major infrastructure projects, the atmosphere and

Table 15.2 Guideline checklist for implementing RM and RIVANS in PPP

ID	RM Procedure	Checklist and Commentary
Soliciting		
a	Awareness and Agreement for the RM Needs	Identify PPP project objectives in terms of economic, environmental and social aspects and overall and network value
		Appoint a Senior Executive responsible for RM
		Analyse the potential benefits from RM and RIVANS
		Identify procurement and operational processes linked to existing procedures and processes
		Identify constraints, initial risks and potential benefits in terms of RM and RIVANS
		Identify resources and skills development requirements in terms of RM and RIVANS
b	Internal Assessment and Improvement	Assess the political will
		Identify strengths and weaknesses
		Develop competencies and recruit suitable staff if needed
		Set up an interdisciplinary team for RM
		Establish initial criteria for private consortium selection
c	Prequalification and Shortlist	Inject relevant relational/cultural items into prequalification (basic) selection criteria
		Develop the common objectives
		Develop an initial RM Statement
		Assess potential partners according to initial criteria
		Incorporate methodology and detailed criteria for evaluating the collaborative capabilities and culture
Contracting		
d	Competitive Dialogue	Evaluate private partner's willingness to form a strong, long-term PPP relationship
		Appraise and refine the common objectives of the PPP project
		Ensure that the common objectives are compatible and align with each other
		Evaluate the relationship in the context of achieving overall common objectives
e	Negotiation	Establish a cross-sector RM team
		Establish contract arrangements
		Establish the RM Statement
		Consider holding joint workshops
		Identify the boundaries of the relationships with guidelines and examples and a Code of Ethics

f	Establish the Project Governance Committee	Build up a set of project-based and relational (e.g. stakeholder satisfaction, innovation, attitude and trust) KPIs for relationship performance evaluation Identify joint processes and protocols to manage knowledge and information flows between groups Establish management structure of a Project Governance Committee The Committee comprises senior managers from the public and private sector partners, to ensure top-down commitment on RM. The main functions of the Project Governance Committee are: to monitor service performance and review relationship performance to be innovative and proactive in RM to ensure partnership to resolve differences swiftly to report on and continuously improve performance
g	Sign off a RM Statement	The Statement should embrace and echo principles and best practices of co-operative partnerships (e.g. as in a long-term 'partnership charter'). It should include the following elements/components: the relationship boundaries with a Code of Ethics shared vision and objectives relational KPIs duties under the Statement succession planning to ensure that the good working relationship is maintained when new staff take-over compliance and legal enforcement where needed (when to revert to contract)

Implementing

h	Relationship Diagnostic	Incorporate a joint programme for reviews, performance measurement and reporting Incorporate measures to monitor and maintain appropriate behaviours Hold separate-sector meetings to discuss/resolve significant issues prior to any formal Committee meeting Incorporate a process for issue resolution at the appropriate levels Invite and incorporate perceptions from each party in terms of the KPIs
i	Relationship Improvement	Evaluate changes in the organisation, project environment, personnel and performance Undertake externally facilitated team-building workshops throughout the life of the project Maintain both formalized communication lines and informal communications Identify potential triggers for disengagement and/or exit strategy

outcomes of many of these events have attracted criticism. Previous studies indicate that opinions and interests of end-users and the public are often not given sufficient consideration and stakeholders may feel that they have been consulted or engaged too late (Akintoye *et al.*, 2003; Majamaa *et al.*, 2008; Ng *et al.*, 2013). Inadequate consideration of end-users needs can erode trust between relevant parties and trigger adversarial relationships. Relationship management (RM), which is critical, as discussed in the previous section, would then be handicapped from the outset.

For proper understanding of end-user needs, AM teams (property/facilities managers, operators and maintenance units, etc.) should be treated as important sources of information and experiential knowledge. This is because AM teams interact closely and continually with end-users, unlike project management (PM) teams (those involved in planning, design and construction), since AM teams deal with various end-user groups in handling complaints, providing assistance, performing maintenance works on facilities, while also monitoring usage patterns, service quality, etc. These position AM teams as critical stakeholders during the project planning phase of a project. Unfortunately, AM teams are often brought on board too late in the process, hence their views are inadequately reflected in the overall facility / built asset design (Wong *et al.*, 2014).

The issues highlighted above are a wake-up call for a revamping of the planning and design process to embrace a broader scope of stakeholders and potential 'co-creators' of value. Co-creation is a relatively new concept, recently surfacing in various industries such as automobile manufacturing, consumer electronics, sports apparel and aviation. The main purpose of co-creation is to build up organisational knowledge through proactively working with customers and partners to seek their feedback and enhance products or designs. Viewing external stakeholders as active partners can inject new ways of thinking, improve interaction and drive innovation in organisations (Payne *et al.*, 2008; Roser *et al.*, 2009).

In co-creation, customers and suppliers play an active role in the product development process. For example, an automobile manufacturer may invite a group of customers to test a prototype vehicle in development and provide feedback on how improvements can be made. Supplier co-creation may involve long-standing suppliers being invited to participate in joint product development meetings to share their expertise, lessons learned and inputs on new technologies (Roser *et al.*, 2009). Co-creation can lead to win-win-win situations where the manufacturer can build brand loyalty, observe reactions to new products from potential customers and foresee potential issues (e.g. problems related to new technologies, logistics, supply-chain management, quality control, etc.); the customers feel involved in the design process, resulting in the development of products that better meet their needs; and suppliers can secure long-term business relationships and consolidate business continuity. The co-creation concept is illustrated in Figure 15.8.

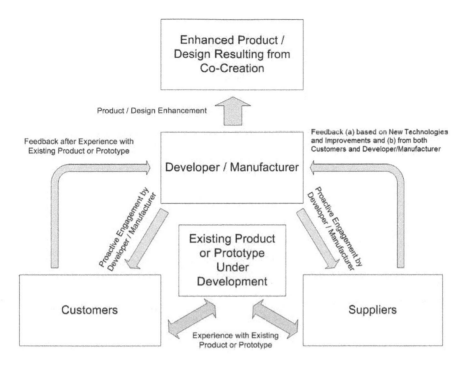

Figure 15.8 Customer and supplier co-creation

Adoption of co-creation in the construction industry remains limited. This can be attributed to the traditional mind-sets of industry players, adversarial mentalities, even 'cut-throat' approaches, hence poor relationships. Furthermore, construction projects are not generally viewed as 'products' in the same way as a car or a mobile phone since they are not something people can touch and feel on a daily basis. Despite this disconnect, co-creation can be very relevant and useful in construction and there are important lessons that can be learned from other industries. In transportation infrastructure projects such as highways and railways, many of which are delivered through PPPs (InTransit BC, 2006; Infrastructure Partnerships Australia, 2008), the public/end-users are in essence the customers. They make use of these infrastructure projects on a daily basis and rely on them for transport, which makes them 'products'. The vision of co-creation adopted in the context of the construction industry where AM teams are recognised as co-creators of value, is illustrated in Figure 15.9.

In PPP schemes whose expected project lifespans usually stretch into decades, delivering greater overall value to end-users is more important than delivering the project at the lowest price. Adequately addressing end-user needs and concerns can address all three sustainability dimensions,

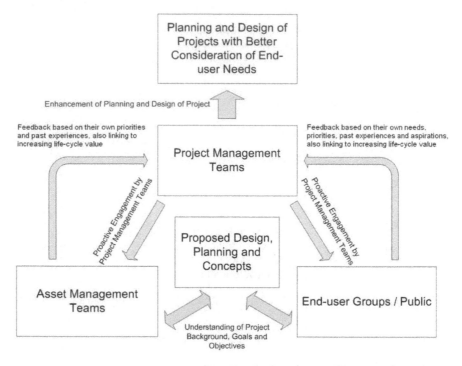

Figure 15.9 Potential for co-creation in the construction industry

i.e. social, economic and environmental, while also contributing to greater social harmony. Furthermore, properly engaging industry stakeholders and working together can lead to the long-term health and sustainability of organisations, allowing them to be more competitive in the market by building up their collective knowledge and expertise. Having recognised AM teams as co-creators of value particularly in PPPs, translating this into win-win-win benefits seems best achieved by integrating them effectively in the RM and RIVANS frameworks for PPPs as formulated in the previous section. Delving deeper, one needs to set up specific mechanisms for feedback and knowledge transfer from AM to PM teams.

Transferring knowledge from asset management teams to the planning and design of PPP projects

Having established the importance of AM teams as co-creators of value in addressing end-user needs, the next step is to create channels for knowledge (expertise, experience and lessons learned) from AM teams to be transferred to the planning and design of PPP projects. In many development projects, PM and AM teams typically work independently of each other. Interaction and links between the two are weak and fragmented. Even within their own domains (i.e. planning, design and construction in PM; operations and maintenance in AM), knowledge and information flows are usually one-way, from one group to the next (effectively, from one phase to the next) with very limited feedback.

As part of a study on extending the RIVANS concept to cover the integration of PM and AM teams (Wong *et al.*, 2014), a workshop held in November 2012 (CICID, 2012) elicited useful views and feedback on this theme from 33 senior-level construction industry experts in Hong Kong. The workshop participants also highlighted mutual benefits in keying into knowledge management (KM) systems and Building Information Modelling (BIM) that are increasingly important in the construction industry. These tools can play a key role in the knowledge exchange between various stakeholders. They can also help mitigate uncertainties through better handling and record-keeping of project information, including in capturing knowledge and lessons learned in a project. Unfortunately, it is still common for different organisations in the same project to maintain their own KM and BIM systems. It was suggested during the workshop that PM and AM teams should jointly establish these systems to save time and resources, as well as to minimise compatibility issues in information exchange. Interestingly, the AM team must be identified and brought on board earlier in a PPP scheme, compared to non-PPP procurement since they should be included in the consortium to bid for the project. The whole PPP team must then plan their activities, resource needs and cash flows together and well in advance to compile a realistic bid. So PM and AM teams in PPPs – compared to those in other project delivery modes – should be more focused in aligning their objectives and more motivated to develop KM and BIM systems jointly.

Three focus areas were recommended by Wong *et al.* (2014) to facilitate the integration of PM and AM teams based on the RIVANS concept. 'Organisational/Management Structure, Procurement Strategies, and Operational Mechanisms' is the first focus area. This may include: (1) establishing a cross-linked organisational structure with common platforms for interaction such as periodic joint workshops and meetings (at both senior management and project levels); (2) identifying appropriate procurement and delivery strategies; and (3) putting in place continuous evaluation and improvement measures such as comprehensive key performance indicators (KPIs). The second focus area is 'Fostering a Culture of Team Building and Providing Additional Means of Communication'. Examples include: (1) making available resources, venues and time for team members to interact within the work environment but outside structured meetings (from coffee breaks to longer brainstorming sessions); and (2) organising group outings or activities outside the work environment such as field trips and sporting events to build strong relationships. 'Information and Communication Technology Tools' is the third focus area. This may be as simple as setting up internet/intranet platforms like basic ftp sites or network drives for file

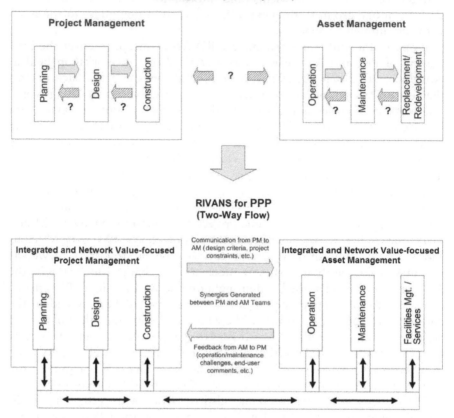

Figure 15.10 Required shift from one-way to two-way information and knowledge
flows (Source: adapted and developed from Wong *et al.*, 2014)

sharing and e-mail/chat groups, or developing more comprehensive systems
including relational databases, KM and BIM-based systems.

In the typical one-way flow scenario, information from PM teams (e.g.
design specifications, drawings, maintenance manuals, etc.) is passed over
to AM teams during project handover. The extension of RIVANS to cover
AM promotes greater interaction, communication and feedback within each
of their respective domains as well as between PM and AM teams (Wong
et al., 2014). The recommended transition from traditional one-way flows
to two-way knowledge transfers is illustrated in Figure 15.10. Under such
a structure, valuable knowledge exchange can take place, especially in PPPs
where AM teams or their representatives are brought on board much earlier in the project. This allows stakeholders from both teams to share and

brainstorm issues such as design and operational challenges, durability and sustainability priorities, lessons learned and previous experiences in these contexts.

The inclusion of operations and/or services conditions is common in PPP schemes, unlike in most non-PPPs. Therefore, the above two-way communication and knowledge exchange potential takes on even greater significance in PPPs. Decisions made in planning and design would have profound implications on the effectiveness, efficiencies and sustainability not only of the built infrastructure, but also of the operations and services therefrom, which being part of the PPP contract are thereby of more than mere altruistic or 'corporate social responsibility'-type interest to the PPP consortium.

Chapter summary

Integrating relationship management and value focus frameworks and mechanisms in RIVANS for PPPs

The foregoing sections have made a sound case for RIVANS for PPPs. Having conveyed the scope and overview in the introduction, the seven sections that followed focused on the specific building blocks of this case. Of course, many of the building blocks are themselves interlocking, hence strengthening the base case – e.g. the 'Importance of value delivery in PPPs' and the 'Importance of relationships in PPPs', provide the foundation for the section on the 'Need for RIVANS in PPPs'.

Next, drilling deeper into relationship management (RM) intricacies in the context of RIVANS and having identified the critical success factors of RM in general the chapters proceeds to develop RM and RIVANS frameworks for PPPs. This application of RIVANS to PPPs dovetails neatly into the recently concluded exercise on RIVANS for TAM (total asset management), given that PPPs usually involve the operational and 'service' phases, and hence need that extra focus on TAM as well.

The two previous sections 'complete the circle' in linking back RM to network value and RIVANS, in the context of the full network, focusing therein on the often neglected asset management teams, end-users and society at large. Even if not ignored, these less visible and less audible stakeholders are rarely paid the attention they deserve. Indeed, it is often alleged that only lip-service has been paid to environmental and societal considerations, apart from maintainability and operability priorities.

Bringing operational and maintenance teams on board early is an obvious prerequisite for more maintainable designs and sustainable built assets. 'Early operator involvement' is what is recommended, now that the benefits of 'early contractor involvement' have been recognised. The latter helps achieve more buildable designs, which also translate to cost-effectiveness and time-savings during the relatively shorter construction project; so 'early operator involvement' should similarly help boost performance levels over the much longer period of beneficial use of the built asset. As mentioned therein, the Hong Kong-based study reported in these two sections also spawned a set of parallel studies in the UK, Singapore and Sri Lanka (CICID, 2012; Ling *et al.*, 2014b; De Silva *et al.*, 2014). These examined relevant domains such as strategies and mechanisms for integrating design and construction and operations and maintenance supply chains, with a focus on extracting much greater overall life-cycle value, for example, through well-structured knowledge-sharing and far better formal and informal relationships, facilitated by excellent relationship management.

Concluding observations

Despite some limitations in combining conclusions from different research streams, the clear evidence and findings in each exercise, as well as the overlap in the researchers and the synergies in the research objectives, justify this over-arching exercise to consolidate the outcomes and benefit from their integration. Specific developments in each stream strengthen the links and boost the synergies in RIVANS for PPPs. For example, expanding PPP to include the fourth 'P' (people) requires stronger RM as well as sharpens the focus on overall life-cycle value, thereby also highlighting the value of RIVANS frameworks in making the best of these. Secondly, extending RIVANS to the operational phase in TAM effectively stretches the supply-value chain, again calling for RM to hold it together better. The latter will be even easier in PPP scenarios, since the formal collaborative structures will be in place, hence the focus on RIVANS for PPPs, while bringing in the fourth 'P' would again make it more likely to succeed and to remain successful for longer.

Apart from the above higher-level and/or broader findings, many specifics also emerge. For example, a strong case is made for 'early operator involvement', but although expected and hence easier in PPPs, it does not usually happen as well as it should, due to the usually segregated teams. This is what is addressed by introducing the RIVANS, RIVANS for TAM and now RIVANS for PPP frameworks. Another example of specifics worth highlighting is the importance of inculcating a supply-value chain focus on life-cycle value for the whole network of stakeholders. This bodes well not just for the sustainability of the built infrastructure, but also of the relationships and the teams themselves.

Acknowledgement

The University of Hong Kong Internal Funding Programme is acknowledged for funding the research study entitled 'Capacity Building of SMEs for Sustainable Housing Development', project number: 201109176151, which contributed to some findings in this chapter.

Reflections

1   RIVANS suggests deeper integration of stakeholders and enhanced ('super-charged') teamworking for PPPs to succeed. What are the key factors for this to be achieved?

2   In their 2014 report, entitled *Private Capital, Public Good: Drivers of Successful Infrastructure PPPs*, the Brookings Institution, Washington DC, makes nine main recommendations, two of which focus on 'active engagement with stakeholders' and 'building an empowered team'. The report contains several supporting case studies and is a very useful source of supporting information when considering the PPP approach. The report can be accessed directly at: http://www. brookings.edu/~/media/Research/Files/Reports/2014/12/17-ppp/BMPP_ PrivateCapitalPublicGood.pdf?la=en.

3   The authors identify a possible fourth 'P' for PPP, as in 'people', in that stakeholders (people) should have greater involvement during the planning and implementation stages of a PPP. This is reinforced by Wisa Majamaa's PhD thesis, entitled *The 4th P – People – in Urban Development Based on Public-Private-People Partnership* from Helsinki University of Technology (2008) where findings confirm that the end-user (people) of the PPP project is often left out of the process until the facility is operational.

4   The issue of 'early involvement' of project stakeholders in construction projects has been discussed by several authors in recent years, particularly as far as 'contractors' are concerned (Bresnan and Marshall, 2000; Li *et al.*, 2005; Mosey, 2009; Riemann and Spang, 2012; Harris and McCaffer, 2013), however, what has been written with regards to the 'early involvement of the operator' in PPPs?

5   The Hong Kong Institute of Surveyors (HKIS) produced a *Practical Guide to PPPs* in 2009, it can be accessed at: http://www.psdas.gov.hk/ content/doc/2005-1-11/PDP%20-%202005-1-11.pdf.

References

Akintoye, A., Hardcastle, C., Beck, M., Chinyio, E. and Asenova, D. (2003). Achieving best value in private finance initiative project procurement. *Construction Management and Economics* 21(5): 461–470.

Anvuur, A. M., Kumaraswamy, M. M. and Mahesh, G. (2011). Building 'relationally integrated value networks' (RIVANS). *Engineering, Construction and Architectural Management* 18(1): 102–120.

Aziz, A. M. A. (2007). Successful delivery of public-private partnerships for infrastructure development. *Journal of Construction Engineering and Management* 133(12): 918–931.

Bresnen, M. and Marshall, N. (2000). Building partnerships: case studies of client-contractor collaboration in the UK construction industry. *Construction Management and Economics* 18(7): 819–832.

Cartlidge, D. (2006). *Public Private Partnerships in Construction*, Abingdon, Routledge.

CCPPP (2014). Models of public-private partnerships. Retrieved 24 August 2014, from http://www.pppcouncil.ca/resources/about-ppp/models.html.

Cheung, Y. K. F. and Rowlinson, S. (2011). Supply chain sustainability: a relationship management approach. *International Journal of Managing Projects in Business* 4(3): 480–497.

Chung, J. K. H. (2012). Adaptive reuse of historic buildings through PPP: a case study of old Tai-O police station in Hong Kong. *The 2nd International Conference on Management, Economics and Social Sciences* (ICMESS 2012). Bali Kuta Resort by Swiss-Belhotel, Indonesia.

CICID (2008). Building RIVANS. Workshop Summary: Boosting Value by Building RIVANS. Centre for Infrastructure and Construction Industry Development (CICID), Hong Kong, 31 May 2008. Retrieved 31 August 2014, from http://www.civil.hku.hk/cicid/3_events/68/68_RIVANS_Workshop_II_Report.pdf.

CICID (2009).Revamping PPPs. Workshop Summary. Centre for Infrastructure and Construction Industry Development (CICID), Hong Kong, 28 February 2009. Retrieved 31 August 2014, from http://www.civil.hku.hk/cicid/3_events/78/78_PPPs_2009_workshop_summary.pdf.

CICID (2012). Relationally Integrated Value Networks (RIVANS) for Total Asset Management (TAM). Workshop Summary. Centre for Infrastructure and Construction Industry Development (CICID), Hong Kong, 3 November 2012. Retrieved 31 August 2014, from http://www.civil.hku.hk/cicid/3_events/124/124_summary.pdf.

Clifton, C. and Duffield, C. F. (2006). Improved PFI/PPP service outcomes through the integration of alliance principles. *International Journal of Project Management* 24(7): 573–586.

Construction Industry Council (1998). *Constructors' Key Guide to PFI*. London, Thomas Telford.

De Silva, N., Ranadewa, K. A. T. O., Kumaraswamy, M. M. and Ranasinghe, M. (2014). Better values and characteristics in 'Relationally Integrated Value Networks' to enhance Total Asset Management. *3rd World Construction Symposium*. CIB and Ceylon Institute of Building, Colombo, 20–22 June.

Grimsey, D. and Lewis, M. K. (2004). *Public Private Partnerships: The Worldwide Revolution in Infrastructure Provision and Project Finance*. Northampton, MA, Edward Elgar Publishing.

Harris, F. and McCaffer, R. (2013). *Modern Construction Management*, Oxford, Wiley-Blackwell.

HDR (2005). Creating effective public-private partnerships for buildings and infrastructure in today's economic environment. Retrieved 28 October 2007, from http://ncppp.org/resources/papers/HDRP3whitepaper.pdf.

Infrastructure Australia (2008). *National PPP Guidelines Volume 1: Procurement Options Analysis*. Australian Government.

Infrastructure Partnerships Australia (2008). CityLink Melbourne Case Studies, Infrastructure Partnerships Australia, Australia.

InTransit BC (2006). Canada Line Traffic Management Strategy, February 2006, Transit BC, Province of British Columbia, Canada.

Jefferies, M. and McGeorge, W. D. (2009). Using public-private partnerships (PPPs) to procure social infrastructure in Australia. *Engineering, Construction and Architectural Management* 16(5): 415–437.

Jefferies, M. C., Brewer, G. and Gajendran, T. (2014). Using a case study approach to identify critical success factors in alliance contracting. *Engineering, Construction and Architectural Management* 21(5): 465–480.

Kumaraswamy, M. M., Rahman, M. M., Palaneeswaran, E., Ng, S. T. and Ugwu, O. O. (2003). Relationally Integrated Value Networks. In (ed.) Anumba, C. *2nd International Conference on Innovation in Architecture, Engineering and Construction.* Loughborough, UK, Loughborough University, 25–27 June, pp. 607–616.

Kumaraswamy, M. M., Ling, F. Y. Y., Anvuur, A. M. and Rahman, M. M. (2007). Targeting relationally integrated teams for sustainable PPPs. *Engineering, Construction and Architectural Management* 14(6): 581–596.

Kumaraswamy, M., Mahesh, G., Anvuur, A. and Chung, J. (2008). Targeting truly integrated 'value networks' for PPPs. In (ed.) Nielsen, Y. *Proceedings of the 5th International Conference on Innovation in Architecture, Engineering and Construction (AEC2008).* Antalya, Turkey, 25–28 June 2008. Middle East Technical University, Turkey.

Kumaraswamy, M. M., Anvuur, A. M. and Smyth, H. J. (2010). Pursuing 'relational integration' and 'overall value' through 'RIVANS'. *Facilities* 28(13/14): 673–686.

Lee, V. (2005). *Public Private Partnerships.* Hong Kong: Hong Kong Legislative Council.

Li, B., Akintoye, A., Edwards, P., and Hardcastle, C., (2005) Critical success factors for PPP/PFI projects in the UK construction industry. *Construction Management and Economics* 23(5): 459–471.

Ling, F. Y. Y., Ke, Y., Kumaraswamy, M. M. and Wang, S. Q. (2014a). Key relational contracting practices affecting the performance of public construction projects in China. *Journal of Construction Engineering and Management* 140(1): 04013034-1–04013034-12.

Ling, F. Y. Y., Toh, B., Kumaraswamy, M. M. Wong, K. K. W. (2014b). Strategies for integrating design and construction and operations and maintenance supply chains in Singapore, *Structural Survey: Journal of Building Pathology and Refurbishment* 32(2): 158–182.

Majamaa, W., Junnila, S., Doloi, H. and Niemisto, E. (2008). End-user oriented public-private partnerships in real estate industry. *International Journal of Strategic Property Management* 12(1): 1–17.

MOF (2004). *Public Private Partnership Handbook.* Singapore: Ministry of Finance.

Mosey, D. (2009). *Early Contractor Involvement in Building Procurement: Contracts, Partnering and Project Management.* Oxford: Wiley-Blackwell.

NCPPP (2014). PPP definitions. Retrieved 23 August 2014, from http://www.ppp-council.ca/resources/about-ppp/definitions.html.

Ng, S. T., Wong, J. M. W. and Wong, K. K. W. (2013). A public private people partnerships (P4) process framework for infrastructure development in Hong Kong. *Cities* 31: 370–381.

Payne, A., Storbacka, K. and Frow, P. (2008). Managing the co-creation of value. *Journal of the Academy of Marketing Science* 36(August): 83–96.

PPIAF (2009). Toolkit for public-private partnerships in roads and highways. Version: March 2009. Retrieved 19 April 2011, from http://www.ppiaf.org/ppiaf/sites/ppiaf.org/files/documents/toolkits/highwaystoolkit/6/pdf-version/toolKit_ppp_roads&highways.zip.

Riemann, S. and Spang, K. (2012). Towards early contractor involvement for infrastructure projects in Germany. *Proceedings of the RICS COBRA Conference*. Las Vegas, September, pp. 519–526.

Roser, T., Samson, A., Humphreys, P. and Cruz-Valdivieso, E. (2009). *Co-creation: New Pathways to Value – An Overview*. London: Promise Corporation, London School of Economics Enterprise.

Schaeffer, P. V. and Loveridge, S. (2002). Toward an understanding of types of public-private cooperation. *Public Performance and Management Review* 26: 169–189.

Sinclair, J. (2001). *Collins Cobuild English Dictionary*. London: Collins.

Smyth, H. and Edkins, A. (2007). Relationship management in the management of PFI/PPP projects in the UK. *International Journal of Project Management* 25(3): 232–240.

Smyth, H. and Pryke, S. (2008). *Collaborative Relationships in Construction: Developing Frameworks and Networks*. Chichester, UK: Wiley-Blackwell.

Wong, K. K. W., Kumaraswamy, M. M., Mahesh, G. and Ling, F. Y. Y. (2014). Building integrated project and asset management teams for sustainable built infrastructure development. *Journal of Facilities Management* 12(3): 187–210.

Yong, H. (2010). *Public-Private Partnerships Policy and Practice: A Reference Guide*. London: Commonwealth Secretariat.

Zhang, J. Q. and Kumaraswamy, M. M. (2013). Developing public-private-people partnerships (4P) for post-disaster infrastructure reconstruction. *Proceedings of Public Private Partnership (PPP) International Body of Knowledge Conference*. Preston, UK, 18–20 March, pp. 281–290.

Zhang, X. Q. and Kumaraswamy, M. M. (2001). Procurement protocols for public-private partnered projects. *Journal of Construction Engineering and Management* 127(5): 351–358.

Zineldin, M. (2004). Co-opetition: the organisation of the future. *Marketing Intelligence and Planning* 22(7): 780–790.

Zou, W. (2012). *Relationship Management in Public Private Partnership Infrastructure Projects*. Unpublished PhD thesis, University of Hong Kong.

Zou, W. W. and Kumaraswamy, M. M. (2009). Game theory based understanding of dynamic relationships between public and private sectors in PPPs. In (ed.) Dainty, A. *Association of Researchers in Construction Management, 25th Annual Conference Proceedings*. Nottingham, UK, September, pp. 197–205.

Zou, W. W., Kumaraswamy, M. M., Chung, J. K. H. and Mahesh, G. (2008). The contribution of value management to developing collaborative working relationships. In (eds) Shen, G. Q. P., Chung, J. K. H., Yu, A. T. W., Au, I. Y. K. and Yang, R. J., *Proceedings of the 9th International Value Management Conference*. InterContinental Grand Stanford Hong Kong, 29 October – 1 November. Hong Kong: Hong Kong Institute of Value Management, pp. 170–176.

Zou, W., Kumaraswamy, M., Chung, J. and Wong, J. (2014). Identifying the critical success factors for relationship management in PPP projects. *International Journal of Project Management* 32(2): 265–274.

16 De-marginalisation of the public in Public-Private Partnership (PPP) projects

Lessons from the e-tolling of Gauteng's freeway in South Africa

P. D. Rwelamila

Chapter introduction

Public-Private Partnerships (PPPs) have emerged internationally since the 1980s as a means to involve the use of private finance for public sector projects (Grimsey and Lewis, 2004; Rwelamila *et al.*, 2014). The exact origin of PPP is hard to track but there are a few examples which suggest that PPP is a way of transforming the delivery of public services to boost economic growth. In Britain, Turnpike Trusts had powers to collect road tolls for maintaining the principal highways as early as the 1560s. Between the 1660s and the 1860s over 1,000 trusts administered around 30,000 miles of road in England and Wales, taking tolls at almost 8,000 toll-gates (William, 1972). The turnpike road development system reduced the rapid deterioration in the condition of main roads and built a network of well-maintained highways that allowed road transport to move more efficiently and reliably. Income from the collected tolls permitted substantial investment in the construction of new engineered structures such as drainages, embankments and bridges to provide faster routes where horse power could be used more efficiently to haul vehicles. The English Turnpike Trusts were replaced by highway boards in the late 1870s. The United States also used private turnpike companies as early as the 1790s to construct essential highways operated as toll roads (US-DOT, 2004). The first turnpike was chartered and became known as the Philadelphia and Lancaster Turnpike in Pennsylvania. Turnpike construction resulted in the incorporation of more than 50 turnpike companies in Connecticut, 67 in New York, and others in Massachusetts and around the country. The use of concessions to provide public services was not limited to developed countries. The development of the Suez Canal, according to Grimsey and Lewis (2004), shows how concessions were widely used to develop major transport links such as canals, tunnels and bridges. The role of the private sector in financing and operating public infrastructure declined during the world war era following opposition from the overtaxed citizens in perceptions of more private gains than public good (Prefontaine *et al.*, 2000).

The recurrence of PPPs has allowed for a more formal transfer of responsibilities from the state to a private sector company. PPPs bring together the public sector (client) and the private sector (supplier) in a medium-to-long-term relationship that allows the parties to blend their special skills to serve the needs and interests of the public (Grimsey and Lewis, 2004). The private sector takes for example responsibilities associated with design, financing, construction and operating the facilities whilst allowing the public sector to perform primary functions in the delivery of the services to the citizen. This also implies a directive to serve the public. Private finance is prompted by the changing market for public services and the need for an intense involvement of the private sector in boosting the capacity of the public sector to provide more and better services.

Based on the seminal PPP work of Henjewele *et al.* (2013) and Rwelamila *et al.* (2014), the aim of this chapter is to provide a critical reflection on the resulting developments in PPPs and to discuss current public protests against PPP initiatives, which are perceived as a siphon rather than a solution to public needs. The chapter will show: how dominant approaches in PPP arrangements have failed to embrace an inclusive approach and have favoured an exclusive approach – marginalised the public; how the 'public' should be central in PPP projects – the need for an appropriate stakeholder management system; and how the process of stakeholder consultation and management in PPP projects should be done in practice. This chapter complements many of the other PPP-based sections of this book and in particular Chapter 17, which also discusses this form of relationship contracting within the context of South Africa.

Theory and practice of PPPs and the twin challenges

Any initiatives to understand and deal with PPPs should be approached through two fronts, identified as the *twin challenges*. These are: the importance of defining the 'public' (dealing with the first 'P') and the implications to PPP arrangements; and the need for de-marginalise the public in PPP projects through multi-stakeholder management.

Dealing with the first 'P'

Public-Private-Partnership suggests a partnership between public and private sectors, which would seem that there are two clearly defined agencies that speak for the partnership. This chapter contests this fact and considers that 'public' could have several different meanings which need to be clarified and it also considers that with these parties defined, the management of such a partnership might become clearer and the outcomes be more successful if all the parties are identified and managed appropriately for mutual benefits and public satisfaction.

PPPs are described as bringing public and private sectors together in long-term partnerships for mutual benefit and include (HM Treasury, 2000):

- the selling of a minority or majority stake to a private partner to give ownership or strategic partnership;
- concessions and franchises given to private companies where the government contracts to purchase services on a long-term basis include 'maintaining, enhancing or constructing necessary infrastructure';
- the use of private sector expertise to exploit the commercial potential of government assets.

PPPs can come in a number of forms such as PFI (Private Finance Initiative), BOOT (Build, Own, Operate and Transfer), DBFO (Design, Build, Finance and Operate) and BOT (Build, Operate and Transfer). Each of these forms has been set up to transfer, fully or partly, risks associated with financing, construction and operating the infrastructure to provide public services which might otherwise not have gone ahead without the leverage of private funds and expertise. Grimsey and Lewis (2005) categorise different types of PPPs as economic or social, hard or soft, so that a hard economic PPP might apply to roads, airports or telecommunications, hard social to schools, prisons or public housing, soft social to social security or community services, and soft economic to vocational training, technology transfer and R&D facilitation.

One of the issues that emerges is whether the public sector actually wants the project. In addition, if the public sector is expected ultimately to pay back the initial financing that is invested in their name then they will also want to be assured of value for money (VFM) for what is actually provided. It also needs to be provided in a scope and form which meets expectations and needs. PPP is controversial because many believe that the private sector has inefficiencies in it which are not properly accounted for, or that provision satisfies those commissioning the project more than those using it.

The majority of risk transferred to the private sector is eventually paid for by the public through taxation and/or user chargers. The public as the ultimate beneficiary needs to be seen as part of the agreement rather than a merely uncontrollable risk (Shen *et al.*, 2006). Amidst the tug-of-war over expectations, PPP projects are also struggling to respond to relentless criticism for poor-to-mixed performance. Studies by Raisebeck *et al.* (2010) in Australia and that by Henjewele *et al.* (2011) in the UK showed that PPP projects may perform well in some aspects such as cost but underperform in others such as time efficiency and variations during operations. One of the issues to emerge is whether the public sector is given due consideration for its role as the ultimate payer of the initial financing and so assured of good value for money (VFM) in return.

In this respect, the chapter is not testing the actual VFM or efficiency of management in other ways e.g. risk management, but as a precursor to this it asks the more fundamental question of who should actually approve the project and provide objectives for it – a question of VFM for whom? For example, the debate in the UK that forced the coalition government to seek to reform the delivery of PFI implies that what constitutes VFM to the procuring authority may not necessarily represent VFM to the public (UK Treasury, 2011). This question can also look at the appropriate scope of the project in view of its cost and utility. It may even have a bearing on the speed at which provision is needed and expected.

The public component in PPPs can have the meaning of taxpayers, users of public facilities such as roads or sewers, the government, the commissioning department such as the prison service or healthcare departments or those who are impacted by the operation of such services – which is broader than the immediate stakeholder groups. It is not easy to satisfy all of them because each group will have quite different and even contradictory requirements. However, more effort could be made to engage the public if it was clearly contextualised and defined. Unfortunately, due to a narrow contextualisation of what constitutes the public and marginalisation of the general public, there are perceptions that the public can be represented by the end-users or groups affected by the project and only consulted on few issues such as social, ecological and economic impacts (El-Gohary *et al.*, 2006; Majamaa *et al.*, 2008). Complaints of exclusion of the public from PPP transactions on grounds of commercial confidentiality are rampant (Pollock *et al.*, 2007). In many cases, this may not be deliberate, but even unconscious exclusion for commercial interests creates dissatisfaction, distrust and conflicts between the general public and the public sector, which negotiates the deals on their behalf.

Figure 16.1, according to Rwelamila *et al.* (2014) represents two positions possible in practice. The artificial position (CI), where the public as direct users and the greater public are marginalised (peripheral) and the real position (CII), where the public as direct users and the greater public are afforded their rightful position as central stakeholders (principal). In the artificial position (CI) the government or public allied organisation $(P_{(a)}i)$ takes the position of both an agent and principal and ignores (peripheral treatment) the public as the principal, while in the real position (CII), the government or public allied organisation accepts its role as an agent and the public as the principal $(P_{(u)}i + P_{(p)}e)$. There are strong indications to suggest that the current PPP stakeholder structure is based on a CI arrangement in which the false assumption is that $(P_{(a)}i)$ as an agent has full mandate from $(P_{(u)}i + P_{(p)}e)$ as principal to deal with all matters of PPP arrangement. Hence $(P_{(a)}i)$ is primarily an agent of $(P_{(u)}i + P_{(p)}e)$ by default.

Embracing CII – the real public (Figure 16.1) – means that the public sector organisation managing the processes of conceiving the PPP project will make sure that direct users of the facility $(P_{(u)}i)$ and the greater public $(P_{(p)}e)$ are managed both in terms of understanding the key decisions taken on their behalf and keeping them in the communication loop during the process of value creation. It should be noted though that the level and intensity of communication between the government or public allied organisation and direct users of the facility and the greater public will differ significantly. Direct users of the facility will be more closely involved (at close distance to

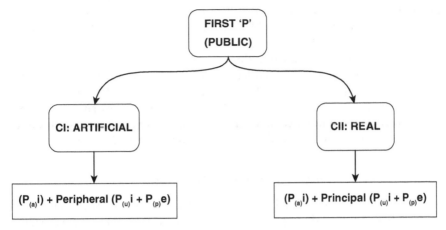

Figure 16.1 Understanding the public in a PPP environment (Source: Rwelamila *et al.*, 2014)

the agent) than the greater public (at a distance to the agent). This should be considered as the ideal though practice brings a different picture as will be discussed later.

Marginalisation of the first 'P'

According to Gutner (2005) and as clearly illustrated in Figure 16.1, the tangle of principal-agent (P-A) relationships and the difficulty of measuring some of the outputs create opportunities for any party to avoid accountability. In a PPP environment the definition of project stakeholders becomes blurred because the chain of stakeholders' command becomes too long and the public which constitutes both internal and external project stakeholders (P(u)i + P(p)e) remain at the bottom and are 'forgotten'.

The chain of stakeholders' command ultimately leads to project managers 'grabbing' more powers while assuming that they have full mandates from the principals to make all decisions across the PPP project life-cycle. The fact that PPP client project managers are agents of the public body loses its meaning and indirectly the parastatal agent becomes both principal and agent. This agency slack (Gutner, 2005) leads into a situation where decisions are made without stakeholders' (P(u)i + P(p)e) knowledge or consent. Through media and other information dissemination mediums stakeholders learn of what is going on around the PPP project, and react accordingly.

The dynamics of a P-A model seem to have pushed the real first 'P' [(P(a)i) + Principal (P(u)i + P(p)e)] into a back seat where the artificial 'P' (Pa) has dominated PPP projects. There is a need to revisit current PPP project management practices and embrace the real first 'P' concept in stakeholders' management as described in the remaining sections of this chapter.

Twin challenges: symptoms from the PPP coal face

A public-centred partnership between the public and private sectors presents a sound base to provide improved efficiency in service delivery. According to Rwelamila et al. (2014), one of the major challenges which have been facing PPP arrangements is public outcry in various countries – the public that PPP projects intend to serve is not accepting the end results of initiatives. Using Figure 16.1 to describe this unacceptable environment, there is an artificial stance across most PPP projects where the first 'P' is considered to be 'artificial' (CI) and where the public sector institution initiating a PPP project [the agent – $(P_{(a)}i)$] takes all responsibilities and keeps the public as direct users of the facility [$(P_{(u)}i)$] at arm's length.

A selection of statements, reproduced below, from public sectors initiating PPP projects, protesters and protest organisations could shed more light on the dominance of the artificial understanding of the public.

In Nigeria, for example, one of the political leaders at the Lekki Toll Road Concession in Nigeria (Iyoghojie, 2011) had this to say:

> I have fears that if nothing is done to halt the toll collection commencement date for now, something very bad may happen. We have seen how the people have complained about the toll gates and now the contractor has said the collection will start next Monday. I cannot imagine what those living in that area will do that Monday. So, that is why I have brought this to the House so that we can do something fast.

In Greece (Smith, 2011), when drivers began refusing to pay road tolls it was seen as a bit of good-natured defiance, born of economic necessity, but when the mayor of Stylida took control of a municipal bulldozer and broke his way through the barriers of a toll booth, civil disobedience on the PPP project took on an altogether different hue. The mayor, who was arrested and charged, stated: 'What I did was within the realm of my duties to defend the legal rights of citizens. I don't regret it and would do it again.'

Reflecting on the above statements, according to Rwelamila et al. (2014), it is possible to have two views. First, there are strong indications to suggest that the opposition arise due to lack of adequate information and a questionable relationship between the two sectors, particularly where such partnerships benefit one side while providing poor value for money to the other. Second, the public seems to have been kept in the dark regarding what has been conceived and constructed and how the facility will be operated. Poor communication seems to make matters worse. Generally, it could be argued with a certain level of certainty that the 'artificial public' (CI) as clearly illustrated in Figure 16.1 seem to dominate PPP project procurement. Hence the agent (government or public allied organisation as the direct custodian of the PPP project) by default seems to assume the role of the principal stakeholder and marginalise other stakeholders.

When the public feels marginalised by the PPP scheme, as clearly demonstrated by the above statements, the repercussions are uncontrollable – as some other examples show in Table 16.1.

The centrality of identifying the 'real public' and an appropriate model to manage it

The twin challenges as argued above are real and appropriate measures need to be taken to make sure that the real first 'P' is identified. Furthermore, an appropriate multi-stakeholder management conceptual model needs to be used to bring a coherent system to the management of a PPP project.

The centrality of the public partnership

As clearly indicated in the preceding analysis, and to reiterate the point made above, the dynamics of a principal-agent (P-A) model seem to have pushed the real first 'P' [($P_{(a)}$i) + Principal ($P_{(u)}$i + $P_{(p)}$e)] into the back seat such that the artificial 'P' (P_a) has dominated PPP projects. There is thus a need to revisit current PPP project management practices and embrace the real first 'P' concept in stakeholder management. It is possible to draw a number of implications for management and management theory regarding the governance of PPPs facing a demanding public base and specifically the dynamics surrounding the marginalisation of the public.

There is sufficient evidence from the literature to suggest strongly that the recurrence of PPPs has allowed for a more formal transfer of responsibilities from the state to private sector companies. In principle, therefore, this supports Grimsey and Lewis's (2005) argument that PPPs should bring together the public sector (client) and the private sector (the supplier) to allow the parties to blend their special skills to serve the needs and interests of the public (in this case, the real principal). In a nutshell, the central focus of any PPP initiative construct should be the public, which is represented by the first 'P'. Experiences of PPP initiatives in various countries as discussed and indicated in Table 16.1, and specifically the work of Yuan *et al.* (2010) and Li *et al.* (2005), do not support the centrality of the public in PPP initiatives,

Table 16.1 Public protest against PPP projects

Country Source	Description
Argentina	Thousands of demonstrators blocked the roads leading to the city of Buenos Aires protesting against a new water and sewer connection fee of US$800 announced by PPP project company Aguas Argentinas. Finally, the government rescinded the concession blaming Aguas Argentinas for not complying with obligations on expansion and quality. Sources: Cuttaree (2008); Santoro (2003)
Australia	Public pressure under the theme 'Power to the people' against a PPP in the New South Wales electricity sector, Sydney Ferries and Parklea Prison becoming a PPP. Sources: The Greens (2012); NSW Treasury (2006); PoNSW (2009)
Bolivia	Absence of an assessment of willingness to pay led to a widespread public opposition to a 40-year concession for a water system in Cochabamba. In October 1998, groups gathered in protests, which led to an outbreak of violence when the Bolivian army killed as many as nine, injured hundreds and arrested several local leaders. Finally, Aguas del Tunari announced its withdrawal from the project. Source: Cuttaree (2008)
Canada	The 14-hospital plan in Ontario province was vigorously opposed by the Ontario Health Coalition. The plan faced extensive protest from the concerned public on grounds that PPP programmes would result in expensive and inflexible contracts, unaffordable health services and a change in public sector ethos. Source: Dobbin (2007)
Denmark	The first Danish PPP in the Farum Municipality failed due to public dissatisfaction with the relationship between the public sector and private sector organisations involved in the projects – linked to fraud and mismanaged financial transactions. Source: Koch and Buser (2006)
Greece	Local drivers outside Athens refusing to pay what they saw as extortionate charges for roads. Sources: Balezdrova (2011); Chrisafis (2011); Smith (2011)
India	Protests against slum clearance for a toll road (the drives were estimated to affect the livelihoods of 500,000 people). Sources: BBC (2006); MoSRTH (2011)

New Zealand	Planned protest in the Octagon, Dunedin against legislation change driven by the Ministry of Local Government, leading to PPP and privatization. Sources: NZP (2011); NZPSA (2011)
Nigeria	The Lekki Toll Road Concession is a 30-year BOT-type partnership between the Lagos State government and the Lekki Concession Company to upgrade, expand and maintain approximately 50 km of an expressway and 20 km of a road. The public, including 74 Lekki estates and 18 villages agitated against the collection of tolls on grounds of unfairness, injustice and affordability for villagers. As a result, the government paid over US$2.5 million in a year in compensation for the suspended toll collection. Sources: All Africa (2011); Iyoghojie (2011)
South Africa	Opposition to the e-tolls N1, R24 and R21 to Pretoria and Johannesburg by trade unions, political parties and other pressure groups. Sources: Delonno et al. (2011) and Zerbst (2011)
United Kingdom	The first major projects such as Skye Bridge were strongly opposed by the public on the grounds of unfair tolls and super-profits for the private sector. The fight against the London underground PPP – proposed 15-year concession. Source: Monbiot (2001)
United States	Between 1977 and 1991 the Arizona hazardous waste facility project experienced a series of public protests against the decision to site incinerators in Yuma County in Arizona. The reasons for opposition were centred on lack of public consultation. As a result of the growing opposition, the state legislature placed a construction moratorium on the facility. Source: Ibitayo (2002)

Source: Rwelamila et al. (2014)

but a distorted definition of the first 'P'. Distortions of the first 'P' are clearly described by using 'Agency theory' as an analysis instrument.

When P-A models or 'Agency theory' are applied to a PPP project environment, it is clear that the first 'P' could be looked at from two distinct positions: the *artificial* position, where the government or public allied organisation dominate and the general public is ignored (treated as a peripheral player) and the *real* position, where the opposite of the *artificial* position prevails – where the public is considered a real principal stakeholder with the users of the PPP facility as internal stakeholders and the rest of the public (citizens) as external stakeholders. A reflection on the two positions strongly suggests that the dominant current scenario in the PPP environment seems to concur with the *artificial* position (Ahadzi and Bowles, 2004; Akintoye *et al.*, 2003; Hodge and Greve, 2005, 2007; Hodge, 2004; Alli, 2013; Domberger and Fernandez, 1999; Holmes *et al.*, 2006). The four factors of the 'Agency theory', as advanced by Lupia and McCubbins (2000) and Brehm and Gates (1997) are important in PPP in promoting the general public. These four factors are:

Factor 1: To recognise a principle and an agent. In a PPP environment this could refer to the relationship between a member of parliament and his or her constituents (members of the public), for example, in which the constituents are the principals and the member is the agent.

Factor 2: The possibility of conflicting interests. An agent could drag his or her heels in making unpopular decisions in a PPP project by ignoring public complaints because he/she believes him/herself to be an expert in the matter and his/her judgement is right while his/her principal requires him/her to listen to public outcry before making any decision on the matter. This could well be applied to PPP projects where there is agent-general public conflict. Alternatively 'dissent-shirking', where an agent does not agree with the principle.

Factor 3: The possibility of asymmetric information. There are strong assumptions that principals are ignorant of their agent's activities. In a PPP environment, the agent may have or assume to have more quality information than the principal, thus the potential to act against his or her principal without fear of being held accountable. For example, it is well acknowledged that requesting information on VFM will be rejected on what is purported to be commercial sensitivity (Henjewele *et al.*, 2011).

Factor 4: The principle may be able to adapt to agency problems. Principals can attempt to solve the problems of delegation in any of three ways: direct monitoring of an agent's activities; accepting without question what the agent says about the activities; attending to a third-party testimony about the agent's actions. Each of the options can provide a principal with valuable knowledge about his or her agent. Each option also has a drawback. Direct monitoring can be very expensive. In a PPP environment, it might take years for a general citizen to understand all the terminologies and documentation used in PPP projects. The drawback of relying on the agent's self-report is that the agent may be reluctant to reveal what he or she knows. For example, if an agent and principal have conflicting interests, then the agent may have no incentive to share information with the principal. In a PPP context, PPP technical experts could develop conflicting interests to the public, and then the agents will have no incentives to share their secrets with principals. Third-party testimony (e.g. from the media) may be unreliable because it is incomplete or prejudiced to certain viewpoints.

The dynamics of the four factors have pushed the position of the greater public to the periphery and governments or public sector organisations have taken up what could be described as a 'position with two hats' (as both principals and agents) (CI: Artificial Position), and thus problems of antinomic delegation and mission creep have significantly contributed to marginalising the public sector as clearly shown in Figure 16.1.

The dynamics of delegation which are primarily based on the principal-agent (P-A) model are real and seem to have contributed significantly to the dilution of the first 'P'. If PPP initiatives are going to take a central position in the provision of infrastructure for the betterment of the public, there is only one way – embracing the 'CII: Real Position' – that the greater public becomes the real principal. This is possible if and only if the concept of stakeholder management is embraced within the true spirit of a PPP project construct where the 'CI: Artificial Position' is tabooed, as will be described in the next section.

It is fundamental to state clearly that in order to maintain a true definition of the first 'P', the way in which PPP stakeholders are managed will be central. Appropriate PPP stakeholder management will ensure that the P-A model challenges are overcome and the 'CII: Real Position' (the real first 'P') is maintained.

In order to maintain the 'CII: Real Position' it is important to argue for a viable partnership with the real public and advocate consideration of transparent communication, good dissemination as well as the normal processes of efficiency. This will need the consideration of additional mechanisms of inclusiveness to ensure the role of the general public. The test for maintaining a true definition of the first 'P' for any PPP initiative as indicated above will be through appropriate stakeholder management, where transparent mechanisms will be needed for deciding on objectives for the public good, sustaining and maintaining these objectives even in stringent circumstances, deciding on the criteria for affordability, resolving conflicting positions and setting accountability by measuring the success in terms of public good (real success). It is important to note that existing mechanisms exist for ensuring objectives are set and met, but if the objectives are not known or agreeable to the public stakeholder, both their acceptance and the long-term success of the project will flounder. It is proposed that additional mechanisms for the real public are added to the existing partnership ones.

These mechanisms, which are described in the next section, are developed to add to (or replace) existing structure to ensure that the real 'P' is taken care of. There is a need to share and confirm the initial objectives of a project with the real principal, then to continue monitoring the implementation of these objectives and also to resolve conflict when objectives or their interpretations differ. These types of issues are sometimes addressed in stakeholder management. The question of ensuring ultimate acceptance and ownership of the project by the general public is an important one otherwise full effectiveness and use of the PPP are at risk or the project may even be rendered obsolete.

An appropriate multi-stakeholder management model

There is enough evidence in the work covered in the preceding sections of this chapter to suggest strongly that the public as a key group of stakeholders for any PPP project has been marginalised. This has happened primarily because a dominant group of researchers, government officials and private consultants seem to have taken a narrow perspective on stake holding. This approach, according to El-Gohary *et al.* (2006) and Rwelamila *et al.* (2014), as described in detail above, has contributed significantly to PPP project failures. Existing stakeholder management models have followed the same narrow definition of stake holding and are thus inadequate for PPP projects.

The need for an appropriate PPP project stakeholder management model therefore has become necessary and fundamental in order to regain the lost strength of Public-Private Partnership as one of the viable dynamic approaches in infrastructure development. El- Gohary *et al.*'s (2006) work should be considered a significant pioneering undertaking in finding an appropriate stakeholder management model in PPP projects. Although grounded on all fundamental aspects of a broader perspective to stake holding, their proposed semantic model, which includes stakeholder involvement processes and stakeholder involvement products, seems too complex to understand and thus articulate in practice. Furthermore, it is important to note that it was primarily developed to capture and incorporate stakeholder input in the design and not across the PPP project lifecycle. Even if the semantic model is critically assessed around its primary focus area of project design, it fails to provide a typical PPP project framework. Generally, the semantic model provides umbrella requirements for stakeholder management and doesn't provide the would-be PPP project manager or practitioner specific requirements in terms of stakeholder management processes, standards or rules on which judgement or decisions can be based (criteria) for each process, tools/techniques to be employed when dealing with stakeholders, and the identification of those responsible with stakeholder management. The model described in Figure 16.2, which was developed by Henjewele *et al.* (2013), provides all the missing parts in El-Gohary *et al.*'s (2006) semantic model and goes beyond the design phase of a PPP project to cover all the PPP project phases.

The model proposed in Figure 16.2 provides a clear framework for managing stakeholders in a PPP project procurement set-up. The model is primarily based on two inter-connected processes: managing a PPP project through four phases (conception, business case development and financial close, design and construction, and operation and maintenance); and stakeholder management activities which are defined through five clusters (identification of stakeholders, prioritisation of stakeholders, building relationships, identifying and managing concerns and conflicts, and managing communication). The five clusters of stakeholder management are appropriately adjusted post the conception phase to reflect the changing nature of the project during the three remaining phases.

The process of stakeholder management through the PPP project lifecycle takes a systematic approach within each phase and under each activities cluster, where a standard on which judgement is based (criteria) is formulated, and appropriate tools or techniques are used to generate information and aid decision making.

A number of tools or techniques are suggested which are necessary in order to allow stakeholders to participate in the process of real ownership of the project. These tools or techniques can be used in conjunction with the more traditional forms of public consultation that are already built into various working systems in non-PPP projects, such as exhibitions and calling for formal written submissions. As clearly indicated in Figure 16.2, prominent possible tools or techniques include: search conferences, brainstorming, mapping, Charrette, dinner parties, workshops, consensus conferences, focus groups, etc. These are proposed under various locations across the project lifecycle based on their relative characteristics, merits and demerits. For lack of space and brevity, detailed particulars of each of these tools or techniques are not provided in this chapter and can be found elsewhere (e.g. Henjewele *et al.*, 2013; NSW Department of Urban Affairs and Planning 2001; Renn *et al.*, 1995; Seargent and Steele, 1988).

Collecting project information and opinions of stakeholders

As a point of departure toward addressing the theme of this paper in the context of the Gauteng e-toll project, it was necessary to develop a rational explanation of events which took place during the e-toll project lifecycle to understand their sequence, speculate about cause-and-effect relationships among them, and draw inferences about the effects of events on the project stakeholders (the public). Hence the historical research method was

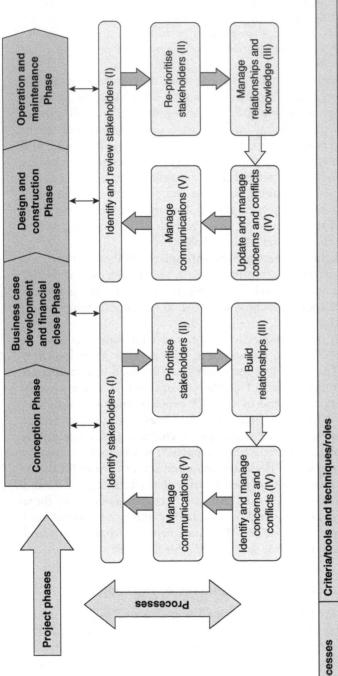

Processes	Criteria/tools and techniques/roles			
Identify/review stakeholders (I)	*Criteria*	Developmental and public needs (brief)	Project objectives and concerns	PPP project scope	Scope and sustainability
	Tool/ Techniques	Search conferences, brainstorming and mapping	Focus groups, brainstorming, review and mapping, Charrette	Focus groups, brainstorming, mapping, Charrette, consensus conference	Focus group or Charrette

		Project needs (impact, interest, power, predictability, etc.)	Project objectives and output specifications (impact, interest, power, predictability, etc.)	Agreed scope (impact, interest, power, predictability, etc.)	Functioning (product and services and contract terms)
Profile and prioritise (II)	*Criteria*	Project needs (impact, interest, power, predictability, etc.)	Project objectives and output specifications (impact, interest, power, predictability, etc.)	Agreed scope (impact, interest, power, predictability, etc.)	Functioning (product and services and contract terms)
	Tool/ Techniques	Search conferences/ meetings, mapping	Charrette, dinner parties, workshops, mapping	Consensus conferences, mapping	Focus group or Charrette
Build/manage relationships and knowledge (III)	*Criteria*	Established needs according to priority groups and empowerment	Established project objectives and output specifications according to priority groups and empowerment	Agreed scope according to priority groups and empowerment	Agreed operational performance - based on roles and empowerment
	Tool/ Techniques	Open consultation and Charrette, training	Charrette and public feedback panels (PFPs)	Charrette, PFPs	Feedback boxes and help desks, training
Update/manage concerns and conflict (IV)	*Criteria*	Impact of established needs according to priority groups	Impact of established project objectives and scope according to priority groups	Impact of agreed scope according to priority groups	Impact of operational performance - based on roles
	Tool/ Techniques	Search conferences, brainstorming, workshops/meetings, negotiations, mediation	Charrette, brainstorming, workshops/meetings, negotiations, mediation	Brainstorming, consensus conferences, surveys, negotiations, workshops, mediation	Focus group or Charrette, mediation
Manage communication (V)	*Criteria*	Effective communication of project brief	Effective communication of business case and adjudication procedures	Effective communication of project parameter and logistics	Effective communication of project operational performance and tariffs
	Tool/ Techniques	Story boarding, search conferences, press releases, posters, project websites	Story boarding, search conferences, press releases, posters, project websites	Mail shots, websites, story boarding, PFPs, press releases, posters, messages through public outlets	Newsletters and regular senior staff bulletins, press releases, multimedia
Responsible party (I –V)		Gov/Agency PM/PR	Gov/Agency PM/PR and SPV	Gov/Agency PM/PR and SPV	Gov/Agency reps/PR and SPV

Figure 16.2 A multi-stakeholder management model for PPP projects (Source: Henjewele *et al.*, 2013)

considered the most appropriate approach to address issues defining the theme of this study. The method was considered appropriate to identify events which took place from conception and planning to construction and ownership in order to interpret the facts around those events.

Every effort was made to find first hand events which took place from inception to ownership of the e-toll project. It was fundamental to ensure the internal validity of the research study, hence triangulation strategy was adopted – through newspaper clippings, websites, documents (from private organisations, government and the South African National Roads Agency) and individuals using the e-toll freeway.

Data sources

In order to find the opinions of stakeholders and their differences on the e-Toll GFIP (e-TGFIP; e-Toll Gauteng Freeway Improvement Project), various methods were used: documentation released to the media, newspapers coverage, TV presentations and presentations at organised public meetings.

Evaluating and interpreting historical data

Once historical data relevant to the theme of the study were located, it was important to decide on what is fact and what is fiction. In other words, the need to determine the validity of data was central to the whole process of evaluating and interpreting it. Taking a leaf from Leedy and Ormrod's (2005) work, two types of evaluation were carried out. First, judging whether documents identified were authentic. Second, deciding, if the items were indeed authentic, what they mean. In these situations, it was fundamental to review the data to determine their external evidence (concerned with the question: Are these articles genuine?) and internal evidence (here dealing with the questions: What does it mean? What interpretations can be extracted from the words?) respectively.

When interpreting historical data, as will be shown later, it was necessary to establish a base-line framework of conceptual constructs for two sets of 'specific lenses' to guide the interpretation. These are described below.

Specific lenses I

These lenses are based on Rwelamila *et al.*'s (2014) seminal work on understanding the public in a PPP environment. Establishing how the first 'P' was embraced – as CI (artificial) or CII (real):

CI: artificial is formulated as: $(P_{(a)}i) +$
Peripheral $(P_{(u)}i + P_{(p)}e)$ (*Equation 1*)

and

CII: real is formulated as: $(P_{(a)}i)$ + Principal $(P_{(u)}i + P_{(p)}e)$ (*Equation 2*)

where:

$P_{(a)}i$ = government or public allied organisation = agent
$P_{(u)}i$ = public as direct users of the facility
$P_{(p)}e$ = greater population public (non-direct users of facility)
i = internal stakeholders
e = external stakeholders
Position (CI): The agent by default assumes the role of the principal stakeholder and marginalises other stakeholders.
Position (CII): The agent and other stakeholders become 'principal stakeholders'.

Data was analysed to establish the position of SANRAL (South African National Roads Agency Limited) and other parties when interacting with stakeholders throughout the e-toll project value creations process.

Specific lenses II

Here 'specific lenses' to guide interpretation of historical data were based on Henjewele *et al.*'s (2013) multi-stakeholder management model. Again, it was important to reflect on historical events which took place across the e-toll project lifecycle.

Through four phases of the e-toll project – conception, business case development and financial close, design and construction, and operation and maintenance – as will be discussed later, it was possible to reflect on how stakeholders were identified and reviewed. The evaluation and interpretation of historical data was done through the following processes:

* identification/review of stakeholders (I/RS)
* profiling and prioritising (P&P)
* building/managing relationships and knowledge (B/MR&K)
* updating/managing concerns and conflict (U/MC&C)
* managing communication (MC).

These lenses ('Specific lenses II') were used to supplement 'Specific lenses I' to assess the process of stakeholder management through the e-toll project lifecycle.

Results, synthesis and analysis

The governance of PPPs in the stakeholder perspective as described above forms the backbone of this chapter. The complexity of the system of relationships in which a PPP is embedded in practice and the challenges which have been experienced across the world naturally led to the adoption of a stakeholder perspective to study it. This perspective traces its roots to a long tradition that professes that an organisation is before all else a coalition of interests in which each actor or category of actors holds conflicting stakes that must be satisfied if that organisation is to be functional (Berle and Means, 1932; Cyert and March, 1963). In the past 20 years, it has been expanded to the study of the relationships among these actors, both internal and external to the PPP project organisation set-up.

Since 2007, when the Minister of Transport indicated that the basis for determining the toll tariffs at each tolling point was 50c/km without discounts, and subsequently SANRAL announced the proposed toll tariffs for the GFIP in 2011 – the full tariff for light vehicles (based on 66c/km; equating to 50c/km in March 2007) – a number of stakeholders have expressed their concerns regarding the public affordability of toll tariffs. These stakeholders (Table 16.2) include: Opposition to Urban Tolling Alliance (BUSA), Road Freight Association (RFA), Retail Motor Industry (RMI), The Southern African Tourism Services Association (SAT SA), Afriforum, Johannesburg Business Chamber (JBC), South African Local Government Association (SALGA), Automobile Association (AA), South Africa Road Federation (SARF), Democratic Alliance (DA), Solidarity, South African Communist Party (SACP), Freedom Front (FF), African National Congress Youth League (ANCYL), Confederation of South African Trade Unions (COSATU), National Taxi Alliance (NTA), Mamelodi Commuter Forum (MCF), South African Commuter Organisation (SACO) and Ekurhuleni Metropolitan Municipality (EMM).

All stakeholders, as indicated in Table 16.2, who were part of the public outcry to the proposed GFIP Toll Tariffs in 2011 formed the study sample.

The two specific lenses described above ('Specific lenses I' and 'Specific lenses II') were used to assess the stakeholders' opinions on e-Toll GFIP.

Using specific lenses ('Specific lenses I')

Using Rwelamila et al.'s (2014) seminal work, it was necessary to establish how the first 'P' was embraced – as a CI (artificial) or CII (real) – in the project. The stakeholders' opinions were as indicated in Table 16.3.

Rwelamila et al.'s (2014) findings are confirmed by opinions gathered from the e-Toll GFIP stakeholders' organisations listed in Table 16.3. With the exception of SANRAL, there is a clear consensus on how real stakeholders have been treated. The dynamics of a P-A construct, formulated by Rwelamila et al. (2014) and described in Figure 16.1, seem to have pushed the real first 'P' $[(P_{(a)}i) + \text{Principal } (P_{(u)}i + P_{(p)}e)]$ into the background, making space for the artificial 'P' (P_a) to dominate the e-Toll GFIP. Among both internal and external e-Toll GFIP stakeholders 94.5 per cent seem to be concerned with the way in which SANRAL handled the project stakeholder management function. Clearly, there is no evidence to support Grimsey and Lewis's (2005) ideal expectations that PPP projects should bring together

Table 16.2 Gauteng freeway e-toll project stakeholders

Stakeholder	Characteristics
Opposition to Urban Tolling Alliance (OUTA)	A voluntary civil action group of business associations and individuals formed in March 2012 to challenge SANRAL's decision to implement e-tolling of the upgraded freeway network in Gauteng, on the basis that it was an irrational decision
Road Freight Association (RFA)	A voluntary association of small and medium-sized trucking companies, including family-owned businesses, owner operators and the largest trucking companies in South Africa
The Southern African Tourism Services Association (SAT SA)	A voluntary association of tourism service providers
Afriforum	A voluntary civil rights organisation linked to the Solidarity trade union
Johannesburg Business Chamber (JBC)	A voluntary and independent, non-political, subscription-based association dedicated to promoting a business-friendly environment
South African Local Government Association (SALGA)	An autonomous association of municipalities with its mandate derived from the Constitution of the Republic of South Africa, as the voice and sole representative of local government
Automobile Association (AA)	A voluntary association offering motoring services, which include: roadside assistance, technical advice, motor-related legal advice, insurance, driver training, as well as exclusive travel packages and travel advice
South Africa Road Federation (SARF)	A voluntary non-political organisation representing all bodies that have an interest in any aspect of the road industry and road administration
Democratic Alliance (DA)	Political party – South African official opposition political party
South African National Roads Agency (SANRAL)	Statutory state-owned entity (SOE): category 3A responsible for managing more than 21,000 km of the national road network
Solidarity	Trade union within the Christian tradition of unionism
South African Communist Party (SACP)	Political party – with an alliance with the South African ruling party – the African National Congress (ANC)
Freedom Front (FF)	Political party – opposition
African National Congress Youth League (ANCYL)	Youth league of the ANC
Confederation of South African Trade Unions (COSATU)	Trade union – with an alliance with the South African ruling party – the African National Congress (ANC)
National Taxi Alliance (NTA)	A voluntary alliance of taxi owners
Mamelodi Commuter Forum (MCF)	A voluntary forum of commuters
Business Unity South Africa (BUSA)	A voluntary principal representative of business in South Africa

N = 18; n = 18

Table 16.3 Specific lenses I: opinion on how the first 'P' was embraced

Stakeholder	Embracing CI (Artificial) $[(P_{(a)}i) + Peripheral (P_{(u)} i + P_{(p)}e]$	Embracing CII (Real) $[(P_{(a)}i) + Principal (P_{(u)}i + P_{(p)}e)]$
Opposition to Urban Tolling Alliance (OUTA)	✓	
Road Freight Association (RFA)	✓	
The Southern African Tourism Services Association (SAT SA)	✓	
Afriforum	✓	
Johannesburg Business Chamber (JBC)	✓	
South African Local Government Association (SALGA)	✓	
Automobile Association (AA)	✓	
South Africa Road Federation (SARF)	✓	
Democratic Alliance (DA)	✓	
South African National Roads Agency (SANRAL)		✓
Solidarity	✓	
South African Communist Party (SACP)	✓	
Freedom Front (FF)	✓	
African National Congress Youth League (ANCYL)	✓	
Confederation of South African Trade Unions (COSATU)	✓	
National Taxi Alliance (NTA)	✓	
Mamelodi Commuter Forum (MCF)	✓	
Business Unity South Africa (BUSA)	✓	

N = 18; n = 18
CI (Artificial): marginalisation of real stakeholders
CII (Real): embracing real stakeholders

the public sector (client) and the private sector (the supplier) to allow the parties to blend their special skills to serve the needs and interests of the public (in this case, the real principal). The central focus of e-Toll GFIP seems to be the public by default. SANRAL's traditional approach of seeking comments through the government Gazette seems to be ineffective and outmoded. All the e-Toll GFIP stakeholders who have been involved in public protest seem to share this sentiment.

SANRAL's official approach in dealing with the e-Toll GFIP stakeholders could be described as orthodox and bordering on arrogance. SANRAL's Chief Executive Officer's (CEO) words are a true testimony of this situation. The CEO had this to say in June (2013) following protests and debate on what he suggested was one of the key questions on the e-Toll GFIP:

> Why are we spending money and why are we asking South Africans – in some selected cases to pay for better roads through tolls?

SANRAL's research showed that most Gauteng freeways had reached their capacity and, therefore, peak hours were extending by 10 to 15 minutes each year. This resulted in congestion and uneconomical use of time, coupled with increased vehicle costs and carbon emissions.

A very sound argument indeed, which could have been raised during the e-Toll GPIF conception phase, but making this pronouncement once the freeway was already built borders on marginalising the first 'P'. He continued to argue:

> For the past two years commuters in Gauteng have reaped the benefits of this project, and although we may continue to disagree on many things surrounding the project, the one thing I think we can all agree on is that it has already helped to improve the lives of Gauteng citizens, and that, we must not forget, was the centre point of this exercise.

Another strong argument, but argued very late when you consider the e-Toll GFIP lifecycle.

Some of the protesters of the e-Toll GFIP approach to e-tolling raised 'moral' arguments (Mtshali, 2014). For example, the South African Moral Regeneration Movement (MRM) chairman stated:'The sentiments of the clergy – who engaged with hundreds of thousands of people weekly – should not be ignored as they wielded significant influence in their communities.' He further argued: 'What started like a very good project, and I really believe it is a good project . . . has become . . . a nightmare. It's causing a lot of bitterness. It's also causing lots of confusion.'

Representing one of the South African trade unions (DeIonno *et al.*, 2011), the Congress of South African Trade Unions (COSATU) made this remark on toll gantries in its memorandum prepared for handing over during its anti-toll protest:

> All the evidence indicates that the revenues from the tolls are going to be enormous and that the loans will be paid off quickly, leaving the private operator to milk the public.

> If more money was put into stopping fraud and corruption, government would easily have enough to fund road construction and maintenance.

> . . .

> They will make it more expensive for the poor to travel by road, and will also increase food inflation by adding to the cost of transporting goods in and out of Gauteng.

> . . .

> Toll roads will further exclude the poor and create divisions.

Writing about developments in financing infrastructure in South Africa, Ilkova (2013) comments on the e-Toll GFIP: 'Government is under pressure to prove that it can enforce unpopular policies as it finally starts electronic tolling on Gauteng's freeways after more than two years of delays and legal challenges to the system.' Looking ahead, she further argued:

> But e-tolling is also a signal to consumers of what is to come. Government plans to spend R4.3 Trillion (US$0.43 Trillion) on infrastructure over the next 15 years. Despite an already steep rise in administered prices, users are expected to fund at least some of that spending.

Study results on 'Specific lenses I' described above should provide fundamental lessons for future PPP infrastructure ventures. Lessons need to be learnt by all public institutions responsible for infrastructure development and specifically those using PPPs as a procurement route. The narrow definition of stake holding is inadequate for any PPP project. Rwelamila *et al.*'s (2014) two positions, as described in Figure 16.1, need to be very well understood by all involved in PPP projects. No public institution involved in procuring infrastructure through PPP can afford to marginalise the public as direct users of infrastructure. Embracing CII is the only option for successful delivery of infrastructure through PPP, where the public is afforded its rightful position as central stakeholder (principal).

Using specific lenses ('Specific lenses II')

In order to interpret e-Toll GFIP historical data through its value creation process, Henjewele *et al.*'s (2013) multi-stakeholder management model described above was used. The stakeholders' opinions were as indicated in Table 16.4.

Table 16.4 summarises descriptive sample information on the e-Toll GFIP stakeholders' opinions on stakeholder management using 'Specific lenses II'. Of the stakeholders 88.9 per cent strongly disagree that a best practices stakeholder management ethos characterised the process of developing the e-Toll GFIP.

The opinions of the e-Toll GFIP stakeholders strongly suggest the process of managing stakeholders was very weak and fell short of dealing with central fundamentals encapsulated within Henjewele *et al.*'s (2013) multistakeholder management model described above. While the four phases of the e-Toll GFIP seem to have a framework of various dynamics relevant to the project, two inter-connected processes, advanced by Henjewele *et al.* (2013), which include the process of managing stakeholders, leave a lot to be desired. Following public presentations, newspaper coverage and TV presentations it is clear that SANRAL paid lip service to the implementation of stakeholder management activities which are defined through five clusters (identification of stakeholders, prioritisation of stakeholders, building

Table 16.4 Specific lenses II: opinion on the process of stakeholder management
through the e-toll project lifecycle

Stakeholder	I/RS	P&P	B/MR&K	U/MC&C	MC
Opposition to Urban Tolling Alliance (OUTA)	1	1	1	1	1
Road Freight Association (RFA)	1	1	1	1	1
The Southern African Tourism Services Association (SAT SA)	1	1	1	1	1
Afriforum	1	1	1	1	1
Johannesburg Business Chamber (JBC)	2	2	2	2	2
South African Local Government Association (SALGA)	1	1	1	1	1
Automobile Association (AA)	1	1	1	1	1
South Africa Road Federation (SARF)	1	1	1	1	1
Democratic Alliance (DA)	1	1	1	1	1
South African National Roads Agency (SANRAL)	5	5	5	4	4
Solidarity	1	1	1	1	1
South African Communist Party (SACP)	3	3	4	3	4
Freedom Front (FF)					
African National Congress Youth League (ANCYL)	3	3	2	2	2
Confederation of South African Trade Unions (COSATU)	1	1	1	1	1
National Taxi Alliance (NTA)	1	1	1	1	1
Mamelodi Commuter Forum (MCF)	1	1	1	1	1
Business Unity South Africa (BUSA)	1	1	1	1	1

N = 18; n = 18
Likert scale: Strongly disagree (1); Disagree (2); Undecided (3); Agree (4); and Strongly agree (5)
I/RS: Identification/review of stakeholders
P&P: Profiling and prioritising
MC: Managing communication
B/MR&K: Building/managing relationships and knowledge
U/MC&C: Updating/managing concerns and conflict

relationships, identifying and managing concerns and conflicts, and managing communication). There are strong indications to suggest that these were completely ignored.

Ignoring stakeholder management activities meant that there was no systematic process of dealing with typical dynamics forces originating from internal and external stakeholders' interactions. Strong indications suggest that there were no standards on which judgement was based (criteria), and thus no evidence of appropriate tools or techniques being used to generate information and aid decision making.

Chapter summary

Whether hospitals in Africa, toll roads in South America or hydroelectric projects in Europe, and including both successful and failed projects, the e-Toll GFIP experience described in this chapter offers a rich and diverse array of lessons for anyone contemplating partnership between the public and private sectors. Indeed, the history of the e-Toll GFIP and the opinions of stakeholders identified a handful of 'lessons learned', which should provide a good base for the future management of PPP projects.

This chapter has attempted to address what are identified as the twin challenges facing emerging economies when involved in PPP projects by putting into context the role of the public as the main beneficiaries of the facilities and allied services and the need to manage multi-stakeholders who are always at the centre of any PPP project.

It is clear from the literature and analysis done in the chapter that the real public represented by the first 'P' is being marginalised through the dynamics of PPP transactions. The main source of this marginalisation could be traced through the delegation dynamics found in the public sector environment. Since the delegation mechanisms within the public sector are influenced by the principal-agent (P-A) theory, it is evident that the dynamics shown in the theory has significantly contributed to the marginalisation. The solution lies in understanding the connection between stakeholder management and the appropriate contextualisation of what constitutes the real first 'P'

From the foregoing it is clear that the future success of PPP initiatives will partly depend on how the first 'P' is defined and how the stakeholders' structure is formulated towards embracing the 'CII: Real Position'. If PPP initiatives are to continue assuming a central position in the provision of infrastructure for the betterment of the public, the chapter recommends that the public as in the 'CII: Real Position' should be embraced within the spirit of a PPP project construct.

What is referred to in this chapter as PPP appropriate stakeholder management, in order to embrace transparent mechanisms will be different from current stakeholder management thinking that doesn't accommodate the true dynamics of PPP structures. Furthermore, the chapter has attempted to demonstrate how PPP stakeholder management approaches will be fundamental building blocks and how they differ from contemporary approaches.

Public protests and sweeping statements from public sector agencies characterise the face of current PPP project dynamics. Current tensions and public protests across Africa, South America, Australasia and Europe with regard to PPP projects, and specifically the e-Toll GFIP, seem to be centred on public stakeholders' marginalisation. The public outcry across continents is characterised by a huge group of people calling for full information on mushrooming PPP projects. The protest theme across most of the PPP projects seems to have a primary message: the public does not know how PPP projects were conceived, and why the private sector is involved in these projects. In short the general public as a central stakeholder-group is demanding its rightful place in the procurement process for PPP projects.

The chapter also takes a closer look at the second challenge facing PPP projects: *the marginalisation of the public*. First, it briefly reflects on the essence of PPP projects, the reason for partnership, social claims, and bad and good examples from practice on issues of stakeholders' involvement in projects. It furthermore takes a closer look at the unique nature of PPP stakeholders and the process of setting the real objectives for PPP projects. In order to find an informed solution to the challenge, the chapter presents a review summary of current theory and practice of stakeholder management. Finally a model that brings the public to the centre of PPP project planning and construction and operation of the facility is formulated and its applicability in practice described.

By tracing the history of the e-Toll GFIP and opinions of its stakeholders, the proposed model presents a thorough representation of multi-stakeholder management processes across the PPP project lifecycle. Each stakeholder management process cluster is described and clear criteria constructs and tools/techniques necessary to make decisions are identified. It is expected that the proposed model will help PPP project managers and other experts create a focused framework to involve the greater public in PPP projects and thus eliminate the element of marginalisation.

Acknowledgements

The author acknowledges his PPP Think Tank team members during his sabbatical at the University of West of England, Bristol, UK, and in particular Peter Fewings and Christian Henjewele. The foundation and framework of this chapter were mooted through debates and research which also led to the publication of Henjewele *et al.* (2013) and Rwelamila *et al.* (2014).

Reflections

1   What are the current challenges facing PPPs in South Africa and what initiatives are being used to overcome them?
2   The Support Programme for Accelerated Infrastructure Development (SPAID) is a partnership between business and government in order to support the achievement of the infrastructure development targets. It produced a report in 2007 that identified key challenges to south African PPPs and can be accessed at: http://castalia-advisors.com/files/12345.pdf.
3   Fombad (2014) identifies several measures that need to be adopted to enhance accountability in South African PPPs if the vision of improving service delivery and realising the national and international developmental projects is to be attained. These include clarifying accountability relations, monitoring measures, transparency, ethical standards, risk transfer and institutional reform.

4   The National Treasury of South Africa established a PPP Unit in 2000 with a mission to facilitate and enhance quality public service delivery by utilising efficient, effective and value-for-money best practice procurement solutions. PPPs are seen by the government as containing key features that are excellent for achieving 'Black Economic Empowerment' (BEE) objectives such as growing black equity and management, providing sub-contract opportunities for black enterprises and creating new jobs amongst the black community. Their website contains policy and guidelines, tender details, project information etc. and can be viewed at: http://www.ppp.gov.za.

5   The World Bank produced a 2014 booklet on PPPs in continental Africa that contains several case studies as well as information on lessons learned from past experience with water PPPs in Africa. This is a useful body of knowledge and can be accessed at: http://ppp.world-bank.org/public-private-partnership/library/water-ppps-africa.

References

Ahadzi, M. and Bowles, G. (2004). Public-private partnerships and contract negotiations: an empirical study, *Construction Management and Economics*, **22** (9): 967–978.

Akintoye, A., Chinyio, E. and Beck, M. (2003). Achieving best value in private finance initiative project procurement, *Construction Management and Economics*, **21** (5): 461–470.

All Africa (2011). Nigeria: issues, protests against Lekki toll plazas. Available from: http://allafrica.com/stories/201112130487.html – viewed: 06 September 2014.

Alli, N. (2013). How Gauteng has reaped benefits of road upgrades, *Sunday Times*, 2 June.

Balezdrova, A. (2011). Greek drivers refuse to pay toll fees, companies concessionaires are looking for a way to collect the lost revenues. Available from: http://www.grreporter.info/en/greek_drivers_refuse_pay_toll_fees_companies_concessionaires_are_looking_way_collect_lost_revenues/3 – viewed: 15 September 2014.

BBC (2006). Delhi protest toll rises to four. Available from: http://news.bbc.co.uk/1/hi/world/south_asia/5366948.stm – viewed: 20 August 2014.

Berle, A. and Means, C. (1932). *The Modern Corporation and Private Property*. New York: Commerce Clearing House.

Brehm, J. and Gates, S. (1997). *Working, Shirking, and Sabotage: Bureaucratic Response to a Democratic Public*. Ann Arbor, MI: University of Michigan Press.

Chrisafis, A. (2011). Greece debt crisis: the 'we won't pay' anti-austerity revolt. Available from: http://www.guardian.co.uk/world/2011/jul/31/greece-debt-crisis-anti-austerity – viewed: 25 August 2014.

Cuttaree, V. (2008). Successes and failures of PPP projects. *The World Bank Presentations – Europe and Central Asia Region*, Warsaw, 17 June. Available from: http://siteresources.worldbank.org/INTECAREGTOPTRANSPORT/Resources/Day1_Pres2_SuccessesandFailuresPPPprojects15JUN08.ppt – viewed: 26 August 2014.

Cyert, R.M. and March, J.C. (1963). *The Behavioral Theory of the Firm*. Upper Saddle River, NJ: Prentice Hall.

DeIonno, P., Hazelhurst, E., Crotty, A. and Enslin-Payne, S. (2011). Gautengers unite as discord takes toll on road plan. *Business Report*. Available from: http://www.iol.co.za/business/opinion/business-watch – viewed: 25 August 2014.

Dobbin, M. (2007). Canada's deadly 'P3' hospital boondoggles. Available from: http://thetyee.ca/Views/2005/10/27/HospitalBoondoggles – viewed: 27 August 2014.

Domberger, S. and Fernandez, P. (1999). Public-private partnerships for service delivery, *Business Strategy Review*, 10 (4): 29–39.

El-Gohary, N.M., Osman, H. and El-Diraby, T.E. (2006). Stakeholder management for public private partnerships, *International Journal of Project Management*, 24: 595–604.

Fombad, M.C. (2014). Enhancing accountability in PPPs in South Africa, *South African Business Review*, 18 (3): 66–92.

Grimsey, D. and Lewis, M. (2004). *Public Private Partnership: The Worldwide Revolution in Infrastructure Provision and Project Finance*. Cheltenham, UK: Edward Elgar Publishing.

Grimsey, D. and Lewis, M. (2005). (eds) *The Economics of Public Private Partnerships*. Cheltenham, UK: Edward Elgar Publishing.

Gutner, T. (2005). Explaining the gaps between mandate and performance: agency theory and World Bank environmental reform, *Global Environmental Politics*, 5 (2): 10–37.

Henjewele, C., Sun, M. and Fewings, P. (2011). Critical parameters influencing value for money variations in PFI projects in the healthcare and transport sectors, *Construction Management and Economics*, 29 (8): 825–839.

Henjewele, C., Fewings, P. and Rwelamila, P.D. (2013). De-marginalising the public in PPP projects through multi-stakeholders management, *Journal of Financial Management of Property and Construction*, 18 (3): 210–231.

HM Treasury (2000). *Public Private Partnerships: The Government's Approach*. London: The Stationery Office.

Hodge, G.A. (2004). The risky business of public-private partnerships, *Australian Journal of Public Administration*, 63 (4): 37–49.

Hodge, G.A. and Greve, C. (2005). *The Challenge of Public-Private Partnerships: Learning from International Experience*. Cheltenham, UK: Edward Elgar Publishing.

Hodge, G.A. and Greve, C. (2007). Public-private partnerships: an international performance review, *Public Administration Review*, 67 (3): 545–558.

Holmes, J., Capper, G. and Hudson, G. (2006). Public private partnerships in the provision of health care premises in the UK, *International Journal of Project Management*, 24 (7): 566–572.

Ibitayo, O. (2002). Public-private partnerships in the siting of hazardous waste facilities: the importance of trust, *Waste Management Research*, 20 (3): 212–222.

Ilkova, E. (2013). Spreading the pain, *Financial Mail*, 29 November – 4 December, p. 31.

Iyoghojie, P. (2011). Lekki tollgates: we will not surrender, LERSA vows. Available from: http://pmnewsnigeria.com/2011/02/24/lekki-tollgates-we-will-not-surrender-lersa-vows – viewed: 25 August 2014.

Koch, C. and Buser, M. (2006). Emerging metagovernance as an institutional framework for public private partnership networks in Denmark, *International Journal of Project Management*, 24: 548–556.

Leedy, P.D. and Ormrod, J.E. (2005). *Practical Research: Planning and Design* (8th edn). Upper Saddle River, NJ: Prentice Hall.

Li, B., Akintoye, A., Edwards, P.J. and Hardcastle, C. (2005). Critical success factors for PPP/PFI projects in the UK construction industry, *Construction Management and Economics*, 23 (5): 459–471.

Lupia, A. and McCubbins, M.D. (2000). Representation or abdication? How citizens use institutions to help delegation succeed, *European Journal of Political Research*, 37: 291–307.

Majamaa, W., Junnila, S., Doloi, H. and Niemistö, E. (2008). End-user oriented public-private partnerships in real estate industry, *International Journal of Strategic Property Management*, 12 (1): 1–17.

Ministry of State Road Transport and Highways (MoSRTH) (2011). Toll barriers on national highways. Available from: http://www.indiantollways.com/category/tolling – viewed: 25 August 2014.

Monbiot, G. (2001). *Captive State: The Corporate Takeover of Britain*. London: Pan.

Mtshali, N. (2014). Moral body portrays e-tolls as a despised evil, *The Star*, 4 September, p. 6.

New South Wales Treasury (NSW Treasury) (2006). The state infrastructure strategy [2006–7 to 2015–16]. Available from: http://www.treasury.nsw.gov.au/__data/assets/pdf_file/0018/5049/part1-pp1-52.pdf – viewed 25 August 2014.

New Zealand Parliament (NZP) (2011). Petition of Penelope Mary Bright and 171 others. Available from: http://www.parliament.nz/en-NZ/PB/Presented/Petitions/e/d/f/49DBHOH_PET2996_1-Petition-of-Penelope-Mary-Bright-and-171-others.htm – viewed: 25 August 2014.

New Zealand Public Services Association (NZPSA) (2011). Prison privatisation: follow the money. Available from: http://www.youtube.com/watch?v=zqTnbK1iClM – viewed: 25 August 2014.

NSW Department of Urban Affairs and Planning (2001). *Ideas for Community Consultation: A Discussion on Principles and Procedures for Making Consultation Work*. A report prepared by Lyn Carson and Katharine Gelber, New South Wales, Australia.

Parliament of New South Wales Legislative Assembly (PoNSW) (2009). *First Session of the Fifty-fourth Parliament Questions and Answers No. 115* (13 March 2009). Available from: http://www.parliament.nsw.gov.au/prod/la/lachapter.nsf/0/C41DA6F1849AAE7FCA2575780011AEA3/$file/115-QA-S.pdf – viewed: 25 August 2014.

Pollock, A., Price, D. and Player, S. (2007). An examination of the UKs Treasury's evidence base for cost and time overrun data in UK value-for-money policy and appraisal, *Public Money and Management*, 27 (2): 127–133.

Prefontaine, L., Ricard, L., Sicotte, H., Turcotte, D. and Dawes, S. (2000). New models of collaboration for public service delivery. Draft report for the Centre for Technology in Government. Available from: http://www.ctg.albany.edu/publications/reports/new_models_wp – viewed: 23 February 2012.

Raisebeck, P., Duffield, C. and Xu, M. (2010). Comparative performance of PPPs and traditional procurement in Australia, *Construction Management and Economics*, 28 (4): 345–359.

Renn, O., Webler, T. and Wiedemann, P. (1995). *Fairness and Competence in Citizen Participation: Technology Risk and Society*. Dordrecht: Kluwer Academic Publishers.

Rwelamila, P., Fewings, P. and Henjewele, C. (2014). Addressing the missing link in PPP projects: what constitutes the public?, *Journal of Management in Engineering*, 35 (5), 164–171.

Santoro, D. (2003). The 'Aguas' tango. Centre for Public Integrity. Available from: http://www.globalpolicy.org/component/content/article/221-transnational-corporations/46887.pdf – viewed: 25 August 2014.

Seargent, J. and Steele, J. (1998). Consulting the public: guidelines and good practice. Policy Studies Institute, London. Available from: http://www.psi.org.uk/consulted – viewed: 25 August 2014.

Shen, L.Y., Platten, A. and Deng, X.P. (2006). Role of public private partnerships to manage risks in public sector projects in Hong Kong, *International Journal of Project Management*, 24: 587–594.

Smith, H. (2011). Greek mayor arrested after bulldozing toll booth barriers. Available from: http://www.guardian.co.uk/world/2011/jan/14/greek-mayor-bulldozes-toll-booth-barriers – viewed: 25 August 2014.

The Greens (2012). NSW not for sale. Available from: http://nsw.greens.org.au/NSW-not-for-sale – viewed: 25 August 2014.

UK Treasury (2011). *House of Commons Treasury Committee – Private Finance Initiative, 17th Report of Session 2010–12*. London: The Stationery Office.

United States Department of Transportation (US-DOT) (2004). Report to Congress on public-private partnerships. Available from: http://www.fhwa.dot.gov/reports/pppdec2004/pppdec2004.pdf – viewed 25 August 2014.

William, A. (1972). *The Turnpike Road System in England, 1663–1840*. Cambridge: Cambridge University Press.

Yuan, J., Skibniewski, M.J., Li, Q. and Zheng, L. (2010). Performance objectives selection model in public-private partnership projects based on the perspective of stakeholders, *Journal of Management in Engineering*, 26 (2): 89–104.

Zerbst, F. (2011). How Gauteng's toll roads will work, *Mail & Guardian*. Available from: http://www.itssa.org/blog/2010/11/09/how-gautengs-toll-roads-will-work – viewed: 25 August 2014.

17 Community-based facilities management (FM) as a form of relationship contracting

Kathy Michell, Hanna Boodhun, Michelle Bunting and Leila Rostom

Chapter introduction

African cities continue to face enormous challenges in terms of rapid urban growth, substantial urban poverty and a multitude of other social problems. In a recent report on the state of African cities, evidence is presented in terms of the reality that African cities are currently in a state of demographic, economic, technological, environmental and socio-political transition (UN-Habitat, 2014). Moreover, African cities are faced with enormous population growth associated with wide-spread poverty across the continent. This report further argues that African cities are too often viewed through a Western paradigmatic approach to urbanism and urban living and what is required is a comprehensive re-think of urban management in an African context in order to attain Africa's vision of sustainable human settlements in terms of human development and prosperity for all. In this context, there is a growing demand from both the public and the private sectors for the need to develop solutions to the creation and management of sustainable African cities (Michell, 2013).

It is clear that a new form of intervention is required in African cities. There is growing evidence that what is required is new ways of managing cities in Africa. The concept of managing public infrastructure and the associated services via the application of facilities management (FM) principles was first postulated by Roberts (2004). Nutt (2004), Roberts (2004), and Tobi *et al.* (2013) argue that the incorporation of facilities management in the management of public infrastructure and its associated services is one of the key elements to the future of public sector facilities management. Hence, it can be argued that facilities management and, more specifically, community-based facilities management has the potential to act as a change agent in the upliftment of the urban poor and a means to addressing the myriad of challenges facing local government. How this is practically implemented remains a challenge due to the scale of the problem and the diverse cultural landscape of different communities within cities in Africa and, more specifically, South Africa. It is argued by some that one way to address the scale of the urban problem is to increase the level of private sector involvement in infrastructure provision and service delivery (Grimshaw *et al.*, 2002,

Ngowi, 2007, Hartmann *et al.*, 2010). UN-Habitat (2014) argues that the 'agents of change' in South Africa are diverse and are likely to require interventions from the private sector in the form of Public-Private Partnerships.

This chapter explores the link between community-based facilities management and relationship contracting, as a form of Public-Private Partnership (PPP) in addressing the challenges facing South African cities. This chapter supports and extends the findings of Chapter 16 by way of an innovative example of community involvement in relationship contracting procurement. The previous chapter traced the history and development of PPPs and also provided a South African contextual case study but in the form of lessons learnt from a more traditional e-toll PPP project.

Contextual landscape of South African cities

It is acknowledged that the percentage of people living in slums and informal settlements in Southern Africa is lower than that compared to the rest of Africa. However, the region faces the same challenges as its counterparts on the continent (UN-Habitat, 2014). These challenges are: high levels of poverty and inequality, urban sprawl, substantial housing backlogs, spatial separation of residential areas defined by race, extensive informal settlements within urban areas, and a lack of adequate infrastructure and services to the urban poor (Robinson, 2008, Michell, 2012, UN-Habitat, 2014). Many of these urban challenges are still a legacy of the apartheid government policies (Southall, 2004, Turok and Parnell, 2009, Michell, 2012). Hence, any intervention on the part of local government will place significant emphasis on ensuring democratic participation, the alleviation of poverty and inequality, improving infrastructure provision and service delivery, overcoming patterns of urban segregation, mitigating xenophobia, addressing civil unrest and achieving interconnection between the formal economy and the informal economy (African National Congress (ANC), 1994, Patel, 2004, Parnell, 2005, Turok and Parnell, 2009, UN-Habitat, 2014).

In this context, it is argued that local government has a central role to play in poverty alleviation, employment creation, infrastructure delivery and service provision (Narayan, 2000, Mufamadi, 2008). More specifically, the central focus of local government ought to be on the adoption of a capacity-focused view of public facilities and the welfare of communities and their ability to self-manage (Michell, 2012). Moreover, the application of community-based facilities management within this urban landscape is likely to provide a level of consistency to the management and operation of public facilities within South African cities.

Community-based facilities management

The emergent concept of community-based facilities management has its philosophical roots in the 'socially oriented' view of facilities management

(Alexander and Brown, 2006, Michell, 2010). This school of thought within facilities management literature argues for a deeper understanding about the relationship between space, place, people and technology and the nature of the interactions between users of facilities and urban precincts, i.e. the socially constructed reality of urban space (Becker and Steele, 1990, van Dommelen et al., 1990, de Bruijn et al., 2001, Grimshaw, 2004, Alexander and Brown, 2006, Price et al., 2009, Michell, 2012). Narayan (2000) contends that, in order to take action at a local level, an understanding of the 'contours of the patterns of poverty' needs to be understood within its local context. This context would take cognisance of the location, the social group, the region and the country.

At the heart of a community-based orientation in alleviating poverty in urban areas is the creation of successful partnerships with communities, thereby providing opportunities that have a direct benefit to the local community (Schriner, 1993). Kasim and Hudson (2006) in establishing the role of facilities management in creating direct benefits to communities, argue for the alignment of public sector facilities management towards the provision of services to support the local community needs and the local economy. Alexander and Brown (2006) define community-based facilities management as:

> the processes by which all the stakeholders in a community work together, to plan, deliver and maintain an enabling environment, within which the local economy can prosper, quality services can be delivered and natural resources protected, in order that citizens can enjoy a quality of life.

Alexander and Brown (2006) further argue that community-based facilities management requires a reorientation from the organisation, workplace, business service and advocacy of the user to the community, neighbourhood, community resources and the citizens in that community. Consequently, the community is seen as a key participant in the planning, delivery, management and maintenance of the facility. Alexander (2008) emphasises the role that community-based facilities management can play as a means of developing social capital and addressing issues around social cohesion in areas of poverty. Furthermore, the case is made by Alexander (2008) that 'community assets include land, the environment and infrastructure owned, managed or impacted on by the community'. Moreover, that the greater the encouragement of community-based options for the use of these assets, the greater the capacity of communities to undertake their own local development is achieved. In providing a framework for the contextualisation of community-based facilities management, Alexander and Brown (2006) argue that it may be seen to encompass facilities management issues from a social and community perspective and comprises four key dimensions:

- the process of engagement and collaboration amongst stakeholders;
- the creation of an enabling environment;
- achieving a balance between social and environmental values; and,
- quality of life for the community.

Michell (2010), in a grounded theory study of public sector facilities management in the townships of Cape Town established that community-based facilities management creates both the platform and the environment for the achievement of the sustainable cities agenda. More importantly, that community-based facilities management has the potential to provide a contextually appropriate vehicle in addressing the challenges facing South African cities. In addition, community-based facilities management is underpinned by four key dimensions/pillars: local economic development, social development, community participation and sustainability, as depicted in Figure 17.1.

In this context, Michell (2010) argues that an Afro-centric view of community-based facilities management requires an acknowledgement of these four pillars and therefore defines community-based facilities management as:

> the integration of processes within a community to plan, develop and maintain the agreed services, in consultation with the community, which support, empower and improve the community in terms of local economic development and social development, in order to achieve sustainable human settlements.

Figure 17.1 Framework for conceptualising community-based facilities management (Source: Michell, 2010)

It was evident from this initial research that the reasons for local governments' failure to manage their physical assets adequately are a lack of capacity and the requisite skills base within local government, a failure to operationalise and manage facilities post construction and a lack of the required operating budget to manage and maintain their facilities sufficiently.

Subsequent research that sought to identify the key components of a facilities management policy that governs the operation of community-based facilities established that relationship building is seen as a key element to the relative success or failure of a facility (Michell, 2012). It was at this point in the evolution of the concept of community-based facilities management, from a wholly theoretical stance toward the implications for practical implementation, that the question can be asked as to the possibility that community-based facilities management could be operationalised by local government in the form of a relational or relationship contract. A relational approach to contracting is seen to encompass a long-term social exchange between parties that is underpinned by mutual trust, interpersonal attachment, a commitment to the partnership, altruism and co-operative problem solving (Walker and Davis, 1999, Duberley and Johnson, 1999). This view is supported by Rahman and Kumaraswamy (2002) who argue that relationship contracting is fundamentally about a recognition of mutual benefits and the creation of sustainable solutions via the development of co-operative relationships. The following section documents the findings of a case study of a facility in the Cape Town metropolitan area where community-based facilities management is being implemented. The aim of this case study was not only to observe the implications of the practical implementation of community-based facilities management, but also to examine the potential for it to be an active agent for change via a relationship contract.

A case study of community-based facilities management in action

The fieldwork comprised a single case study of a public facility within a socially and economically marginalised community (Khayelitsha) in the greater Cape Town metropolitan area. The aim of the case study was to build on the conceptual framework for community-based facilities management proposed by Michell (2010) (see Figure 17.1). The particular case documented hereunder was selected because of its unique setting and is a current prime example of a functional community-based facility. The case study in this sense is an *instrumental case study* with the primary interest being in how community-based facilities management occurs in practice. It is unique in its genre in terms of how the facility is managed and for this reason only a single case study approach could be adopted. Closer investigation of the case in question revealed that it is the first facility displaying the characteristics of community-based facility management in the City of Cape Town.

More importantly, further upgrading is occurring to the facility under a partnership agreement between local government, a foreign European bank, a non-government organisation and the local community. It should be noted at this point that this particular initiative uses an 'area-based approach' and therefore the partnership agreement is effectively managing several facilities within this chosen locale of Khayelitsha. Semi-structured interviews were undertaken with a variety of stakeholders, including local government officials, non-government organisations, community leaders, local residents and the private sector.

History of Khayelitsha

Khayelitsha is situated on the outskirts of Cape Town and is one of the largest marginalised townships in South Africa (Goodlad, 1996). The township came into existence as a result of an urgent need to move 'black' people from the massively overcrowded neighbouring townships of Langa, Nyanga and Gugulethu (Ndingaye, 2005). As a result, Khayelitsha (meaning 'new home' in IsiXhosa) is an area where apartheid political strategies lead to the segregation of 'black' people and it was common practice that the lowest income earners, usually migrant workers, would settle in Khayelitsha to reduce transport costs to their place of work. This history has resulted in Khayelitsha being characterised by high levels of unemployment and poverty. Residents of Khayelitsha have been living in drastically poor conditions since the apartheid regime, with a lack of road infrastructure and access to basic public services (water, sanitation and electricity). In addition, Khayelitsha is considered to be prone to violence and crime (Nleya and Thompson, 2009). The inherent contextual landscape of Khayelitsha makes it a suitable case in terms of being representative of the urban challenges facing South African cities.

Local community context and dynamics

Previous research has identified a direct relationship between the facility, its surrounding area and more specifically the dynamics of the local community within which the facility is located (see Michell, 2012). Some relevant observations about the local area within Khayelitsha where the facility is located can be used as a means of analysing how the community and its dynamics affect the implementation of community-based facilities management. The demographics, the politics, the socio-economic concerns and the rate of violence and crime in the community create internal dynamics that affect the structure of the community. These identified characteristics were shown from the findings to generate positive elements in the community resulting in a high sense of civic duty leading to volunteerism as was demonstrated by the presence of many community-based organisations in the area and voluntary security patrollers. This sense of civic duty also contributes to a high sense of belonging of the local people to their community and, hence, a willingness to take 'ownership' of any project/facility in their area. However, these inherent characteristics of the community also create negative features such as conflicts of interest among people with different agendas and diverging political alliances. In addition, these negative characteristics mean that the restricted resources in the community result in the local community having a low level of education and a high unemployment level. Hence, the local community possesses limited skills, experience and managerial abilities in order to be in a position to manage and operationalise a facility in their community.

These dynamics are also major contributors to the abuse of law and order systems in place as well as a high rate of crime, particularly violent crime in the area. Moreover, these inherent attributes of the community also impact on its external environment. It was evident from the case that the community, in terms of its needs and socio-economic concerns, has a two-way relationship with the local government via a project undertaken by a non-government organisation in Khayelitsha. The latter has set its objectives so that, through its upgrading programme, it can redress the negative features that result from the core characteristics of the community. As explained by different participants when referring to various conflicts and territorial attitudes met within the local area, this urban upgrading project has had to work with and around these.

The objectives of the project, for the area under concern are to: build safe and integrated communities and overcome the four types of exclusion: economic, cultural, social and institutional. Political controversies, though inevitable, are minimised by avoiding any show of favouritism to any local group. The project seeks to empower the community by promoting participation through training and skills development thereby opening up channels for future employment. In addition, this project makes it an imperative for the community to be consulted both at the onset of any intervention and to integrate the community in the management of any facility that is developed. However, through this community participation, the non-government organisation and local government have to deal with both the positive and the negative aspects of the community dynamics. The field data highlighted the relational interface role played by the non-government organisation in ensuring that the facility meets the needs of the local community.

It was evident from the case that the four pillars on which the implementation of community-based facilities management rests (local economic development, social development, community participation, sustainability) are the driving force for the success of this facility. Moreover, this replication of the four pillars can be seen from the formal approach that the local government has been trying to adopt through the concept of 'community delivery of services'.

The facility

McShane (2006) argues that for community-based facilities management to exist, there has to be physical facilities to be managed. Therefore, the facility, on its own, is the axial point of the implementation of the community-based facilities management concept. The analysis of the facility, as a standalone element, provides the researchers with the opportunity to understand how its facilities management is a function of its design, operation, size, location and rate of usage. This is essential for the development of a context-specific understanding of the integration of people, place, process and technology for this facility. Community needs define the functionality of the building be it commercial, retail or community centres and this, in turn, determines the design of the building and how the facility operates. Furthermore, from the data collected, it was found that the size, location and the rate of usage of the facility has a direct impact on the different roles and responsibilities of a potential facility manager. These responsibilities, in the form of portfolios, would reflect the different components of an effective facilities management policy, e.g. financial, administrative, marketing and maintenance.

The emergent data from the case revealed that the local facility is directly affected by the non-government organisation project, the local community (*read community dynamics*) and the facility management structure established by the local government. Moreover, there is a direct relationship between the community's needs and its use of the facility and the level of activity that takes place there. There exists a dynamic relationship between the effectiveness of the facility and the operationalisation and management of the facility, i.e. the performance of the facility is reliant on the degree to which the function of the building aligns with the objectives of the users.

Emergent themes for a community-based facilities management policy framework

The thematic analysis of the data allowed for the emergence of a number of themes which represent the key components for a community-based facilities management policy framework. It is within this framework that the relationship contract acts as the vehicle for the implementation of community-based facilities management. Each of these themes are briefly outlined below together with a few selected verbatim quotes from the interview transcripts in order to highlight key points in the evolution of the emergent themes. The project stakeholders that were part of the interview process are identified using the following *Confidentiality Codes*:

- *LCP* = Local Community Person
- *CNSX* = Consulting Organisation Employees
- *LG* = Local Government.

Emergent theme 1: Facilities management committee

The case under investigation highlighted that one of the emergent themes intrinsic to the implementation of community-based facilities management was the establishment of a facilities management committee that is responsible for the operationalisation and management of the facility. In this regard the facilities management committee can be seen as a subset of the wider local community, in that it is treated as a separate entity that acts with the sole purpose of engaging in the facilities management of the facility. As stated by one of the respondents:

> CNSX5: So, they [facilities management committee] do the mainte-nance and everything else but the direction and decisions making and everything else and policy input about what happens there is actually working within that facilities management committee . . . they make their own money through their own fundraising and that money goes back to the facility . . . basically, the City builds these facilities and it hands them over to the community although it belongs to the city and then hands over but then everything is then 100 per cent controlled and owned.

Hence, the facilities management committee emerged to fulfil the objective of local community participation in socio-economic activities and the civic affairs of their local community to improve service delivery. In this regard, the facilities management committee sought to have representativeness of the whole community by selecting various members of different local organisations. This is seen as a key issue that needs to be considered when creating a facilities management committee, especially in an area which has different local groups already operating in it. It is further noted that this representativeness will be inversely proportional to the cohesion of the group forming the facilities management committee. The more varied from each other the agendas and aims of the different facilities management committee members are, the lower will the alignment of their objectives be in terms of the facilities management committee.

Here again, the dynamics in the community thus impacted on the structure of the facilities management committee and the interaction between the committee members where, for example, diverging agendas of the different people in the committee led to clashes and conflicts. It was also observed that the limited skills and low education level of the community influenced the scale at which the facilities management committee could operate. From the data that was collected, it could be inferred that the most important element in how community-based facilities management can be implemented is via the creation of a locally representative committee. In the same manner that the facility, on its own, was identified as being the foundation upon which community-based facilities management relied for its very existence, the implementation of the concept requires that a committee of local people be created to manage the facility.

Emergent theme 2: Formal conflict resolution mechanisms

The second emergent theme was formal conflict resolution mechanisms that need to form a part of the policy framework that would guide the implementation of community-based facilities management. This theme has been identified directly from this case study, as its absence in the case under investigation was a major contributor to the ineffectiveness of the facilities management committee. As stated by two respondents:

> CNSX6: And so you had, competing groups in the area, literally, legally at loggerheads and we were trying to find a community group to manage the thing peacefully. So, we basically said look, if this group manages it, this group will burn it down, this group manages it, and this group will burn it down.

> CNSX4: When there were elections, some people wanted to be in certain positions. And also when they were not elected then they started challenging that I want to be this . . . There were people in the neighbourhood who wanted to see themselves as 'The People' [indicated by respondent] in the Case X management.

Though the internal breakdown of the facilities management committee, through conflicts of interests, could be observed by the representatives of the non-government organisation and local government representatives, there was no formal mediation that took place between the members to work through these problematic areas. Although there exist reference meetings through which these dissensions could be addressed, there needs to be formal structured channels through which these disputes can be highlighted, discussed and settled.

Emergent theme 3: Training and mentoring

The third emergent theme is that of training and mentoring. In the context of urban challenges facing South African cities, the majority of the community facilities requiring direct interventions on the part of local government are in, and will be built in, marginalised communities where there are restricted resources. More importantly, the majority of the local community, who form the foundation of a community-based intervention have limited skill sets. Clearly sufficient resources are required within a community in order to manage and operationalise the facility successfully:

> CNSX6: But none of the groups we work with are ready to do that. And we've tried to get them to take on those roles, but it became clear that

they are not ready . . . Well, we gave them the responsibilities, won't you manage the finances? But they couldn't do that . . . But they're not able to do the . . . they don't have people with the time and the infrastructure and resources.

CNSX2: Yes training. Lots of training . . . Training skills.

This reinforces the need to train the community members prior to appointing them to the facilities management committee, in order to provide them with the necessary tools that they will need to manage a facility. To cater for the ability of participants to take on responsibilities regarding all the components that make up the discipline of facilities management, training as a tool is crucial. Training is especially important to overcome the inverse relationship that may exist between project efficiency and the degree of community participation, particularly, in an area with identified dynamics similar to the case study. Training would thus allow for a reversal of such a relation so that a directly proportional correlation can be established between the two. In as much as training is a vital element in community-based facilities, it is a short-term instrument that is provided at one specific point in time at the onset of the facilities management committee creation. There is clearly a need for a separate measure to ensure that community members continue to learn and build on the skills that they have acquired through the training programmes. An element that could be implemented is that of mentoring: 'So, we just decided that let's go a little more slowly, advisory first, as they are able to take on more then, let's explore that' (CNSX6).

Providing a formal mentoring programme to the facilities management committee members offers long-term guidance and the ability to assess their progress and abilities. It is a means of gauging any possible weaknesses or additional skills that are required throughout the operation of the facilities management committee. Mentoring is a two-way communication path that provides the beneficiary of the programme a constant support if facing any problem until such time as the facilities management committee is self-sustaining.

Emergent theme 4: Accountable financial system

The fourth emergent theme is that of the need for an accountable financial system. Depending on the function of the facility, the facilities management committee holds different portfolios. If one of them is concerned with the handling of income that may be generated from the facility or any other form of financial management, an accountable financial system is one of the essential tools in the creation of the facilities management committee. At this stage there is no financial system in place for the operation and management of the facility:

LG2: The City of Cape Town is carrying all these costs so we actually don't see the water usage and the electricity usage . . . it could be that, for example, have a water leakage and I do not see it because we don't have access. We don't even know where the water meter is so that we could even just write it down just to see the average consumption.

This is significant in terms of the trust and confidence that exist between the non-government organisation, local government and the facilities management committee. Such an element of trust underpinned by this accountability contributes to the sustained authority of the facilities management committee in managing the facilities. This accountable system can be enforced through mechanisms such as voluntary informal annual general meetings where the facilities management committee can provide a review of financial accounts to the non-government organisation, local government and the local community.

Emergent theme 5: Effective community-related incentives

The fifth emergent theme, that in this particular case could be seen to have a high level of resonance within the community, was discovered to be the identification of the most effective incentives for the community to participate in such a project and its facilities management. This theme was identified as being one of the main issues that was found to drive the community members to offer their services. The philosophy of the non-government organisation involved was that the participants should participate in the programmes as volunteers. Moreover, that the reward would be their influence on the facility, and therefore a tacit form of ownership of the facility. Hence, incentives, judged to be effective in the long term, were discovered to be less so in the short term. It was found that in this particular case where the non-government organisation, through its volunteerism philosophy, on which the community-based facilities management framework relied and which underpinned its whole training programme, was dependent on incentives with long-term benefits. On the other hand, the community people need motivations that are more beneficial in the short term:

CNSX6: At the heart of the Programme X approach is saying [mimicking voice] 'if you guys volunteer for things, there's a whole of good things that will start to happen. I mean you will gain experience, you will get new networks, you will meet government officials, you'll get skills that could lead to jobs; we know of you, we could recommend you for things.' So, for us . . . but it is not formalized, but we're saying that everything starts with volunteering, not with being paid.

LCP3: To, to be on the-the FMC? We did volunteer; there was no employ-ment there, as from 2007 till this year. There was no payment. We did try maybe there will be a payment after, if the FMC did go the right way as I was saying they were having, we did form our bank account, so everybody who were going to participate there were thinking in future.

Thus, the community-based facilities management framework should cater for the provision of long-term and short-term incentives so as to increase the effectiveness of the latter in ensuring community participation throughout the life cycle of the facility. Consequently, this will ensure that the concept of encouraging the community to engage in the facilities management of their local facilities can be implemented effectively.

Emergent theme 6: Formal channels of communication

The last emergent theme highlighted from the research is formal commu-nication structure/channels between the facilities management committee, the representatives of non-government organisation and local government:

LCP3: Yah, he [NI3] was reporting to us if there is anything there . . . yah, he [NI3] did do the reports and he was a part of the facility management at that time . . . CNSX2 also did do the reports to us . . . no . . . we claimed to him that, for the reports of this facility you need to come direct to us before you can report to your supervisor who is CNSX3.

CNSX3: I tend to differ in that in the sense that we have got a Community Facilitator that we have to-to communicate with him and then he is the one who is supposed to . . . to take that to the people. And then if they have got something, they actually communicate through him and he will bring it to the Operations and Maintenance without us being involved there. It has to stay purely community people in that faction. And then they must take that mandate from the people through the community participation.

A breakdown in communication was identified as a serious challenge faced by the stakeholders involved in the implementation of community-based facilities management in the case in question. It is interesting to note that Doloi (2009) identified communication as the most significant factor impact-ing on the success/failure of a relationship contract. Trust and confidence are also seen as having an impact on the efficacy of the relationship between the parties in a relationship contract (Ngowi, 2007, Doloi, 2009).

Earlier in the chapter it was argued that a relationship contract could provide the vehicle for the practical implementation of community-based facilities management at the grassroots level. Moreover, that community-based facilities management provides a contextually appropriate vehicle for

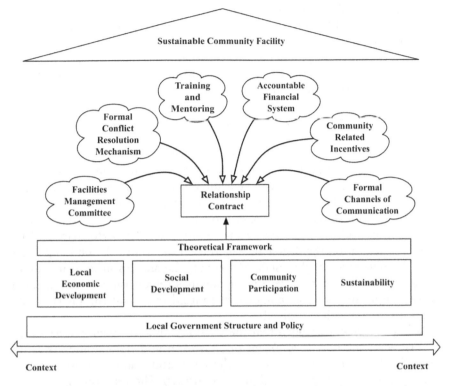

Figure 17.2 Key components of a relationship contract within the context of
community-based facilities management

capacity and skills development, local economic development, community
empowerment, community development and ultimately community control
over the management and operation of the facilities within their locale. The
case study has identified a number of emergent themes that relate directly
to the practical reality of the implementation of community-based facilities
management in a marginalised community. In seeking to align the concept
of community-based facilities management with relationship contracting,
it is clear at this point that these emergent themes would need to form the
very foundation of any relationship contract that local government would
be seeking to implement, as depicted in Figure 17.2.

In other words, the long-term sustainability of the relationship contract
would need to address specifically the following:

- the formation of a facilities management committee, comprised primarily
  of local community representatives that is responsible for the day-to-day
  management and operationalisation of the facility;
- conflict resolution mechanisms;

- training and mentoring strategies for the local community representatives involved in the facilities management committee;
- an accountable financial system for the 'partnership agreement' or relationship contract;
- effective and transparent community-related incentives; and
- clearly defined formal channels of communication between all parties involved in the relationship contract.

Conclusion

The emergent themes identified in the case study are intrinsic to the successful implementation of community-based facilities management and therefore it is likely that they have a direct bearing on the long-term sustainability of public facilities. It is clear from the case that there exists a dynamic relationship between the effectiveness of the facility and the operationalisation and management of the facility, i.e. the performance of the facility is reliant on the degree to which the function of the building aligns with the objectives of the users. Moreover, the successful implementation of community-based facilities management is reliant on a commitment to partnership, effective communication, mutual trust and the interpersonal relations of all parties involved.

While it is acknowledged that the partnership agreement between local government, the non-government organisation and the community is not termed a 'relationship contract' in the case study under investigation, it is in essence a form of a relationship contract. The relationship contract will allow local government to adopt a capacity-focused view of community assets and an increased focus on the well-being of communities and their capacity to self-manage. In addition, the application of a relationship contract will provide a level of coherence to the management and operation of public facilities within local government. The use of a relationship contract in the implementation of community-based facilities management in the management, maintenance and operation of infrastructure and public facilities in marginalised communities provides a means of acting as an agent for change in addressing some of the urban problems associated with African cities.

Reflections

1   A number of themes emerged from the case study project presented in this chapter and they represent the key components for a community-based facilities management policy framework. Can you summarise them and in doing so investigate other project-specific community-based innovations in construction procurement?

2   Community involvement and regional focus are also evident in another relationship-based contract in the form of indigenous social housing

in Australia's Northern Territory. Further discussion can be found in Jefferies M.C., Rowlinson, S. and Schubert A. (2012). 'The procurement of Indigenous social housing in Australia: a project alliance approach'. *Proceedings of CIB W070, W092 and TG72 International Conference*. Cape Town, South Africa. The paper can be accessed directly at: http://hdl.handle.net/1959.13/1050846.

3   South Africa's Construction Industry Development Board (CIDB) was established to provide leadership to stakeholders and to stimulate sustainable growth, reform and improvement of the country's construction sector. Their website is a useful resource for industry policy and guidelines.

4   The Netherland's Ministry of Foreign Affairs, through its Department of Policy and Operations Evaluation (IOB) commissioned a report into *PPPs in Developing Countries*. The report was published in 2013 and a significant part of the report focuses on sub-Saharan Africa. The report is available at: http://www.government.nl/files/documents-and-publications/reports/2013/06/13/iob-study-public-private-partnerships-in-developing-countries/iob-study-no-378-public-private-partnerships-in-developing-countries.pdf.

5   Another important resource within the field of FM is that of Alexander, K. and Price, I. (2012). *Managing Organizational Ecologies: Space, Management and Organization*. New York: Routledge. The book explores how space and the built environment are managed and constructed in policy and practice within the broader context of FM.

References

African National Congress (ANC) (1994). *The Reconstruction and Development Programme: A Policy Framework*. Johannesburg: Umanyano Publications.

Alexander, K. (2008). Managing community assets for urban sustainability. *International Conference on Urban Sustainability*. Hong Kong.

Alexander, K. and Brown, M. (2006). Community-based facilities management. *Facilities*, 24: 250–268.

Becker, F. and Steele, F. (1990). The total workplace. *Facilities*, 8: 9–14.

De Bruijn, H., Van Wezel, R. and Wood, R.C. (2001). Lessons and issues for defining 'facilities management' from hospitality management. *Facilities*, 19: 476–483.

Doloi, H. (2009). Relational partnerships: the importance of communication, trust and confidence and joint risk management in achieving project success. *Construction Management and Economics*, 27: 1099–1109.

Duberley, J. and Johnson, C. (1999). Contracting in local authorities. *Public Management: An International Journal of Research and Theory*, 1: 531–554.

Goodlad, R. (1996). The housing challenge in South Africa. *Urban Studies*, 33: 1629–1645.

Grimshaw, B. (2004). Space place and people: facilities management and critical theory. In (eds) Alexander, K., Atkin, B., Bröchner, J. and Haugen, T. *Facilities Management: Innovation and Performance*. Abingdon: Spon Press, pp. 15–32.

Grimshaw, D., Vincent, S. and Willmott, H. (2002). Going privately: partnership and outsourcing in UK public services. *Public Administration*, 80: 475–502.

Hartmann, A., Davies, A. and Frederiksen, L. (2010). Learning to deliver service-enhanced public infrastructure: balancing contractual and relational capabilities. *Construction Management and Economics*, 28: 1165–1175.

Kasim, R. and Hudson, J. (2006). FM as a social enterprise. *Facilities*, 24: 292–299.

Mcshane, I. (2006). Community facilities, community building and local government: an Australian perspective. *Facilities*, 24: 269–279.

Michell, K. (2010). *A Grounded Theory Approach to Community-based Facilities Management: The Context of Cape Town, South Africa*. Unpublished PhD Thesis, The University of Salford, UK.

Michell, K. (2012). FM as a social enterprise. In (eds) Alexander, K. and Price, I. *Managing Organizational Ecologies: Space, Management and Organization*. New York: Routledge, pp. 167–177.

Michell, K. (2013). Urban facilities management: a means to the attainment of sustainable cities? *Journal of Facilities Management*, 11 (3).

Mufamadi, S. (2008). Foreword. In (eds) Van Donk, M., Swilling, M., Pietserse, E. and Parnell, S. *Consolidating Developmental Local Government: Lessons from the South African Experience*. Cape Town: UCT Press, pp. v–vii.

Narayan, D. (2000). *Voices of the Poor: Can Anyone Hear Us?* New York: Oxford University Press for the World Bank.

Ndingaye, X.Z. (2005). *An Evaluation of the Effects of Poverty in Khayelitsha: A Case Study of Site C*. Master of Arts in Development Studies, University of the Western Cape, South Africa.

Ngowi, A.B. (2007). The role of trustworthiness in the formation and governance of construction alliances. *Building and Environment*, 42: 1828–1835.

Nleya, N. and Thompson, L. (2009). Survey methodology in violence prone Khayelitsha, Cape Town, South Africa. *IDS Bulletin*, 40: 50–57.

Nutt, B. (2004). Infrastructure resources: forging alignments between supply and demand. *Facilities*, 22: 335–343.

Parnell, S. (2005). Constructing a developmental nation: the challenge of including the poor in the post-apartheid city. *Transformation: Critical Perspectives on Southern Africa*, 58: 20–44.

Patel, Y. (2004). New urban realities: overview of urban challenges facing South Africa. *World Urban Forum*. Barcelona: United Nations Habitat.

Price, I., Ellison, I. and Macdonald, R. (2009). Practical post-modernism: FM and socially constructed realities. *European Facility Management Conference*. Amsterdam.

Rahman, M.M. and Kumaraswamy, M.M. (2002). Joint risk management through transactionally efficient relational contracting. *Construction Management and Economics*, 20: 45–54.

Roberts, P. (2004). FM: New urban and community alignments. *Facilities*, 22: 349–352.

Robinson, J. (2008). Continuities and discontinuities in South African local government. In (eds) Van Donk, M., Swilling, M., Pietserse, E. and Parnell, S., *Consolidating Developmental Local Government: Lessons from the South African Experience*. Cape Town: UCT Press, pp. 27–50.

Schriner, J.A. (1993). Communities and companies. *Facilities*, 11: 7–11.

Southall, R. (2004). The ANC & black capitalism in South Africa. *Review of African Political Economy*, 31: 313–328.

Tobi, S.U.M., Amaratunga, D. and Noor, M.N.M. (2013). Social enterprise applications in an urban facilities management setting. *Facilities*, 31: 238–254.

Turok, I. and Parnell, S. (2009). Reshaping cities, rebuilding nations: the role of national urban policies. *Urban Forum*, 20: 157–174.

UN-Habitat (2014). *The State of African Cities: Reimagining Sustainable Urban Transitions*. Nairobi: United Nations Human Settlements Programme (UN-Habitat).

Van Dommelen, D., Noordegraaf, R. and Buma, H. (1990). Decision making in a strategic approach to facilities management. *Facilities*, 8: 12–16.

Walker, B. and Davis, H. (1999). Perspectives on contractual relationships and the move to best value in local authorities. *Local Government Studies*, 25: 16–37.

Index

accountability: in Asian PPPs 177, 179, 184–5, 186, 190; in community-based facilities management (FM) 336–7
additionality in PFI projects 44
adversarial relationships 147, 168, 286, 287
Africa *see* community-based facilities management (FM)
Agency theory 306–7
alliances 13–14, 15, 23–4, 26–31
alternative technical concepts (ATC) 110–11
annual PPP project list publications 259–61, 262
arbiter's role in London Underground's PPP 59, 60, 61, 62
Argentina 304
artificial first 'P' (public) in the PPP environment 300–1, 302, 306, 307, 312–13, 314
artificial neuronal networks 78
Asian financial crisis (1997) 181, 187, 217, 236, 237
Asian Infrastructure Investment Bank (AIIB) 191
Asian PPPs: accountability and transparency in 177, 179, 184–5, 186, 190; governance issues 178–9, 184, 186, 187, 188, 189; Hong Kong 178, 180, 183–5; infrastructure types 179–80; international ventures 185, 190; Japan 178, 180, 185–6, 188; risk management 181; Singapore Sports Hub PPP project 201–11; South Korea 178, 180, 186–8, 235–44; *see also* Chinese PPPs
asset creation and ownership 23–6
asset management (AM) teams 283–91

Australian PPP research think tanks 77
Australian PPPs: and Asian contractors 185, 190; average procurement timeline 263; characteristics of 156–8; fiscal drive for 182; market 76, 153–5, 188; origins of 155–7; public protests against 304; social infrastructure projects 158–60; *see also* Top Ryde PPP project (Australia)
authority (Cialdini's persuasion theory) 201

bid (upfront) costs and reimbursement 60–1, 66, 164
bid rigging in Japan 185, 186
bidding procedure for Chinese PPPs 221–3
Bolivia 304
Brazil 247
Build-Operate-Transfer (BOT) 217, 220, 221, 223, 237, 259, 266
Build-Own-Operate (BOO) 2, 237
Build-Own-Operate-Transfer (BOOT) 2
Build-Subsidise-Operate-Transfer (BSOT) 220–1
Build-Transfer (BT) 222
Build-Transfer-Lease (BTL) 237–8
Build-Transfer-Operate (BTO) 237, 238
Building Information Modelling (BIM) 289
building procurement, definition 2
Bundle 401 Project (Oregon) 103–4, 109
Busan-Geoje Fixed Link (South Korea) 242

Canada 181, 257, 304
Chengdu No. 6 Water Plant (China) 223, 224

Milton Keynes UK
Ingram Content Group UK Ltd.
UKHW021637071024
449327UK00020BA/1332

9 780367 001186